ELECTROCHEMICAL SCIENCE
AND TECHNOLOGY OF POLYMERS—1

ELECTROCHEMICAL SCIENCE AND TECHNOLOGY OF POLYMERS—1

Edited by

R. G. LINFORD

School of Chemistry, Leicester Polytechnic, Leicester, UK

ELSEVIER APPLIED SCIENCE
LONDON and NEW YORK

ELSEVIER APPLIED SCIENCE PUBLISHERS LTD
Crown House, Linton Road, Barking, Essex IG11 8JU, England

Sole Distributor in the USA and Canada
ELSEVIER SCIENCE PUBLISHING CO., INC.
52 Vanderbilt Avenue, New York, NY 10017, USA

WITH 29 TABLES AND 47 ILLUSTRATIONS

© ELSEVIER APPLIED SCIENCE PUBLISHERS LTD 1987

British Library Cataloguing in Publication Data

Electrochemical science and technology of
polymers.
 1. Addition polymerization
 I. Linford, R. G.
 547.8′4   QD281.P6

Library of Congress Cataloging in Publication Data

Electrochemical science and technology of polymers.
 Bibliography: p.
 Includes index.
 1. Polymers and polymerization—Electric properties.
 I. Linford, R. G.
 QD381.9.E38E37 1987   620.1′9204297   86-19660

ISBN-13: 978-94-010-8026-2   e-ISBN-13: 978-94-009-3413-9
DOI: 10.1007/978-94-009-3413-9

The selection and presentation of material and the opinions expressed are the sole responsibility of the author(s) concerned

Special regulations for readers in the USA

This publication has been registered with the Copyright Clearance Center Inc. (CCC), Salem, Massachusetts. Information can be obtained from the CCC about conditions under which photocopies of parts of this publication may be made in the USA. All other copyright questions, including photocopying outside of the USA, should be referred to the publisher.

All rights reserved. No part of this publication may be reproduced, stored in a retrieval system, or transmitted in any form or by any means, electronic, mechanical, photocopying, recording or otherwise, without the prior written permission of the publisher.

Preface

Polymers are normally thought of as insulators. In the last few years, however, a rapidly advancing and changing field has developed which exploits the ability of certain polymers to conduct charge, in some cases electronically and in others by means of ions.

Certain electrochemical processes of major present-day industrial importance depend on the presence of polymeric materials for their efficient operation. The chlor-alkali industry is a prime example. Exciting new power sources, in which polymers replace conventional electrodes and/or electrolytes, are being intensively developed. Remarkable advances in the understanding of electrochemical processes and the development of a range of sophisticated sensors and other devices have been made possible by the use of polymer-coated electrodes.

The impact of polymers on the electrochemical field is still in its initial growth phase. The results of a rapidly escalating volume of industrial and academic research are being applied in many contexts, especially in the information technology field. In certain areas, the use of polymers is only just beginning to show its impact. In the next year or so, the use of polymerised Langmuir–Blodgett films as a substitute for conventional E-beam resists in electronics can be anticipated. By the end of the decade, polymerised mono- and multi-layers may be incorporated in very large-scale integrated circuits.

It is the aim of the present volume to review some of the areas in which the application of polymers in electrochemical systems has reached a comparatively mature stage. In order to guide the reader who may not be immediately familiar with both polymeric and electrochemical concepts, two introductory chapters have been included to set the scene. The following chapters then each emphasise a particular aspect of present-day scientific or technological importance.

Future volumes will cover the materials, applications and techniques of study of this rapidly developing interface between polymers and electrochemistry.

R. G. LINFORD

Contents

Preface	v
List of Contributors	xi

Chapter 1 Ionic and Electronic Transport ... 1
ROGER J. LATHAM and ROGER G. LINFORD
1 Introduction ... 1
2 Electrochemical reactors ... 4
3 The electrode/electrolyte interphase ... 7
   3.1 Reversible electrodes and the standard electrode potential ... 7
   3.2 Dependence of electromotive force on concentration ... 10
   3.3 Dependence of electromotive force on temperature ... 10
   3.4 Overpotential ... 11
   3.5 Summary ... 13
4 Solid state cells ... 13
   4.1 Mechanisms of ionic conduction in the solid state ... 15
   4.2 Polymeric electrolytes ... 16
5 Measurement of conductivity ... 17
   5.1 The complex plane representation ... 18
   5.2 Electronic conductivity ... 19
References ... 21

Chapter 2 Polymer Structure and Conductivity ... 23
MALCOLM D. GLASSE and ROGER G. LINFORD
1 Introduction ... 23
2 Macromolecular structure ... 24
3 Amorphous polymers ... 29
   3.1 Flexibility and flow ... 29

	3.2 The glass transition		30
4	Polymer crystallinity		32
	4.1 Crystallisability		32
	4.2 Crystallinity		34
	4.3 Crystallisation		37
5	Spherulites		38
6	Implications		40
7	Conclusions		42
	References		43

Chapter 3 Ion Conducting Polymers 45
JOHN R. OWEN

1. Introduction . 45
2. The chemistry and structures of polymer complexes and solutions . 49
3. The kinetics of ion transport . 52
   - 3.1 Free volume theory . 53
   - 3.2 Application to ion conduction . 55
   - 3.3 Conclusions regarding kinetics . 57
4. Electrical measurement techniques . 58
5. Current research directions . 61
   - References . 64

Chapter 4 Organic Polymers as Electroactive Materials . 67
ALAN G. MACDIARMID and MACRAE MAXFIELD

1. Introduction . 67
2. p- and n-doping of polyacetylene . 69
3. Basic concepts of electrochemistry . 71
4. Polyacetylene as an electrode material . 72
5. The polyacetylene cathode . 74
   - 5.1 Use of p-doped (oxidized) polyacetylene . 75
   - 5.2 Use of n-doped (reduced) polyacetylene . 80
6. The polyacetylene anode . 86
7. Batteries employing polyacetylene anodes and cathodes . 89
8. Batteries using other conducting polymers . 90
9. Present status of batteries employing polymer electrodes . 96
10. Conclusions . 97
    - Acknowledgements . 98
    - References . 98

Chapter 5 Polymer Modified Electrodes: Preparation and Characterisation . 103
A. ROBERT HILLMAN

1. Introduction . 104
2. Preparation . 106
   - 2.1 Pre-formed polymers . 107
   - 2.2 Polymers formed/coated simultaneously . 145

3	Electrochemical characterisation .	180
	3.1 Objectives and limitations .	180
	3.2 Simple models of polymer films on electrodes and general approach .	181
	3.3 Thermodynamic parameters	182
	3.4 Kinetic parameters	190
4	Spectroscopic and other non-electrochemical characterisation techniques	209
	4.1 Macroscopic observations and properties .	210
	4.2 'Monomer'/redox centre observation	211
	4.3 Sub-molecular units	216
	4.4 Atoms	217
	4.5 Other techniques .	225
5	Conclusions	225
	References	226

Chapter 6 Reactions and Applications of Polymer Modified Electrodes 241
A. ROBERT HILLMAN

1	Introduction .	241
2	Theoretical treatments of mediated charge transfer.	242
	2.1 Basic models and concepts .	242
	2.2 Analysis of mediated charge transfer	244
	2.3 Experimental tests of the analysis	250
	2.4 Other mechanistic studies	264
	2.5 Summary.	266
3	Applications .	266
	3.1 Introduction .	266
	3.2 Electrochemical synthesis	266
	3.3 Sensors	273
	3.4 Corrosion protection of metals .	276
	3.5 Semiconductor electrochemistry .	276
	3.6 Photogalvanic effects	278
	3.7 Immobilisation of particulates	279
	3.8 Display devices	279
	3.9 Electronic and electrochemical devices	281
4	Conclusions	283
	References	284

Chapter 7 Perfluorinated Ionomer Membranes for Use in the Production of Chlorine and Caustic Soda 293
PETER J. SMITH

1	Introduction .	293
	1.1 Historical perspective to membrane cell technology	293
	1.2 Current commercial chlorine technology .	296

2	Perfluorinated sulphonate ionomers	301
	2.1 Synthesis of perfluorosulphonate polymers	301
	2.2 Commercial exploitation of perfluorosulphonate ionomers	302
	2.3 Physical properties of perfluorosulphonate ionomers	305
3	Perfluorinated carboxylate ionomers	309
	3.1 Synthesis and polymerisation of perfluorinated carboxylate monomers	311
	3.2 Fabrication of perfluorinated membranes	313
	3.3 Surface topography of perfluorinated ionomers	315
	3.4 Physical properties of perfluorocarboxylate ionomers	318
4	Mixed perfluorinated sulphonate/carboxylate ionomer membranes	319
5	Structure/transport relationships in perfluorinated ionomer membranes	320
	5.1 Basic structural features of perfluorinated ionomers	320
	5.2 Experimental correlations between polymer structure and transport properties	324
	5.3 Quantitative relationships involving ionic transport	328
6	Present and future trends in perfluorinated ionomer membrane cell technology	329
	Acknowledgement	329
	References	329
Index		335

List of Contributors

MALCOLM D. GLASSE
  School of Chemistry, Leicester Polytechnic, PO Box 143, Leicester LE1 9BH, UK

A. ROBERT HILLMAN
  School of Chemistry, University of Bristol, Cantock's Close, Bristol BS8 1TS, UK

ROGER J. LATHAM
  School of Chemistry, Leicester Polytechnic, PO Box 143, Leicester LE1 9BH, UK

ROGER G. LINFORD
  School of Chemistry, Leicester Polytechnic, PO Box 143, Leicester LE1 9BH, UK

ALAN G. MACDIARMID
  Department of Chemistry, University of Pennsylvania, Philadelphia, Pennsylvania 19104, USA

MACRAE MAXFIELD
  Corporate Research Center, Allied-Signal Inc., Morristown, New Jersey 07960, USA

JOHN R. OWEN
Department of Chemistry and Applied Chemistry, University of Salford, Salford M5 4WT, UK

PETER J. SMITH
General Chemicals Business Group, Imperial Chemical Industries PLC, PO Box 13, The Heath, Runcorn, Cheshire WA7 4QF, UK

CHAPTER 1

Ionic and Electronic Transport

ROGER J. LATHAM and ROGER G. LINFORD
School of Chemistry, Leicester Polytechnic, UK

1. INTRODUCTION

A major aim of this chapter is to introduce the reader who is relatively unfamiliar with electrochemistry to a number of definitions, concepts and principles that are commonly used in electrochemical science and technology.

For our purposes, the materials with which this book is concerned may be divided into two broad categories: *conducting polymers* in which charge is carried efficiently enough for them to be used in an appropriate device, and *insulating polymers* in which the passage of charge is prevented sufficiently to permit their use as containers and other barriers to conduction. Traditionally polymers were thought of as insulators and, as pointed out in Chapter 3, conduction was normally regarded as an undesirable attribute.

Charge may be carried by ions or by electrons. Ions may be positively charged (*cations*) or negatively charged (*anions*). If ions within a polymer attract each other too strongly then *ion-pairs* can be formed; as discusssed in Chapters 3 and 7 these are neutral and do not conduct charge under the influence of an electric field. In some cases such as certain of the polymer–salt complexes mentioned in Chapter 3, ions associate into larger groupings such as *triple ions* e.g. A_2B^+ or AB_2^-, and these are charge carriers. Sometimes, in an *electronically conducting polymer*, the charge is carried not by electrons (*n-type conduction*) but by electron holes (*p-type conduction*). Examples of both types are covered in Chapter 4.

Although some esoteric polymers such as polyacetylene and polypyrrole are naturally electronically conducting, other more common polymers may be converted to electronic conductors by *doping* with substantial quantities of graphite or other electronically conducting material. For ionic conduction, doping with an ionic salt is necessary. Whereas some polymers contain ionic groupings, these are attached to the polymer and are not free to move. A range of examples of ionically conducting polymers is given in Chapter 3; the best studied systems are those based on poly(ethylene oxide) complexed with alkali metal salts.

Typical electrochemical devices contain electrodes and electrolytes. An *electrode* is the region where ions are formed or consumed by oxidation or reduction of an electrochemically active material. The electrode at which oxidation, e.g.

$$Cu = Cu^+ + e$$

occurs is called the *anode*; anions migrate towards the anode and cations migrate away from it. The opposite process, reduction, occurs at the *cathode* which is the electrode towards which the cations migrate. It is necessary for electrodes to be efficient *mixed conductors*, i.e. to conduct both ions and electrons (or electron holes). It may be necessary to add *ion carriers* (ionically conducting material) and/or *electron carriers* to the electrochemically active electrode material to achieve satisfactory conduction performance.

In an electrochemical device such as a cell, it is necessary to separate the pathways used for the passage of ions from those used by electrons. The material through which ions travel within the cell from one electrode to another is called the *electrolyte*. This obviously needs to be a good conductor of ions and a poor conductor of electrons and/or holes. It is common for solutions of metal salts in water and most non-aqueous solvents to have both of these attributes.

Apart from the obvious need to satisfy mechanical, thermal and other requirements concerning stability and performance characteristics, the most important property of an electrolyte is its conductance. The *conductance*, G, is the reciprocal of the resistance, R, and is nowadays measured in Siemens, symbol S; this unit is identical to the reciprocal ohm, Ω^{-1}. The conductance varies inversely with sample thickness (path length), l, and directly with sample area, A, and it is therefore conventional to characterise electrolytes by the proportionality constant, σ, in the equation $G = \sigma A/l$. This constant is called the conductivity, and is normally expressed in $S\,cm^{-1}$, sometimes written as $(\Omega\,cm)^{-1}$. For a given material it may be expressed as the sum of ionic and electronic contributions.

It is important to note that the path length in cells containing thin film solid electrolytes is very low and consequently acceptable conductance is obtained even from materials whose conductivity would seem to be very low. It is because of this that polymeric electrolytes are useful; typically their conductivities, even at 100°C, are $\simeq 10^{-4}\,S\,cm^{-1}$, but they can be cast as films of thickness $1-100\,\mu m$, so that the conductance of $1\,cm^2$ can be as high as $1\,S$.

The ionic conductivity, σ_i, of the majority of ionic solids is too low to be of use in electrochemical devices; for solid NaCl at 25°C, for example, σ is nine orders of magnitude lower than for the value for $1\,mol\,dm^{-3}$ aqueous solution. A comparatively small number of ionic materials, representing a variety of types (as will be discussed later), have sufficiently high ionic conductivity to be useful in devices, and these are called *superionic* or *fast-ion* conductors. These fall into two classes, the mixed conductors mentioned above in which electronic conductivity, σ_{el}, is significant, and the *solid electrolytes* which have low values of σ_{el}.

The essential features of a typical polymeric electrolyte cell, the Harwell battery, are shown in Fig. 1. The *current collectors*, which transport electronic current to and from the electrodes in an electrochemical cell, must have properties that are the reverse of those needed for an electrolyte. They must have low ionic and high electronic conductivity; metal wires, foils or grids, or SnO_2-coated glass find frequent application.

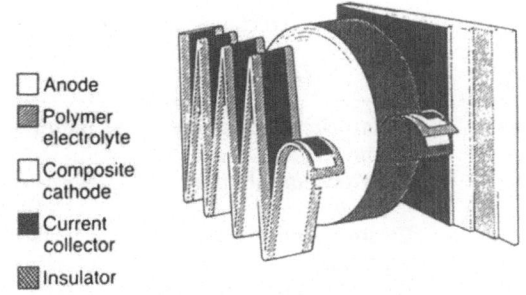

FIG. 1. The configuration of a typical polymeric battery.

There are two main types of electrochemical cell. In a *galvanic cell* the cell reaction between the electrode components produces electrochemical energy, whereas in an *electrolytic cell* voltage is applied externally and causes a reverse cell reaction to take place. Examples of the former can be found in Chapters 3 and 4 and of the latter in Chapter 7.

2. ELECTROCHEMICAL REACTORS

An electrochemical reactor can be a producer or consumer of energy. Polymers are used in many aspects of such devices. Of the types of electrochemical reactor outlined in Table 1, the primary and secondary batteries are probably the most commonly encountered everyday examples. A battery is an electrochemical cell or a bank of such cells. A *primary battery* is a galvanic cell and a *secondary battery* is a battery that can be recharged. On discharge it is a galvanic cell and on recharge it is an electrolytic cell; the lead–acid car battery and the lithium polymeric battery shown in Fig. 1 are both examples of secondary batteries.

Any battery may be simply thought of as a three-compartment system and polymers find application in each of the compartments. In addition to taking an active role in the electrochemistry of the battery, polymers are also used as passive components. A few representative examples[1] of the use of polymers in the lead–acid secondary battery are given in Table 2.

TABLE 1
Types of Electrochemical Reactor

Reactor	*Description*	*Representation*
Primary battery	Electrochemical energy convertor	Battery → Electricity
Secondary battery	Electrochemical energy storer	Electricity → Battery → Electricity
Fuel cell	Electrochemical energy producer	Chemical fuel → Fuel-cell → Electricity
Electrolysis cell	Electrochemical substance producer	Electricity → Cell → Substance

TABLE 2
Polymeric Components in the Lead–Acid Battery

Component	Function	Polymers used[a]
Casings	Containment of electrodes and electrolyte	PP, ABS, SAN, HIPS
Electrode plates	(a) Binder in pasted plates	PTFE, PTFE/PE blend
	(b) Walls of tubular plates	PVC, woven polyester
	(c) Grid to support paste	PP/lead
	(d) Current collector	PE/carbon
Electrode spacers	Location of electrodes	PE, PP, PVC
Separators	(a) Microporous, electrolyte-retaining barrier	PVC, LDPE, PTFE, PF
	(b) Semi-permeable, electrolyte-retaining barrier	Cellophane, LDPE–MAA
Absorbers	Absorption of electrolyte	PP, PA, PE
Gelling agents	Absorption of electrolyte	Substituted cellulosics

[a] Abbreviations: PP, polypropylene; ABS, acrylonitrile-butadiene–styrene; SAN, styrene–acrylonitrile; HIPS, high-impact polystyrene; PTFE, polytetrafluoroethylene; PE, polythene; PVC, poly(vinyl chloride); LDPE, low-density polyethylene; PF, phenol–formaldehyde; MAA, methacrylic acid; PA, polyamide.

A process of *charge-transfer* occurs at each of the electrode/electrolyte interfaces. It is common practice in electrochemistry to refer to these interfacial regions as *interphases* in order explicitly to acknowledge the three-dimensional nature of what might otherwise be seen as a flat, two-dimensional interface. The interphases are in many ways the regions that have the greatest influence on the behaviour of a working battery.

In a working electrochemical device, the net overall reaction arises from the combination of the individual *half-cell reactions*; these are the oxidation and reduction processes occurring at the anode and cathode respectively. In a galvanic cell (e.g. a primary battery or a secondary battery under discharge), the anode is the electrode or plate from which electrons leave and is therefore the *negative electrode*; the cathode is the *positive plate*. In an electrolytic cell (e.g. a secondary battery under recharge), the negative plate is the electrode at which electrons are introduced into the cell; the electrons reduce the species present and the negative electrode is the cathode here. The + and − symbols remain as such on the battery casing (Fig. 2) but the plates

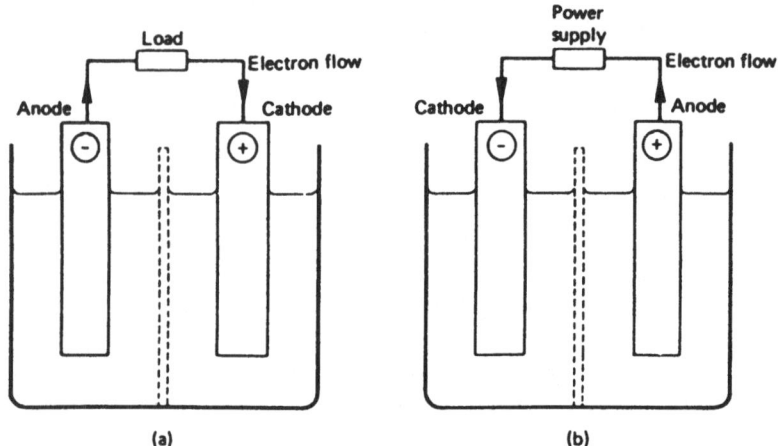

FIG. 2. The role of the electrodes in a secondary battery under (a) spontaneous discharge and (b) charge conditions.

reverse in role. For example, in the lead–acid battery:
At the negative electrode

$$Pb + SO_4^{2-} \xrightleftharpoons[\text{charge}]{\text{discharge}} PbSO_4 + 2e^-$$

At the positive electrode

$$PbO_2 + 4H^+ + SO_4^{2-} + 2e^- \xrightleftharpoons[\text{charge}]{\text{discharge}} PbSO_4 + 2H_2O$$

As with any chemical system, the maximum work available is when the system is poised at equilibrium; in the case of a battery this is when there is no flow of current through the external or internal circuit. When the system is made to work and therefore to depart from equilibrium, *polarisation* develops such that in a battery the usable voltage will be less than the thermodynamically expected voltage of the cell. The contributing factors to the overall polarisation include:

(i) voltage loss due to ohmic effects, i.e. drop of voltage across a resistance, related to the physical construction of the cell;
(ii) charge (or electron) transfer processes as the oxidation/reduction reactions proceed;

(iii) mass transport effects including diffusion of products and/or reactants to and from the electrode surface;
(iv) lattice building due to the removal or incorporation of atoms from the ordered lattice of the electrode material.

An important consideration in a working battery is the possibility of dimensional changes taking place at the electrodes as a result of the chemistry of the cell reaction. Polymeric electrolytes are particularly suited to applications in which this is a major problem, because they are deformable and can accommodate such changes.

In an ideal battery the electrolyte will allow only an ionic, and not an electronic, current to pass. Some solid materials that would otherwise be candidates as solid electrolytes are in fact mixed conductors. In a cell incorporating such material, the cell chemistry would be able to proceed without useful external work being carried out. This situation is called *self-discharge* and it reduces the anticipated *shelf life* of a battery. Most ionically conducting polymeric materials are in fact rather good electronic insulators, and consequently the problem of self-discharge that occurs in many solid state batteries is usually absent in polymeric systems.

In summary, a battery consists of a three-compartment device in which the interphases between the electrode and electrolyte compartments are particularly important. In a solid-state battery, contact at the interphase may be a problem, but the use of polymeric materials may often help to maximise the contact area as a result of the favourable thermomechanical properties.

3. THE ELECTRODE/ELECTROLYTE INTERPHASE

3.1. Reversible Electrodes and the Standard Electrode Potential

When an electrode is acting *reversibly,* as for example in the case of a metal electrode in equilibrium with its ions in solution, the processes occur with no net current being taken from the cell; the potential difference between such electrodes is the *electromotive force* (*emf*) of the cell.

The value of the emf is determined by the cell reaction. It is not always equal to the *open circuit voltage, ocv,* which is the voltage actually detected externally across the electrodes of a practical battery using a measuring instrument of very high input impedance (resistance), i.e. under conditions where virtually no current flows.

Differences between the emf and the ocv can arise from such factors as the following.

(i) The occurrence of corrosion and other side reactions within the total assembly of battery, current collectors and encapsulants. These act as additional electrochemical cells, the emfs of which combine with the emf of the main cell reaction, thus modifying its value.

(ii) the *ohmic or iR drop,* caused by the fact that it is not possible to carry out real measurements under conditions where absolutely no current flows. The measured voltage, V_{meas}, is given by

$$V_{meas} = E_{net} - V_{drop}$$

where E_{net} is the expected emf of the cell, taking account of all the reactions actually occurring, and V_{drop} is given by *Ohm's law, $V = iR$*, R here being the *internal resistance* of the cell and i the current.

Since a complete cell must consist of an oxidation process and a reduction process at the respective electrodes, the value of the cell emf can be obtained from the correct combination of the two *half-cell potentials*. As it is impossible to measure a single half-cell potential, all half-cells are described in relation to a common reference half-cell which has been defined as having zero electrode potential. This is in fact the *standard hydrogen electrode (SHE)*, in which there is equilibrium between hydrogen gas at 1 atmosphere and hydrogen ions at unit activity, using a surface activated 'platinised' Pt electrode. It is not uncommon to quote half-cell potentials relative to other standards which are themselves defined in terms of the SHE; for polymeric lithium batteries, for example, it is often convenient to work relative to the Li/Li$^+$ electrode.

Half-cells are usually described as reduction processes in association with the reference system. When a half-cell which involves unit activity in solution is combined with the standard hydrogen electrode in this way, the resulting cell potential is called the *standard electrode potential*, $E^\ominus$, for that reduction half-cell. Thus for example in the well-known Daniell cell, both

$$Cu^{2+} + 2e = Cu \qquad E^\ominus = +0.34 \text{ V}$$

and

$$Zn^{2+} + 2e = Zn \qquad E^\ominus = -0.761 \text{ V}$$

mean that values +0·34 V and −0·761 V will be observed on a meter when the appropriate electrode is combined with the standard hydrogen electrode.

Closely related to the standard electrode potential is the *standard Gibbs free energy*, $\Delta G^\ominus$. This represents the maximum net amount of work available and thermodynamic considerations show that

$$\Delta G^\ominus = -zFE^\ominus$$

where F is the *Faraday*, i.e. the amount of charge corresponding to one mole of electrons.

When two half cells are combined, the resulting $\Delta G^\ominus_{cell}$ which is obtained must be negative in order for the cell reaction, as written, to proceed from left to right. Thus, for the examples above, the correct combination will be:

$$Cu^{2+} + 2e = Cu \qquad \Delta G^\ominus_1 = -2FE^\ominus_1$$
$$Zn = Zn^{2+} + 2e \quad \Delta G^\ominus_2 = +2FE^\ominus_2$$

where the second half-cell is now taken as an oxidation reaction and the sign of the free energy is therefore reversed.

Hence the overall cell reaction is

$$Zn + Cu^{2+} = Zn^{2+} + Cu$$

and

$$\Delta G^\ominus_{cell} = \Delta G^\ominus_1 + \Delta G^\ominus_2$$
$$-2FE^\ominus_{cell} = -FE^\ominus_1 + 2FE^\ominus_2$$
$$-2FE^\ominus_{cell} = -2F(E^\ominus_1 - E^\ominus_2)$$
$$E^\ominus_{cell} = (E^\ominus_1 - E^\ominus_2)$$
$$= +0\cdot34 - (-0\cdot761)$$
$$= 1\cdot101 \text{ V}$$

By implication, since $\Delta G^\ominus$ must be negative for the cell reaction to proceed spontaneously, $E^\ominus_{cell}$ *must be positive*. Usually the approach is shortened by simply dealing with standard electrode potentials, as the zF term will always cancel.

Thus, for any combination of half-cells,

$$E^\ominus_{cell} = E^\ominus_1(\text{red}) - E^\ominus_2(\text{red})$$

or alternatively

$$E^\ominus_{cell} = E^\ominus_1(\text{red}) + E^\ominus_2(\text{ox})$$

3.2. Dependence of Electromotive Force on Concentration

For the general case of a half-reaction of the type

$$\text{ox} + ze = \text{red}$$

a thermodynamic treatment using chemical potentials gives the well-known *Nernst equation* where the emf is a logarithmic function of the activities of the species involved:

$$E = E^\ominus + \frac{2 \cdot 303 RT}{zF} \log\left(\frac{a_{\text{ox}}}{a_{\text{red}}}\right)$$

Thus for the Daniell cell the reduction potentials of the two half-cells become:

$$E_{\text{Zn}} = E^\ominus_{\text{Zn}^{2+}/\text{Zn}} + \frac{2 \cdot 303 RT}{zF} \log\left(\frac{a_{\text{Zn}^{2+}}}{a_{\text{Zn}}}\right)$$

and

$$E_{\text{Cu}} = E^\ominus_{\text{Cu}^{2+}/\text{Cu}} + \frac{2 \cdot 303 RT}{zF} \log\left(\frac{a_{\text{Cu}^{2+}}}{a_{\text{Cu}}}\right)$$

The overall cell potential, obtained from $E_{\text{Cu}} - E_{\text{Zn}}$, becomes

$$\begin{aligned}E_{\text{cell}} &= E^\ominus_{\text{Cu}^{2+}/\text{Cu}} - E^\ominus_{\text{Zn}^{2+}/\text{Zn}} \\ &\quad + \frac{2 \cdot 303 RT}{zF} \log\left(\frac{a_{\text{Cu}^{2+}}}{a_{\text{Zn}^{2+}}} \cdot \frac{a_{\text{Zn}}}{a_{\text{Cu}}}\right) \\ &= E^\ominus_{\text{cell}} + \frac{2 \cdot 303 RT}{zF} \log\left(\frac{a_{\text{Cu}^{2+}}}{a_{\text{Zn}^{2+}}}\right)\end{aligned}$$

since by convention a_{Zn} and a_{Cu} are unity. It is frequently assumed that for dilute solutions, the activity coefficient of the electrolyte approaches unity and the Nernst equation is then given as:

$$E = E^\ominus + \frac{2 \cdot 303 RT}{zF} \log \frac{[\text{ox}]}{[\text{red}]}$$

3.3. Dependence of Electromotive Force on Temperature

The Gibbs–Helmholtz equation

$$\Delta G = \Delta H + T \left(\frac{\partial \Delta G}{\partial T}\right)_p$$

where H is enthalpy, can be expressed in an electrochemical form

$$-zFE = \Delta H - TzF\left(\frac{\partial E}{\partial T}\right)_p$$

and by analogy with the relationship

$$\Delta G = \Delta H - T\Delta S$$

where S is entropy, the *temperature coefficient of the emf* can be obtained

$$\Delta S = zF\left(\frac{\partial E}{\partial T}\right)_p$$

3.4. Overpotential

When a cell departs from equilibrium, as in the case of an electrolysis (electrolytic) cell such as a secondary battery on charge, or a galvanic power cell, the difference in the potential measured between the electrodes will not be the thermodynamic emf, E. The cell is then said to be in a state of *polarisation*. Polarisation being caused by the presence of an ohmic potential drop arising from the internal resistance, as mentioned in Section 3.1. An alternative cause is the existence of *overpotential* (*overvoltage*) associated with the electrode processes.

The overpotential, η, is the difference between the working potential for a certain current density, E_w, and the thermodynamic reversible potential.

$$\eta = E_w - E$$

Since, at an electrode/electrolyte interphase, electron transfer has to occur and the concentrations of reactants and products of the electrode process must also change, the two main contributions to the total overpotential will be *activation* (*or charge-transfer*) *overpotential* and *concentration overpotential*.

For a polarised cell, the voltage, V, is given by

$$V = E + iR + \eta_A + |\eta_C|$$

when the cell is either an electrolysis cell or a secondary cell on charge; and

$$\dot V = E - iR - \eta_A - |\eta_C|$$

when the cell is giving power: η_A and η_C are the total overpotentials at the two polarised electrodes, anode and cathode.

3.4.1. Activation overpotential

When an electrode system is at equilibrium, the rates of the forward and reverse processes are considered to be equal and opposite. Several treatments are available for predicting the behaviour when the equilibrium is disturbed. However, the important result in all cases is the *Butler–Volmer equation*

$$i = i_0[\exp(-\alpha zF\eta/RT) - \exp((1-\alpha)zF\eta/RT)]$$

where

(i) α is the *charge-transfer coefficient*, a constant in the range 0 to 1·0, which relates the way in which both the forward and reverse processes are disturbed from equilibrium;
(ii) i_0 is the *exchange current density*, which is a measure of the rates of the forward and reverse processes at equilibrium.

The Butler–Volmer equation shows that the net current will be the difference between the contributions due to the forward and reverse processes depending upon the polarity of the electrode (Fig. 3). This equation also shows that the amount of polarisation due to charge-transfer will be small for a given working current density when the exchange current density is relatively high.

In practice, measurements are often made for high overpotentials (>100 mV) and one of the processes becomes dominant. In this case, the Butler–Volmer equation can be simplified to the well-known *Tafel relationship*, allowing values for i_0 and α to be obtained. For *cathodic polarisation*, i.e. when the charge-transfer process at the cathode is suppressed relative to that at the anode, the Tafel relationship, obtained by dropping the second term on the right-hand side of the Butler–Volmer equation, is

$$\eta = \{\log i_0 - \log i\} \times 2\cdot303RT/\alpha zF$$

and for *anodic polarisation*, when the cathode process dominates,

$$\eta = \{\log i - \log i_0\} \times 2\cdot303RT/(1-\alpha)zF$$

3.4.2. Concentration overpotential

Concentration polarisation arises because the supply of electroactive species to the electrode surface must be maintained. Species will arrive from the bulk of the electrolyte either as a result of diffusion due to the presence of a concentration gradient or, if ionic species are involved, as a result of migration in an electric field. This can lead to the establishment of a *limiting current density,* i.e. the process cannot proceed any faster. Application of Fick's first law of diffusion gives

$$i_{\lim} = zFDc/\delta(1-t)$$

where D is the diffusion coefficient of the electroactive species under consideration, t is the *transport number* of that species (i.e. the fraction of total charge carried) and δ is the thickness of the diffusion layer.

The term allowing for electromigration $(1-t)$ is omitted when only diffusion of the electroactive species at the electrode surface is important, as is the case when a supporting electrolyte is used.

3.5. Summary

Standard electrode potentials and related emfs are applicable for reversible electrodes and cells. In a working cell, polarisation develops because of the presence of overpotentials associated with the electrode processes. In certain cases one of the overpotential contributions may be dominant and the overall electrode process can then be described as being limited, for example by charge-transfer.

4. SOLID STATE CELLS

Conceptually, the simplest cell configuration consists of two parallel plate electrodes made of metal or similar dimensionally stable materials, immersed in a liquid electrolyte held in an inert container such as a beaker. In such a case, the dimensional *cell constant, l/A,* is well defined and constant; l is the distance between the electrodes and A is the cross-sectional area of the smaller electrode. Provided that the liquid electrolyte wets the solid electrodes, there are no contact problems. In addition, the electrolyte acts as a relatively large, often virtually infinite, reservoir for both the reactants and the products of the cell reaction. Material transport can take place not only by diffusion but also by convection. Many practical battery systems in everyday use, such as the lead–acid accumulator, broadly conform to this model.

In certain systems intended for future large-scale use, the liquid/solid interphases are retained, but the nature of the components is reversed. In the sodium–sulphur battery, for example, which is designed to operate at 350°C, the electrolyte is solid β-alumina and the electrodes are molten.

Major disadvantages shared by all battery systems incorporating liquid components are:

(i) spillage and leakage, leading to loss of electrolyte and possibly to corrosion;
(ii) relative frailty, because the electrodes and electrolyte do not form a system with mechanical integrity.

These can be overcome by incorporating the liquid within a suitable semi-solid matrix. In the conventional dry Leclanché cell, which is used domestically as torch batteries, etc., the electrolyte is mixed with a filler to form a paste. In other systems a gel is used. In the sealed lead–acid batteries, which are now replacing nickel–cadmium cells as a convenient, rechargeable and small domestic power source, the electrolyte is supported in a glass or polymeric fibre mat.

None of these 'dry' cell systems is truly solid state in nature. The electrolyte, although macroscopically dispersed, still forms a continuous phase in which long-range structural order is lacking, diffusion and conduction is facile, and convection is to some extent possible.

In a *solid state battery*, the electrodes and the electrolyte are solid. The electrolyte may be in the form of a slice of a single or of a polycrystal, a pellet of compacted powder, or a thin layer of a conducting glass or polymer. Convection is precluded and diffusion and conduction are relatively slow processes at room temperature. In consequence, many such systems are designed to operate at elevated temperatures. The handicaps of solid state batteries include the following.

(i) Transport of reactants to, or products from, the charge-transfer interphase is limited and may be rate-determining.
(ii) If the cell reaction product is a poor ionic conductor, as is often the case, then the ohmic drop inevitably increases as discharge proceeds. This is because the reaction product cannot be distributed in a dilute fashion throughout the electrolyte but has to reside within the interphase at which it is formed.
(iii) Contact at the interphase is poor because solid surfaces are not flat. It is not unusual for the actual area of contact to be only about 1% of the apparent surface area. Consequently it is

difficult to ascribe a meaningful value to the cell constant (although it is conventional to use the apparent area), and the true current density at points of contact may exceed the apparent value by a factor of 100.

(iv) If dimensional changes occur during cell discharge or recharge, as is the case with *intercalation cathodes* such as TiS_2 or V_6O_{13} for example, then cell rupture and/or loss of electrode/ electrolyte contact can result.

By way of compensation, solid state batteries have certain inherent advantages that make them particularly suitable for specialised applications. They are:

(i) robust;
(ii) spill-proof;
(iii) leak-proof in general;
(iv) easy to make in a compact configuration.

These properties make them particularly suitable for such applications as heart pacemaker batteries, back-up power sources for computer memory and self-powered very large-scale integrated (VLSI) circuits.

4.1. Mechanisms of Ionic Conduction in the Solid State

In essence, there are three main mechanisms: defect conduction, highly disordered sub-lattice motion and amorphous region transport.

Defects such as vacancies (unoccupied lattice sites) and interstitials (ions occupying positions between lattice sites) can be formed *intrinsically* due to temperature effects or *extrinsically* due to deliberate or accidental impurity dopants. *Defect conduction* can involve anions and/or cations; the very low ionic conductivity possessed by alkali halides is caused by the presence of defects. In high temperature doped oxide conductors such as ZrO_2/Y_2O_3 and also in halide ion conductors including fluorite lattice materials such as CaF_2 and $SrCl_2$, conductivity increases substantially with temperature, but there are no sharp discontinuities in the value of the activation energy. There are usually no first-order solid–solid phase transitions affecting the conductivity of these materials, although they are sometimes said to undergo a Faraday transition.[2] The ions are located at lattice defect sites characterised by deep potential energy wells, and ionic motion takes place by a hopping mechanism between adjacent sites. The site residence time considerably exceeds the flight time during which the ions are in motion.

Certain materials such as AgI and CuI have not inconsiderable conductivity at room temperature, and when they are heated through the so-called β–α first-order phase transition, the value of σ rises by several orders of magnitude. The very high conductivity in the α phase, which approaches that of $0 \cdot 1\,\mathrm{mol\,dm^{-3}}$ NaCl in aqueous solution, is usually described in terms of *disordered sub-lattice motion*. In AgI above 146°C, when the material is in the α phase, the I$^-$ ions form an ordered, body-centred cubic, 'host' sub-lattice. The Ag$^+$ ions are distributed over so many lattice sites that their radial distribution function is more reminiscent of that of a liquid, leading to the statement of O'Keeffe[3] that such materials have liquid-like sub-lattices. In fact, the enthalpy of the β–α transition of AgI is roughly equal to that of the solid–liquid melting process at 555°C. The sum of the two enthalpies is comparable with the enthalpy of melting of an alkali halide, supporting O'Keeffe's suggestion that the Ag$^+$ sub-lattice melts at the β–α transition.

Mobile ions within materials of this type 'reside' in sites characterised by shallow potential energy wells and the flight time between sites is comparable with the residence time. During the past 20 years, it has been found that the addition of substantial proportions of inorganic or organic ionic solids to AgI or CuI produces substantially enhanced room temperature ionic conductivity. Examples include electrolytes such as Ag_2HgI_4, Ag_4RbI_5, $Ag_7[NMe_4]I_8$ and pentacupro-(4-methyl-1,4-oxathianium)hexaiodide.[4]

The third type of solid electrolyte, involving *amorphous region transport*, includes many glassy and virtually all polymeric electrolyte systems, and will be discussed in terms of the latter.

4.2. Polymeric Electrolytes

As discussed in Chapter 2, ionic motion takes place in the amorphous, rather than in the crystalline, phase. It is clear that a defect-oriented conduction mechanism is inappropriate for such materials and it becomes a moot point as to whether a disordered sub-lattice model, or a liquid phase conduction model, is more appropriate.

Polymeric electrolytes were developed as a result of the pioneering work of Wright[5] on complexes of poly(ethylene oxide) (P(EO)) with alkali metal thiocyanates and iodides. This was followed by Armand's[6] development of P(EO):LiClO$_4$ electrolytes for batteries using intercalation electrodes. As discussed in Chapter 3, present-day interest is largely focused on complexes of a variety of polymers with lithium salts, although alternative systems based on complexes of P(EO) with divalent metal salts[7-9] also appear promising for certain applications.

Two features of the elecrochemical behaviour of polymeric electrolytes are worthy of special note. The first is that the cationic transport number is substantially less than unity in many cases,[10] showing that the anions also participate in the charge carrying process. This is in contrast with the majority of solid electrolytes, in which ionic charge is solely carried either by anions or by cations. The second is that the impedance spectrum is different from that of other types of solid or liquid electrolyte; this is discussed in more detail in Section 5.1.

5. MEASUREMENT OF CONDUCTIVITY

In electrochemistry, conductivity is a very important property, especially for electrolytes. It is not possible to measure the conductivity of an isolated electrolyte. Instead, a test cell has to be used, in which the electrolyte is sandwiched between a pair of contacting electrodes. The property that is then measured pertains to the whole assembly, i.e. to the electrolyte plus the attendant instrument leads, electrodes and interphases.

The electrodes can be *blocking,* i.e. incapable of acting as a source or sink for the ions that traverse the electrolyte, or *non-blocking,* in which case ions can cross the interphases. For blocking electrodes, the current is carried by electrons along the leads to and from the instrument into the electrodes; through the electrolyte, the charge is carried by ions. The link at the interphases between the two types of separated charge carrier is the *double layer capacitance.*

In practice, direct current measurements are rarely used because the influence of a constant applied potential sets up a concentration gradient of ions. When equilibrium is reached, the chemical potential caused by this gradient is exactly sufficient to nullify the effect of the applied voltage and no current flows. The test cell is then said to be *concentration polarised.*

Two types of a.c. techniques are normally employed. Most older studies used a *constant frequency* technique to overcome the effect of concentration polarisation. The problem with this approach is that there is no real means of separating the contributions of the electrolyte from those of the other components in the test cell. In consequence, it is now more common to carry out measurements at *variable frequency*.

The current passing through the test cell and the voltage across it are both measured as a function of frequency. Since the test cell contains capacitative (and perhaps also inductive) components as well as resistances, the *impedance, Z,* rather than the resistance, *R,* is obtained. The impedance is given by

$$Z = R - \frac{j}{\omega C} + j\omega L$$

where C is the capacitance and L is the inductance, ω is the angular frequency and j is the square root of -1. Because complex numbers are involved, a convenient way to present the data is by means of a *complex plane representation*.

5.1. The Complex Plane Representation

In test cells involving polymeric electrolytes it is often possible to ignore the inductive components and to represent the data in the form of a plot of *real* (x-axis) against *imaginary* (y-axis) terms, where the real contributions arise from the resistances and the imaginary terms from the capacitances. A plot of the form shown in Fig. 3 is usually found. By comparing the plot obtained experimentally with that produced by calculation of the frequency dependence of the real and imaginary contributions of an *equivalent circuit,* it is possible to deconvolute the electrolyte characteristics from the rest of the test cell components.

Although it is possible to explain a plot consisting of a semi-circle followed by a spike in terms of the simple equivalent circuit shown in Fig. 4, to account for the flattening of the semi-circle and the tilting of the spike requires the use of *constant phase elements*.[11] Physically, these terms replace the capacitors in the circuit; they can be thought of as having characteristics intermediate between a capacitor and a resistor, i.e. a 'leaky' capacitor. The full mathematical representation of these would be out of place here but, as with the simple circuit, the electrolyte resistance can be obtained by extrapolating the spike to the point of intersection with the x-axis. The electrolyte conductivity is then obtained from the measured electrolyte resistance by means of the measured cell constant.

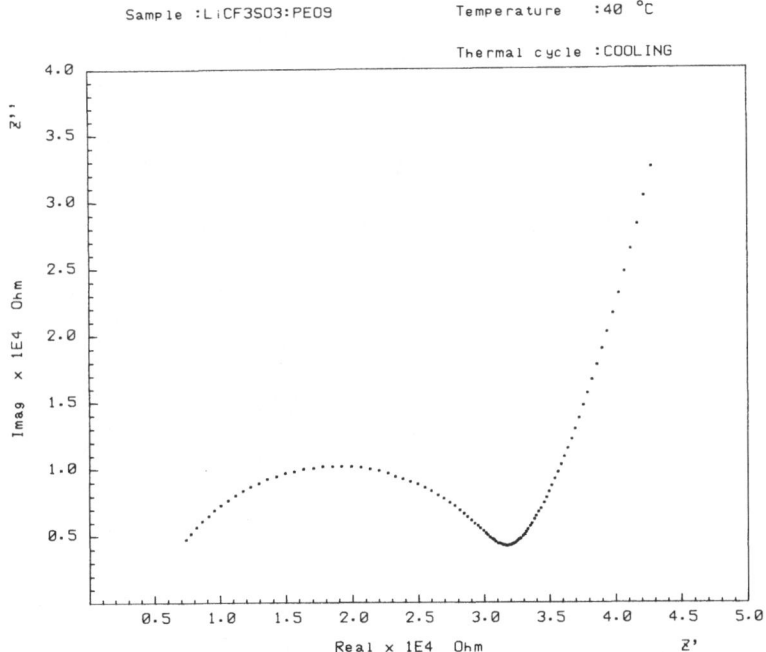

Fig. 3. Typical impedance plot for a polymeric electrolyte.

5.2. Electronic Conductivity

It is important for electrolytes that their electronic conductivity, σ_{el}, is small, and in practice this is usually the case for polymeric electrolytes. The value of σ_{el} can be measured by two methods.

The first is to construct a *Wagner polarisation cell* with one blocking and one non-blocking electrode. An external voltage source is then connected, the polarity being chosen so that conducting ions are made to flow from the end of the cell containing the blocking electrode. The electrode cannot act as a source for these ions and so the region near the blocking electrode interphase becomes severely depleted of ions; concentration polarisation therefore offsets the effect of the applied emf. After equilibrium is reached (typically two days), any residual current flowing can be ascribed to electronic rather than ionic carriers. This method depends on the assumption that the ionic current is

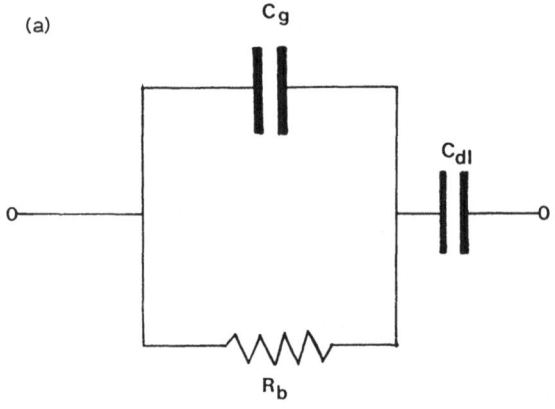

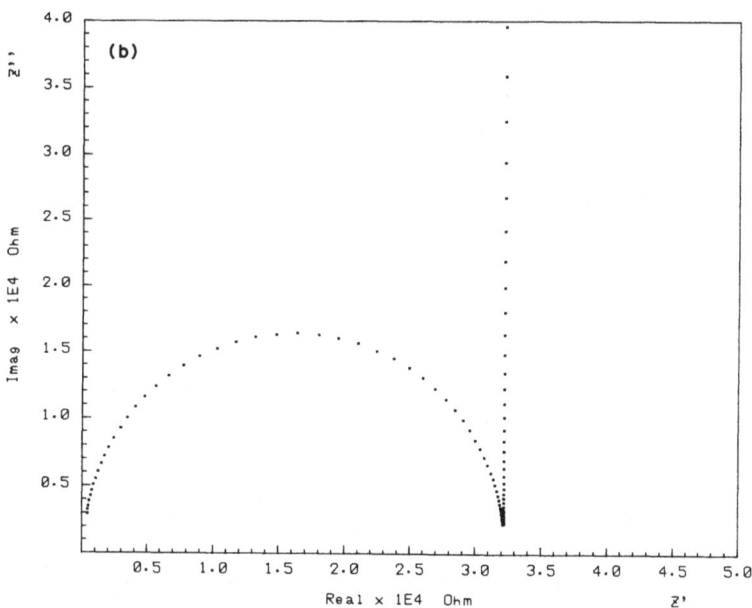

Fig. 4. Simple equivalent circuit (*a*) that produces an impedance plot (*b*) of a semi-circle and a spike.

carried solely by one species, which makes it rather unsuitable for measurements on polymeric electrolytes.

The second method is to measure the self-discharge characteristics of a working cell containing the chosen electrolyte. The cell, which is made in a form such that there is very little material present in one of the electrodes (*anode* or *cathode limited*), is maintained under infinite external load until the galvanic voltage produced descends to a small fraction of its original value. It is then possible to calculate the value of σ_{el} from the cell current and voltage as a function of time.[12]

REFERENCES

1. P. J. Fydelor and K. V. Lovell, Royal Military College of Science, Shrivenham, Physics Dept, Technical Note PD/30/82, 1982.
2. R. G. Linford and S. Hackwood, *Chem. Rev.*, 1981, **81**, 327.
3. M. O'Keeffe and B. G. Hyde, *Phil. Mag.*, 1976, **33**, 219.
4. C. Johnson, R. J. Latham and R. G. Linford, *Solid State Ionics*, 1982, **7**, 331.
5. D. E. Fenton, J. M. Parker and P. V. Wright, *Polymer*, 1973, **14**, 589.
6. M. B. Armand, *Solid State Ionics*, 1983, **9/10**, 745.
7. L.-L. Yang, R. Huq, G. C. Farrington and G. Chiodelli, *Solid State Ionics*, 1986, **18/19**, 291.
8. A. J. Patrick, M. D. Glasse, R. J. Latham and R. G. Linford, *Solid State Ionics*, 1986, **18/19**, 1063.
9. T. M. A. Abrantes, L. J. Alcacer and C. A. C. Sequeira, *Solid State Ionics*, 1986, **18/19**, 315.
10. P. R. Sørensen and T. Jacobsen, *Electrochemica Acta*, 1982, **27**, 1671.
11. P. H. Bottelberghs, in *Solid Electrolytes. General Principles, Characterisation, Materials, Applications*, ed. P. Hagenmuller and W. Van Gool, Academic Press, New York, 1978, p. 145.
12. C. Johnson, A. M. P. Jelfs, R. J. Latham and R. G. Linford, *J. Power Sources*, 1982, **13**, 115.

CHAPTER 2

Polymer Structure and Conductivity

MALCOLM D. GLASSE and ROGER G. LINFORD
School of Chemistry, Leicester Polytechnic, UK

1. INTRODUCTION

The earliest electrochemical devices contained both solid and liquid conductors. Both of these condensed phases have had a role to play ever since, and both have been called upon, under appropriate circumstances, to be electronic, ionic or mixed conductors. Polymers may behave as liquids or solids, or as composites of both of these phases, depending on their nature. The structure of polymers is necessarily more complicated than that of solids or liquids composed of small molecules. However, despite the fact that polymers have only been recognised as such for the last 50 years, the understanding of their behaviour is now sufficient to allow their properties to be related to their structure.

The term *polymer* describes many natural and synthetic materials. A broad definition will include any material which consists of large molecules. These *macromolecules* consist of many units, each being the size of a conventional small molecule. These units are linked to each other by conventional chemical bonds. This definition allows us to divide the properties of such materials into two categories:

(i) Those properties which are independent of molecular size. They therefore depend on the nature of the atoms present and their bonding. They are entirely predictable without any need for a knowledge of polymer science.
(ii) Those properties which are created or affected by the large molecular size.

The latter are our concern here. Most of them centre around the visco-elastic, or semi-solid, behaviour of polymers. However the consequences of this behaviour lead to a wide variety of structures and properties.

The polymeric materials to be considered here are principally those based upon chain structures. In the liquid state the chains are randomly coiled and, even when apparently solid, at least part of the material is *amorphous*, i.e. the randomness persists. This therefore excludes from our scope substances which are entirely crystalline, such as crystalline proteins, and also inorganic polymeric network structures such as zeolites which have symmetrically arranged units with bonding in three dimensions. The latter category includes many materials which are useful as solid electrolytes and intercalation electrodes, but these are not classed as polymeric for the purposes of this book.

2. MACROMOLECULAR STRUCTURE

In a polymer chain, the individual units are often identical to each other, although not necessarily so. They are also usually divalent, having one linkage at each end. A chain of divalent units is called a *linear polymer*, where *linear* does not imply that the line should be straight (Fig. 1a). The units at the ends of any polymer chain must be somewhat different from those in the centre; but such differences can usually be ignored unless specific chemical groupings, *end-groups*, have been introduced to create particular chemical effects. The average number of units in the polymer chain is known as the *degree of polymerisation*. The length of the chains is usually measured and reported as the average *relative molar mass* (*rmm*), previously called *molecular weight*, which will be proportional to the degree of polymerisation and to the relative molar mass of the units.

Degrees of polymerisation may range from 10 to 10^6 with relative molar mass values being about 100 times these figures. They tend to be used in theoretical studies, being derived from a knowledge of the particular polymerisation process, or calculated from the relative molar mass. The molar mass can be determined by several experimental techniques, some (e.g. osmometry) being absolute whereas others (e.g. viscometry) are empirical. The results are averages. For some techniques the average is the conventional arithmetic mean, and the result is known as the *number-average rmm*. Other techniques yield a

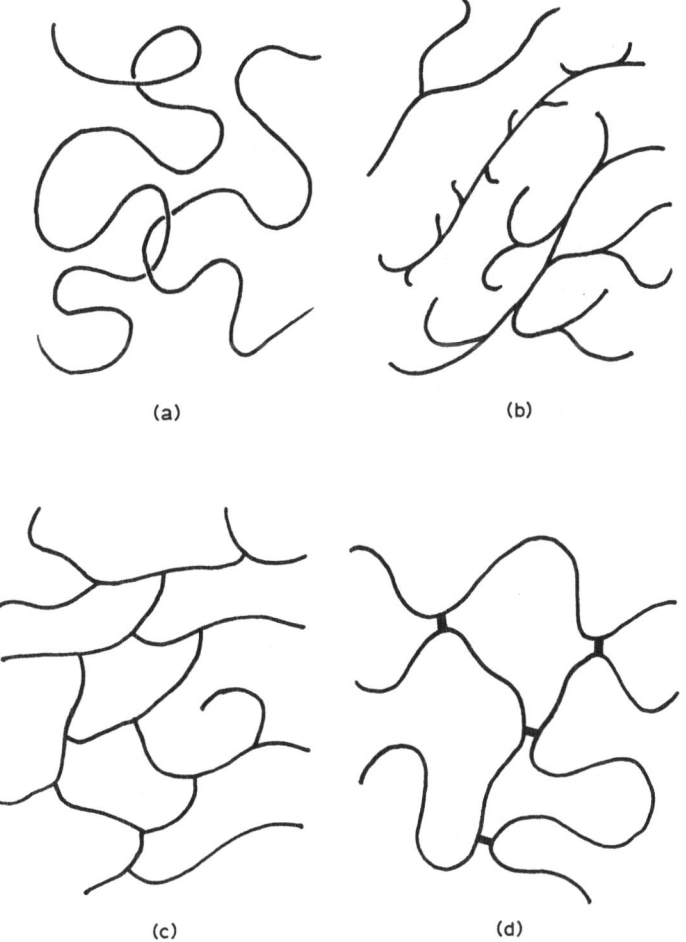

FIG. 1. Macromolecular architecture of polymers: (*a*) linear; (*b*) branched; (*c*) network; (*d*) crosslinked.

higher *weight-average rmm*. The difference between these arises because polymerisation methods often lead to a very wide spread of molecular sizes, covering several orders of magnitude.

Shorter chains may be referred to as *oligomers*. The individual units are sometimes called *monomer units,* but the term *structural repeat unit* (*sru*) is now preferred as the units often differ from the parent monomer. An older term for a structural repeat unit is a *mer*. The nature of the end group is likely to be more significant for shorter chains.

An occasional tri- or tetra-valent unit in a polymer chain will produce a *branched polymer* (Fig. 1b). The presence of many such units will lead to *networks* (Fig. 1c). These can also be created by *crosslinking* linear polymers (Fig. 1d). Linear oligomers with reactive end-groups may be crosslinked by added tri- or tetra-functional reagents. Longer chains require reactive sites, such as the double bonds in natural rubber, or units added during copolymerisation. Polymers without apparent reactivity can often be crosslinked by high-energy gamma rays or neutron radiation.

The structure of a polymer often reflects the chemical mechanism involved in its formation. Most polymerisation processes fall into one of two categories: *step-growth* or *chain-addition*.

Step-growth polymerisation exploits condensation reactions in which two molecules become linked across mutually interacting groups, represented by x and y in Fig. 2a. Each such linking represents one step in the process. The monomer molecules must have two such

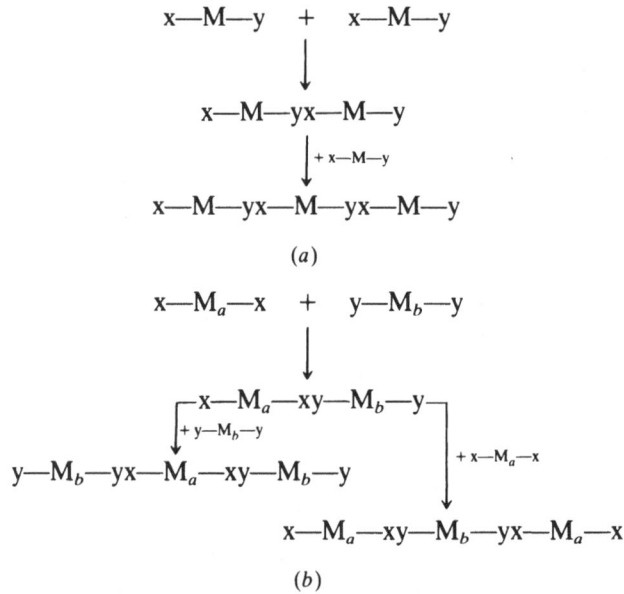

FIG. 2. Step-growth polymerisation: (a) one-monomer type; (b) two-monomer type.

groups, either different (as in x—M—y) or the same (when two monomers are required, x—M_a—x and y—M_b—y as in Fig. 2b). Successive steps in the process build up dimers, trimers, tetramers and so on, any size reacting with any other size according to availability.

The links, which remain in the chain as evidence of the step-growth mechanism, will usually be esters (—CO—O—), amides (—CO—NH—) or urethanes (—NH—CO—O—). It is very difficult to produce very long chains by this process, but the end-unit will usually retain the potential to react further.

In chain-addition polymerisation the word 'chain' refers to the process and *not* to the product. A *chain reaction* regenerates the active species with each addition. The active species, such as a free radical, needs to be generated slowly by *initiation* (Fig. 3). The *propagation* usually takes only a few seconds to add many thousands of units; *termination* destroys the reactivity irretrievably. Like a catalyst, the initiator is required only in small amounts but, unlike a catalyst, it is consumed in the process. The total process may take some hours with the monomer in bulk, or solution, or in an emulsion. However, the rapid growth of individual chains means that a few full-sized chains are produced from the beginning and chain length is a function of reaction conditions rather than time.

For chain-addition polymerisation, the monomer molecules may contain either a double bond or a ring. This is opened out to produce the divalent structural repeat unit and so the chain produced does not

$$I \rightarrow 2R^* \qquad (I)$$

$$R^* + M \rightarrow R—M^* \qquad (I)$$

$$R—M^* + M \rightarrow R—M—M^* \qquad (P)$$

$$R—M—M^* + M \rightarrow R—M—M—M^* \qquad (P)$$

$$\text{and so on} \qquad (P)$$

until

$$R—M_i^* + {}^*M_j—R \rightarrow R—M_{i+j}—R \qquad (T)$$

FIG. 3. Reaction scheme for chain-addition polymerisation. The symbol * indicates that the species is a free radical or else an anion or cation. The initiation steps are shown by I, the propagations by P and the termination by T.

contain those links which are characteristic of step-growth polymerisation. The chains can also be much longer and the nature of the end-groups can usually be ignored.

Ethylene oxide and other cyclic monomers may be polymerised by chain-addition polymerisation and the relative molar mass of the product is typically of the order of 10^5 or 10^6. This type of monomer may also undergo a polymerisation which is related to the alternative step-growth mechanism. This process is initiated by an alcohol and catalysed by acids or bases. At each step, another ring is opened and linked to the chain, leaving a terminal —OH group. The ends remain active, but only to monomer, not to each other. Two such ends may be obtained from a dihydric alcohol initiator. The shorter chains (maximum relative molar mass $\sim 10^4$) obtained by this mechanism are often called poly(ethylene glycol)s, to draw attention to their end-groups.

The necessary changes in bonding that accompany polymerisation sometimes lead to two or more stereochemical (or structural) possibilities in the repeat units. If only one such possibility is present in a sample polymer, it is said to be *stereoregular*. If two or more structures are incorporated into the chain at random, then it will not be regular.

A parallel situation exists if two or more monomers are blended before being polymerised. It often happens that their units are incorporated at random into the chain. The product is then a *random copolymer* (Fig. 4a), the prefix 'co' indicating more than one type of unit in the chain. A different type of polymerisation may cause monomer molecules to combine only with others of a different type; this then yields an *alternating copolymer*. It can be seen from Fig. 4(b)

∿∿ABAABABBBAABA∿∿ (a)

∿∿ABABABABABABABA∿∿ (b)

AAA∿∿AAABBBB∿∿BBBAAAA∿∿AAAA (c)

∿∿AAAAA∿∿AAAAA∿∿AAAAAA∿∿ (d)
    B   B
     B   B
      B   BBB∿∿BBBB
    B
   B
  B
BBBB∿∿BBB

FIG. 4. Copolymer arrangements: (a) random; (b) alternating; (c) block; (d) graft.

that such a material may be regarded as a polymer of the single unit —AB—. Special synthetic processes may be used to build *block* (Fig. 4c) and *graft* copolymers (Fig. 4d). When it is necessary to emphasise that a polymer contains only one type of unit, the term *homopolymer* may be used.

Stereoregular homopolymers and alternating copolymers are potentially crystallisable. Irregular homopolymers and random, block and graft copolymers normally lose this capability and remain amorphous.

Copolymers tend to have properties which are intermediate between those of the two 'parent' homopolymers. Blends of homopolymers and also block and graft copolymers usually exhibit some phase separation at the microscopic level. Their properties therefore tend to be a combination of those of the parents, rather than a compromise or average.

3. AMORPHOUS POLYMERS

3.1. Flexibility and Flow

Polymers which are not crystalline, i.e. which do not show long-range order, are described as *amorphous*. In this respect they are liquid-like, but they do not behave as normal free-flowing fluids because the large molecular size reduces flow and creates visco-elastic behaviour. Amorphous regions are present in almost all polymers. Some are incapable of crystallisation in any case; in samples of others the crystallisation process may be retarded or inhibited. Even those polymers that can crystallise readily never do so completely. In this section, attention will be focused on polymers that are entirely amorphous, although many of the comments apply equally to the amorphous component of semi-crystalline polymers.

At high temperatures most linear polymers are flexible. At the molecular level the flexibility arises from rotation about the single bonds in the polymer backbone (Fig. 5). Molecular flexibility confers bulk flexibility. If forces are applied upon two adjacent chains, but in opposite directions, then the random thermal motions of sections of the chains will eventually permit wholesale translational motion. Such movement is a prerequisite of flow.

Flow may occur in polymers which either have short chains or which have sufficient thermal motion. Flow is therefore more apparent at high temperatures. Very long chains become heavily entangled and

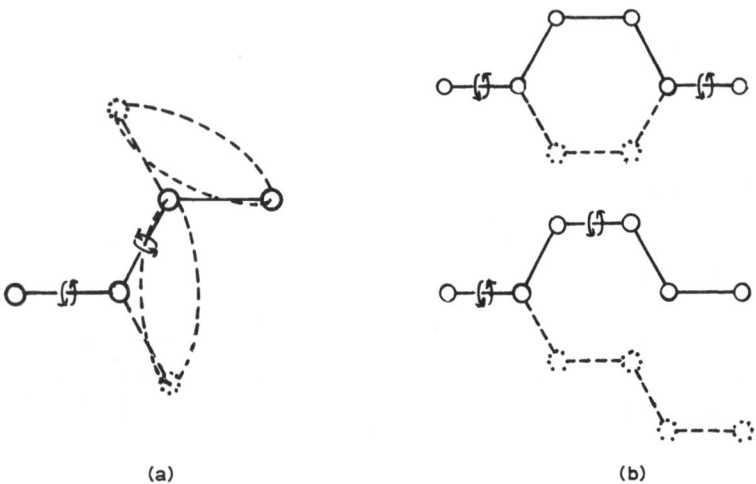

Fig. 5. (a) Bond rotation within a polymer backbone. (b) Conformations which may result.

indeed can become effectively knotted. Such polymers behave as though they have been lightly crosslinked. They can be deformed but they do not flow perceptibly in the short term. Slow, long-term flow is known as *creep*. The long sections of chain between entanglements or crosslinks adopt a statistically random conformation because this maximises entropy. When any deforming force is removed, restoration of this random state will cause an overall return to the original shape in the bulk material. This is the origin of *rubber-like* behaviour.

Liquid-like and rubber-like behaviour are therefore related. They represent the two extremes of visco-elastic behaviour, both being caused by the motion of short segments of chain. Intermediate behaviour is often undesirable; few serious uses have been found for 'bouncing putty', a material that is viscous under sustained load but elastic under rapid deformation. Rubber-like, flexible, viscous amorphous polymers can be considered as analogous to liquids for many purposes, including aspects of ionic conduction.

3.2. The Glass Transition

Viscosity increases with chain length. As fluid short-chain polymers are cooled they become more viscous, until they will no longer flow under gravity, but only under greater applied stresses. They behave more like putty or chewing gum.

Longer-chain polymers are much less fluid. They are *elastic*, i.e. their strain depends on applied stress; if first deformed by an applied stress, they will instantaneously recover their original shape and size when the stress is removed. As elastic long-chain polymers are cooled, the speed of recovery from deformation becomes slow and they are said to become *leathery*. This leathery state is similar to the putty-like state achieved in the cooled short-chain polymers; properties in this state hardly depend upon chain length. In general therefore, if a particular sample or type of amorphous polymer does not crystallise on cooling, it will eventually reach a leathery state.

The increasing stiffness on cooling reaches a limit at which the amorphous polymer becomes rigid. In this state it is *glassy* and usually brittle. Short sections of polymer chains will be capable of vibration and perhaps more adventurous oscillation, but they will not be able to move far enough to displace other similar sections.

The temperature below which such liquid-like cooperative movement becomes impossible is known as the *glass transition temperature*, T_g. Below this temperature physical properties such as density, viscosity, diffusion and conduction become less sensitive to temperature. Above the glass transition, properties which depend on molecular motion change more quickly.

The glass transition is more like a second-order than a first-order transition, in the Ehrenfest classification of phase transitions. This means that there is no discontinuity in entropy, enthalpy or volume at the transition temperature, but that there is a change in such properties as the heat capacity and compressibility. Experimentally T_g manifests itself as a small step or a change in gradient of a differential scanning calorimetry (DSC) trace. It can also be seen as a more substantial feature in a thermomechanical analysis (TMA) trace. This is a technique in which the degree of penetration of a weighted probe into the sample is monitored as a function of temperature.

The value of T_g depends strongly on the nature of the structural repeat units. Unhindered rotation between backbone atoms leads to a low T_g but steric hindrance by side groups and/or polar interactions can raise its value. For applications such as plastic glazing or students' rulers, T_g must be well above room temperature, whereas for rubbers and silicone fluids it must be well below the operating temperature. As discussed in detail in Chapter 3, a low T_g is desirable in polymeric solid

TABLE 1
Glass Transition Temperatures of some Common Polymers

Polymer	Symbol	sru	T_g (°C)
Poly(ethylene oxide)	P(EO)	⌇CH$_2$—CH$_2$—O⌇	−65
Natural rubber	NR	⌇CH$_2$—C(CH$_3$)=CH—CH$_2$⌇	−73
Polythene (polyethylene)	PE	⌇CH$_2$—CH$_2$—⌇	−125
Polystyrene	PS	⌇CH$_2$—CH(C$_6$H$_5$)⌇	+100
Poly(methyl-methacrylate) ('Perspex'/'Oroglass' /'Plexiglass')	PMMA	⌇CH$_2$—C(CH$_3$)(CO·O·CH$_3$)⌇	+105

electrolytes. Some typical values are listed in Table 1. The addition of salt to a polymer such as poly(ethylene oxide) [P(EO)] to form a polymeric solid electrolyte usually raises the T_g value substantially.[1,2]

4. POLYMER CRYSTALLINITY

4.1. Crystallisability

Linear polymers are crystallisable if all the structural units are identical. This will be inevitable for homopolymers such as polyethylene and poly(ethylene oxide) for which the structural repeat units are simple. Alternatively, the structural repeat units may contain a chiral centre, i.e. an asymmetric atom. In this case, the configurations about this centre will determine whether the polymer is stereoregular and hence whether its units may pack together in a regular crystalline fashion.

With a polymer such as polypropylene, one of the carbon atoms in the structural repeat unit is asymmetric, one out-of-chain bond being to hydrogen and the other to a methyl group. As shown in Fig. 6, the polymer is in the *isotactic* form when all the carbon atoms bearing the methyl groups are of the same chirality (i.e. for the conformation shown in Fig. 6a, all the methyl groups are on the same side of the chain), *syndiotactic* when they regularly alternate and *atactic* otherwise. Atactic material is not crystallisable, because of the irregular shape of the chain. It remains as an amorphous, deformable (i.e. *plastic*) mass. Atactic polypropylene has no obvious uses, being too close at room temperature to its T_g to have the properties of a

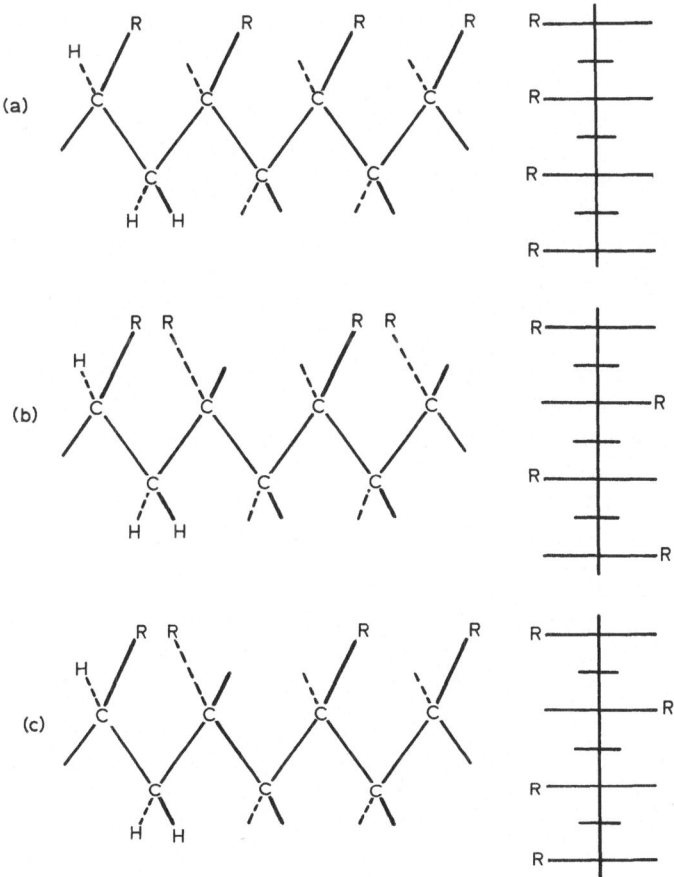

FIG. 6. Tacticity: (a) isotactic; (b) syndiotactic; (c) atactic. The projections on the right-hand side of the figure are viewed from above.

rubber. The commercially available form of polypropylene is the semi-crystalline isotactic form.

Poly(propylene oxide) has the potential for structures similar to those possible with polypropylene. The ether oxygen in the chain confers more flexibility than in the hydrocarbon analogue and hence lowers the T_g compared with polypropylene. The commercially available material is atactic. The low T_g makes it a suitable candidate for combination with salts to form polymeric solid electrolytes but the relatively large methyl side-groups may impede ionic motion.

Other types of isomerism may also be created during polymerisation. When conjugated dienes are polymerised they may produce *cis* or *trans* structures or a mixture of both. Some monomers have two different 'ends' to the molecule, either of which may be capable of adding to a growing polymer chain. In practice, the structural differences usually produce energetic factors which favour one mode of addition only.

Hot polymers are viscous fluids in which the backbone bonds may rotate and adopt a variety of conformations. As they cool, the lower-energy positions come to be preferred. Stereoregular chains will then favour regular zigzag or helical shapes. At the same time, these regular sections of chain may pack together to form regions of crystallinity.

This can only happen below a theoretical *melting point*, T_m. The value of the theoretical T_m may not be detectable in practice. Structural defects lower the observed melting point and supercooling can also occur, causing non-equilibrium crystallisation.

4.2. Crystallinity

The simple *fringed micelle* model of polymer crystallinity is presented in Fig. 7. Here straight lines represent the zigzag or helical sections of chain. This model has now been superseded, but it includes some useful features and explains several phenomena:

(i) The 'micelles' are crystalline regions which are smaller than the polymer chains. One chain may pass through several micelles, and/or re-enter the same one.

(ii) Sections of chain become trapped in uncrystallisable positions between the micelles. A certain amorphous content is therefore inevitable.

(iii) Around each micelle is a 'fringe', or region of doubtful order which might be assignable as either amorphous or crystalline depending on the method of study.

This model is compatible with several observations and is likely to be adequate to interpret the properties of some polymeric electrolytes. X-ray diffraction patterns of crystalline polymers show:

(a) a ring pattern, similar to that characteristic of crystalline powders, which arises from the randomly oriented crystalline regions and is superimposed on:

(b) a diffuse halo or haloes characteristic of liquids.

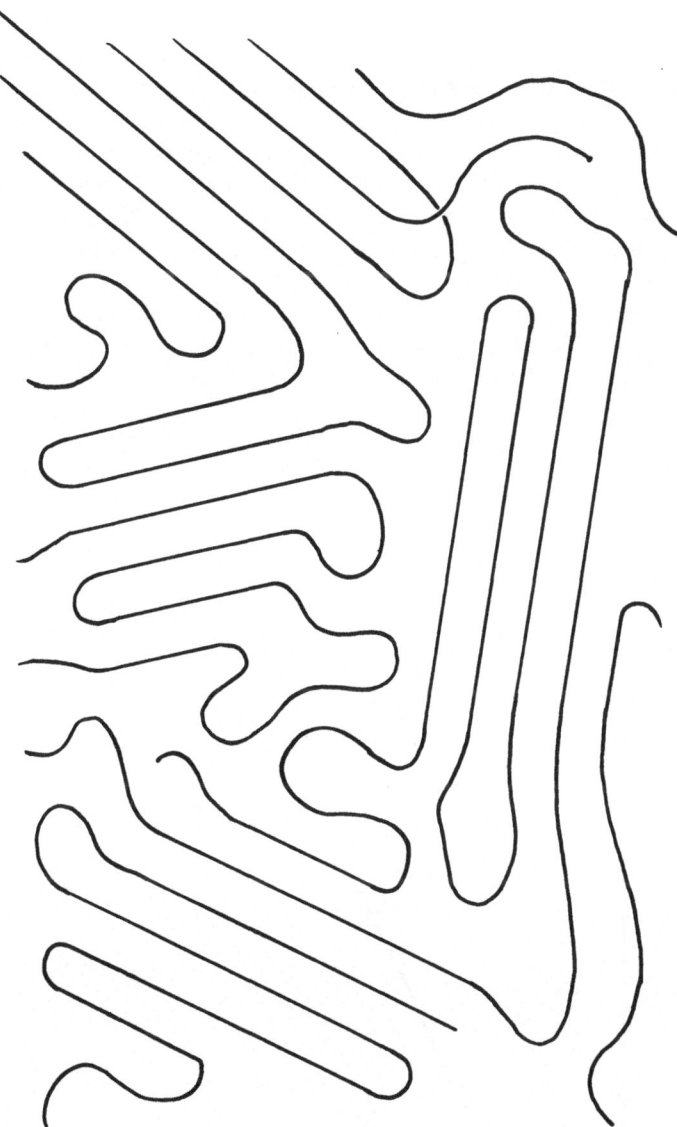

FIG. 7. The fringed micelle model of polymer crystallinity.

Measurement of the amounts of scattering of X-rays from these two regions can lead to a value for the *extent of crystallinity*, χ.

Densities of semi-crystalline polymers lie between the amorphous and crystalline densities. The former is obtained by extrapolating the liquid density, as a function of temperature, to below the T_m value of the crystalline region. The latter is obtained from X-ray and unit-cell data. It is therefore possible to calculate the extent of crystallinity from density measurements. Any difference from the value of χ obtained from X-ray measurements is said, on the basis of the model, to be caused by the intermediate density fringes, which appear as amorphous to X-rays.

Differential scanning calorimetry shows that the melting is spread over a wide temperature range. At high resolution, the same technique may reveal the change in heat capacity that accompanies the glass transition. It is sometimes possible to measure the degree of crystallinity by comparing the melting peak area obtained from the sample with that obtained from a sample of the same material for which χ is known.

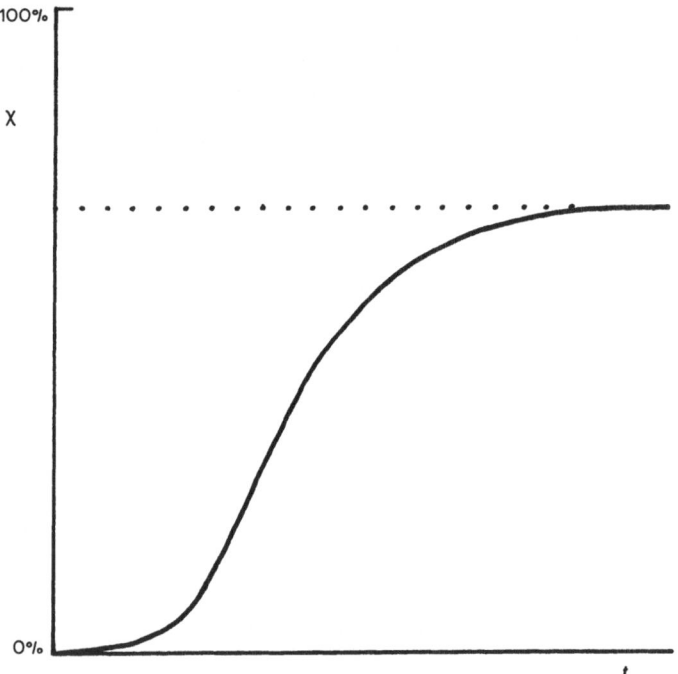

FIG. 8. The variation of the degree of crystallanity, χ, as a function of time, t.

4.3. Crystallisation

Crystallisation is accompanied by contraction. An early technique in the study of polymer crystallisation was dilatometry. The sample is surrounded by a fluid. The volume change of the sample plus fluid that accompanies crystallisation is monitored by measuring the length of the fluid column in an attached capillary tube. The polymer is normally melted first and then cooled and thermostatted. This technique reveals that the degree of crystallinity develops sigmoidally with time (Fig. 8).

The observed acceleration indicates the need for crystalline material to grow on pre-existing similar material or to be nucleated. This requirement is not inherent in the fringed micelle model and led to the development of the spherulite model discussed below.

Dilatometry also indicates the temperature dependence of crystallisation rate, which is at a maximum between T_g and T_m, falling to zero at these points (Fig. 9).

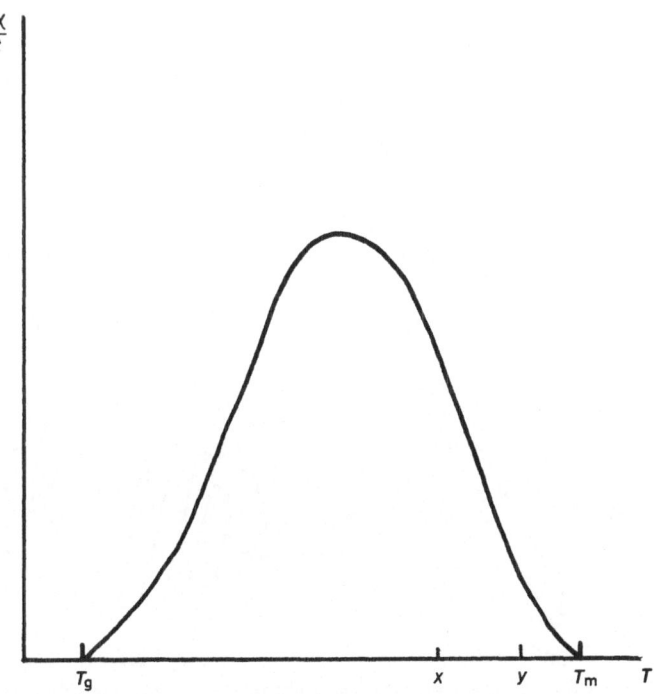

FIG. 9. The variation of the crystallisation rate, $d\chi/dt$, as a function of temperature, T.

Studies such as these show that the extent of crystallinity, and the temperature range over which the crystalline material melts, can be affected by the treatment of the material. The *thermal history* of a sample of polymer may affect its properties. For example, if room temperature is near T_g, then annealing at point x on Fig. 9 will give a high degree of crystallinity, whilst annealing at y will give a higher and narrower melting range and perhaps better thermal stability. Alignment of polymer chains by stretching a sample may enhance the crystallinity. This principle is exploited in the drawing of textile fibres.

5. SPHERULITES

Microscopic examination of crystalline polymers indicates that the fringed micelle model is an oversimplification. Optical microscopy and scanning electron microscopy (SEM) both reveal roughly spherical semi-crystalline regions. They are usually seen in section as discs which grow in area as crystallisation proceeds. Under special laboratory conditions (extremely dilute and hot solutions) SEM shows that small single crystals can be grown. These have no direct practical application, but their structure (Fig. 10) has provided evidence of chain

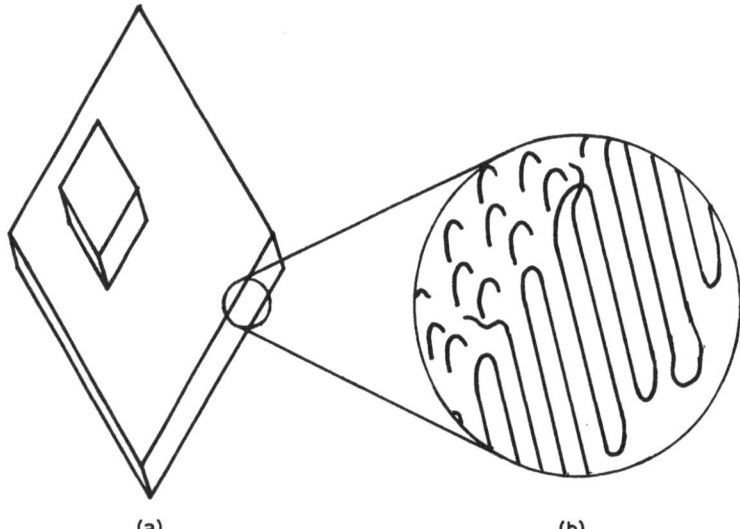

FIG. 10. A polymer single crystal: (*a*) overall view; (*b*) magnified portion of lamella to show chain folding.

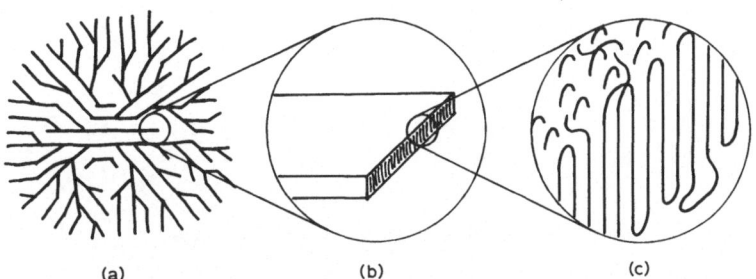

FIG. 11. A spherulite: (a) overall view; (b) magnified view of a lamella; (c) highly magnified view of folded chains forming the lamella.

folding to produce *lamellae*, which are of significance. The ostensibly straight (i.e. zigzag or helical) chains pass through the thickness of the lamellae, fold at the surface and return through them. There may be irregularities at the surface, and multiple lamellae may grow upon each other. The thickness is controlled by the temperature of growth, being of the order of 10 nm.

Similar lamellae are detectable in *spherulites* grown from melted bulk polymers. They take the form of *fibrils* which radiate from the centre, branching to fill most of the available space. On their surfaces and between the fibrils, the amorphous component is trapped (Fig. 11). Under SEM at high magnification, lamellar fibrils are seen on the film surface, or after fracture. This also reveals the 'wheatsheaf' appearance at the centre from which they grow.

Optical microscopy is usually performed with a thin film viewed between crossed polarizers. Amorphous material appears black, but the spherulites can be seen to grow from spots of light at their nuclei into discs. The polarisers cause a spherulite to appear as a pattern of light and dark regions. The pattern often takes the form of a 'Maltese Cross' (Fig. 12), centred on the centre of the spherulite. The orientation of the cross is a function of the position of the polarisers, not the sample.

A hot-stage on the microscope permits potentially valuable studies, for example of the effect of annealing on polymer structure. Annealing may be carried out deliberately, to change the crystal morphology, extent of crystallinity, or melting range. Annealing or quenching may also occur accidentally if a polymer is heated during service. In either case, this may change the conduction properties of polymeric electrolytes made from polymer–salt complexes.[3,4]

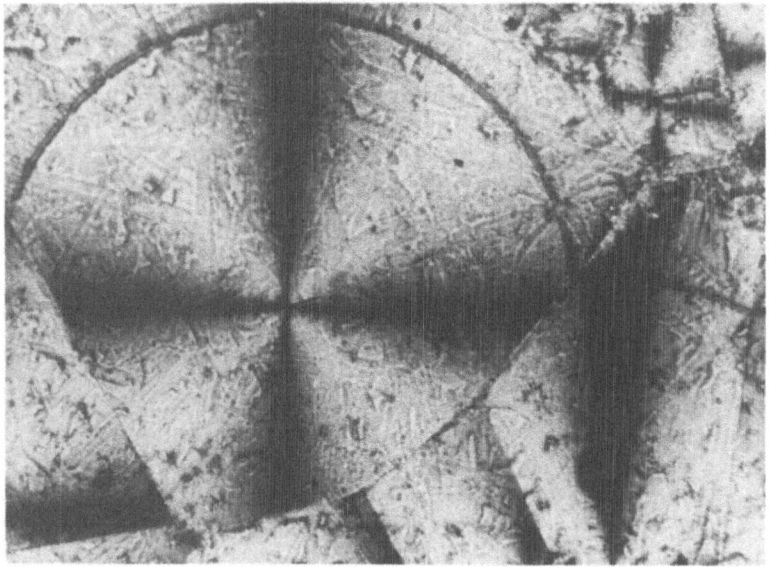

FIG. 12. Maltese Cross pattern obtained by viewing a thin cast film of P(EO)$_7$:LiClO$_4$ through crossed polarisers.

6. IMPLICATIONS

The breakthrough by Armand[5] in 1978, discussed in Chapter 3, had immediate implications for structure–property relationships. Ionically conducting complexes of salts with poly(ethylene oxide) were seen as analogous to crown ether/ionic salt adducts.[6,7] Polymer helices were thought to coordinate cations down their centres in the way that crown ethers coordinate at the centre of their rings. As the complexes were clearly at least partially crystalline, a solid state model was adopted, and it was supposed that these helical paths were responsible for the cation conduction.

Figures 7 and 11 have shown that polymer scientists adopt a two-phase model of polymer crystallinity. For them, the question as to which phase was responsible for ionic transport remained an open one.

The increase in conductivity which accompanies melting tends to indicate that ionic motion is facile in the amorphous, rather than the crystalline, phase. The melting process in a polymer may be spread over a not inconsiderable temperature range which makes the practice of quoting activation energies in the region of the melting point a suspect one. In some materials, such as $P(EO)_n$; $LiClO_4$ when $n \neq 7$, more than one crystalline phase is experimentally observed and more than one melting point is revealed by DSC studies.[3,4,8] The crystalline region can be thought of as 'locking up' some of the ionic carriers, and consequently the optimum conductivity is not obtained until the material is heated above its highest melting point.

The recognition that ionic conduction takes place in the amorphous phase makes crosslinking an attractive way of enhancing conductivity. If crosslinks are introduced into a molten polymer, they will affect the subsequent recrystallisation in two ways:

(a) They will inhibit the movement necessary for polymer chains to organise themselves into crystalline structures.
(b) The links themselves will be structurally different from other structural repeat units. They will therefore behave as impurities which cannot fit into the crystal lattice.

For both of these reasons the crystallinity will be reduced or even eliminated. The resulting polymer, lacking the stiffness imparted by that crystallinity, would be more likely to flow; but here the crosslinking helps by adding to the structural integrity.

The crystalline component in crystalline polymers provides sufficient rigidity for them to be classified as 'solid' electrolytes. The amorphous component both provides the conducting phase and imparts sufficient flexibility for the composite material to accommodate the volume changes which may occur in electrode materials during discharge. There may be the need, and the possibility, to adjust the crystalline content, as suggested at the end of Section 5. This makes it necessary to have a means of assessing such changes.

Differential scanning calorimetry has frequently been used to study polymer crystallinity, especially in polymer–salt complexes. This technique has shown that melting endotherms can sometimes be shifted up or down in temperature, or changed in size, by annealing, quenching or other changes to a cooling profile. If similar experiments are carried out under a polarising microscope, this type of information can be further enriched.

The implications of the Gibbs phase rule are often ignored when considering polymer morphology. For a one-component system under constant pressure, two true condensed phases can only coexist at a single temperature, i.e. the melting point that pertains to that pressure. It is found in practice that amorphous and crystalline regions co-exist over a wide temperature range and this implies that the amorphous region consists of material trapped in a metastable, but virtually permanent, state within the spherulites.

A P(EO) salt complex contains two components and consequently two condensed phases may simultaneously be present. For $P(EO)_n$:$LiClO_4$ for example,[3] two crystalline regions, one melting at ~60°C and the other at >100°C, are observed over the composition range $4 \leq n \leq 7$, in conjunction with amorphous material. The two crystalline regions can be seen as two visually distinguishable spherulite types under the polarising microscope, and SEM/energy dispersive analysis of X-rays (EDAX) studies show the salt contents of the two types to be substantially different. This means that a polymer electrolyte film is often a mixture of regions of three different compositions and the nominal specified overall composition is in fact an average.

7. CONCLUSIONS

Polymeric electrolytes are usually classified as 'solid' electrolytes. As suggested in Chapter 1, however, it may be better, from the point of view of understanding their behaviour, to regard them as liquids. Even in semi-crystalline polymers, the amorphous phase seems to be responsible for the transport of ions.

Solid-like behaviour is usually required for structural stability. This may be conferred on polymers by:

(i) very high chain length;
(ii) high glass transition temperature;
(iii) crosslinking;
(iv) crystallisation.

Completely amorphous polymers with a sufficiently high relative molar mass can be mobile enough to permit the facile movement of ions but yet be virtually incapable of flow. Well above their T_g, such materials are rubbers. The very long chains become entangled and so flow is almost eliminated. The small amount of residual flow is measured as creep and is temperature-sensitive. Poly(propylene oxide), PPO, falls into this category.

Poly(propylene oxide) is also a polymer which has a fairly high glass transition temperature. In this case the T_g is raised further towards room temperature by added salts. This restricts chain movement and so makes such materials seem more solid. Unfortunately, the high T_g also reduces ionic mobility.

Crosslinking can be used almost to eliminate flow or creep. If carried out to excess, it may raise the T_g and be harmful; but such levels of crosslinking should not be necessary if an elastic behaviour is acceptable. The disadvantage lies in the need to insert cross links *in situ*, or at least in a preformed electrolyte. Chemically induced crosslinking involves further additives and often needs a polymer specially modified to produce sites for crosslinking. Radiation-induced crosslinking brings its own special requirements. The risk of damage to the ionic component may perhaps be eliminated if it can be made to diffuse into the polymer in a later process.

Crystallisation considerably reduces deformation in polymers. It may be seen as a stiffening framework. Unfortunately this framework does not itself conduct. Possibilities exist for modifying the crystalline component, but melting, whether followed by subsequent re-crystallisation or not, may negate these changes and lead to profound alterations in properties.

REFERENCES

1. A. J. Patrick, M. D. Glasse, R. J. Latham and R. G. Linford, *Solid State Ionics*, 1986, **18/19,** 1063.
2. J. J. Fontanella, M. C. Wintersgill and J. P. Calame, *J. Polym. Sci. Polym. Phys. Ed.* 1985, **23,** 113.
3. R. J. Neat, A. Hooper, M. D. Glasse and R. G. Linford, *Solid State Ionics*, 1986, **18/19,** 1088.
4. R. J. Neat, A. Hooper, M. D. Glasse and R. G. Linford, in *Transport–Structure Relations in Fast Ion and Mixed Conductors,* eds F. W. Poulsen, N. Hessel Anderson, K. Clausen, S. Skaarup and O. Toft Sørensen, Risø National Laboratory, Roskilde, 1985, p. 341.
5. M. B. Armand, J. M. Chabagno and M. J. Duclot, in *Fast Ion Transport in Solids* eds P. Vashishta, J. N. Mundy and G. K. Shenoy, North-Holland, Amsterdam, 1979, p. 131.
6. B. E. Fenton, J. M. Parker and P. V. Wright, *Polymer,* 1973, **14,** 589.
7. J. M. Parker, P. V. Wright and C. C. Lee, *Polymer,* 1981, **22,** 1305.
8. M. B. Armand, *Solid State Ionics*, 1983, **9/10,** 745.

CHAPTER 3

Ion Conducting Polymers

JOHN R. OWEN

Department of Chemistry and Applied Chemistry, University of Salford, UK

1. INTRODUCTION

Until recently, electrical conduction in polymers was generally regarded as a technologically undesirable phenomenon responsible for the breakdown of polymer components in electrical insulators. Therefore we find, in the older literature, studies of the 'intrinsic' conductivity—a residual conductivity which was ascribed to loosely bound protons in a structure where the concentration of any other ions in the polymers was scrupulously minimized.[1] This chapter arises from a new interest in the opposite goal—optimization of the ionic conductivity. In this case a large concentration of ions is deliberately introduced into the polymer structure, and that structure is designed to give a high mobility for ions. The ionic conductivity is proportional to the product of ion concentration and mobility, and is thus maximized.

The classification of Table 1 shows some of the different ways in which ions can be incorporated into polymers. First, we have the *gel polymer electrolyte*; this is simply a polymer swollen with a solution of a salt in a liquid solvent. The salt is dissolved in the liquid, which in turn is in solution in the polymer. Next, we have the *polymeric salt* concept, in which anionic (e.g. —$CF_2SO_3^-$) or cationic (e.g. —R_3N^+) groups bound to the polymer chain act as counter-ions to small, unbound and potentially mobile ions. However, it appears that the mobility of the latter is severely restricted by strong coulombic attraction if there is no solvent present. Upon addition of liquid solvent the unbound ions become mobile, as in the examples of

TABLE 1
Classification of Ion-Containing Polymers

Type	Composition	Mobile species	Examples[a]
Gel polymer electrolyte	Polymer, salt and solvent	Cations, anions and solvent	PVF_2, PC + $LiClO_4$
Ionomer or polyelectrolyte	Polymeric salt	None, unless wet	Nafion
Solvating polymer	Polymeric solvent + salt	Cations and anions	PEO + $LiClO_4$
Solvating ionomer	Polymeric solvent/salt	Cations or anions	See Section 5

[a] Abbreviations: PVF_2, Poly(vinylidine fluoride); PC, propylene carbonate; PEO, poly(ethylene oxide).

ionomers (crosslinked, and therefore insoluble, solvent-swollen polymers used as ion exchange resins) and *polyelectrolytes* (soluble polymers, used as surfactants). The third class is the *solvating polymer*.[2] Here the polymer itself has the ability to dissolve certain salts and support ionic mobility. Solvating ability is an essential prerequisite for fast ion conduction in a dry polymer. This chapter is concerned mainly with the solvating polymers, which can be classed as genuine solid electrolytes since they do not require the addition of a liquid supporting solvent.

Poly(propylene oxide) (PPO) was reported to dissolve lithium perchlorate as early as 1966. A strong interaction between the solute and the ether oxygens of the polymer chain segments was postulated to explain the volume contraction, 70°C change in glass transition temperature, and anomalous mechanical behaviour.[3] A dramatic increase in the viscosity and solubility of poly(ethylene oxide) in methanol on addition of inorganic salts was explained by postulating a metal ion/polymer complex.[4] Having already been singled out as having anomalously high intrinsic conductivities,[1] the polyethers were found to be much better ionic conductors upon deliberate salt addition[5] and after a more detailed study they were proposed as potentially useful electrolytes.[6]

By analogy with liquid solvents like tetrahydrofuran, the dissolution of the salt is not considered to be due to a high dielectric constant, but to coordination of the cation by electron-donating groups of the polymer as shown for example in Fig. 1. Again, by analogy with liquid

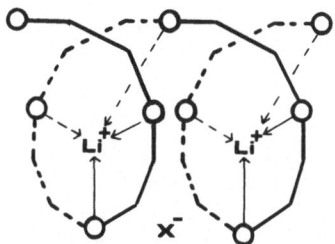

FIG. 1. Proposed structure for a poly(ethylene oxide)$_4$:LiX complex.

solvents, crystalline complexes may be formed at high salt concentration, with fixed ratios between the number of cations and coordinating groups, e.g. poly(ethylene oxide)$_4$: LiClO$_4$. Phase diagrams such as that of Fig. 2 have a direct parallel with solubility curves and eutectic behaviour when we identify the amorphous phases with liquids, and the crystalline phases with solids. However, polymers do not always obey the phase rule, and we often find examples of permanently metastable states in which the expected crystals are presumably prevented from forming by the inherent disorder and slow structural reorganization kinetics.

Given that the proliferation of research into ion conducting polymers began in the solid electrolyte field, it is not surprising that the conductive properties were initially assigned to the crystalline complexes, according to the solid state vacancy mechanism. Models for the

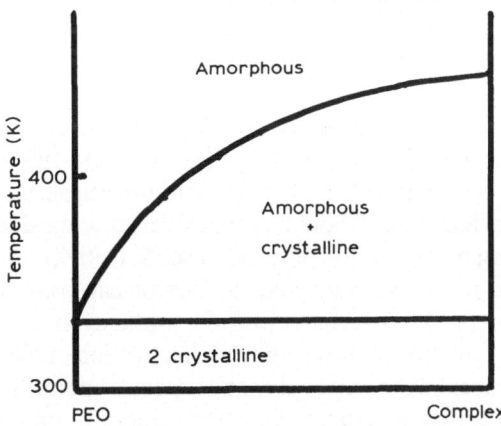

FIG. 2. Phase diagram for the PEO/(PEO)$_{3.5}$LiCF$_3$SO$_3$ system.[13]

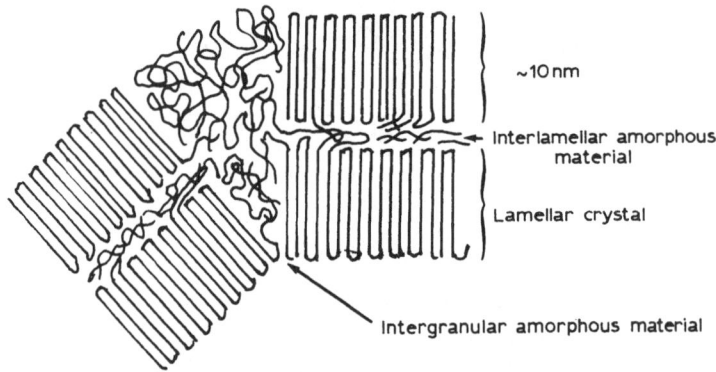

Fig. 3. Example of a microstructure in a semi-crystalline polymer.

conduction process were based on the hopping of cations between vacant sites in a regular helical polymer sheath. However, this model was soon rejected in the light of a knowledge of the detailed microstructure of semi-crystalline polymers (Fig. 3) which shows that a conduction path through the crystal phase would be extremely tortuous. Moreover, totally amorphous samples were equally, if not more, conducting; also the anions were found to be mobile without the benefit of vacant sites.

At the time of writing there is little evidence for any substantial conduction in the crystalline fraction so that the traditional 'solid state' description is inappropriate. Conduction is mostly in the amorphous phase, even in predominantly crystalline samples.

Applications for ion conducting polymers as solid electrolytes are being considered in many fields, for example solid state batteries, sensors, and electrochromic displays. In most cases the figure of merit for the electrolyte material is not the conductivity, σ, but the conductance, $\sigma A/l$. This expression illustrates the superiority of thin polymer films over ceramic materials, and also suggests that mechanical strength in film form should be considered alongside the conductivity in ranking materials for potential use. Selectivity of solvation and conductivity is another important technological consideration. In a sensor this may be the principal criterion, while even in a battery it can be important in order to avoid concentration polarization;[7] the mobility of more than one charged species permits variation of concentration across a sample, which is generally undesirable as the conductivity is often a sensitive function of concentration.

TABLE 2
Relation of Technological Requirements with More Basic Parameters of Polymers

Technological requirement	Polymer parameter
Mechanical strength	High rmm or crosslinked, rather high T_g, filled or crystalline.
Conductivity	High salt concentration and dissociation (i.e. good solvent properties), low T_g, amorphous.
Cation specificity	Bound anion group.
Anion specificity	Bound cation group.

The following sections will attempt to show how polymers can be designed to optimize specific properties such as those shown in Table 2.

2. THE CHEMISTRY AND STRUCTURES OF POLYMER COMPLEXES AND SOLUTIONS

With the recent upsurge of interest in polymer electrolytes, a large number of 'complexes' have been reported between polymers and ionic salts. Strictly, some of these ought to be classified as solutions of salts in polymeric solvents, the term 'complex' being reserved for stoichiometric adducts. However, in either case the polymer is acting as a nucleophilic ligand (Lewis base) and the cation as an electrophile (Lewis acid) in a manner similar to the formation of coordination complexes and solutions with small ligands.[8] In the polymer case, the differentiation and separation of a pure complex from the remaining solution is more difficult, and the material studied is often a polymeric solution containing some stoichiometrically complexed regions.

The preparation of polymer complexes is generally very simple. The usual method is to make an intermediate form of the complex in which both ions are very mobile, e.g. the molten,[9] liquid solution,[10] or pre-crosslinked[11] state, by physical mixing of the polymer and salt, aided by heat or solvent. Then the desired form is obtained by cooling, solvent removal or crosslinking respectively. A disadvantage of the melt method is thermal decomposition, whereas the solvent methods tend to introduce impurities unless the solvent is very pure. Removal of solvent can be difficult and is best done after forming the polymer into a thin (e.g. 100 μm) film.

Polymers are usually non-equilibrium phases. Therefore an equilibrium similar to that between a stoichiometric salt hydrate and its saturated solution is not always attained in a polymer. However, it often turns out that distinct solution and stoichiometric complex phases can be identified, for example by optical microscopy, revealing crystal anisotropy, or by DSC revealing a heat of melting.[12] Solid state nuclear magnetic resonance (NMR) can also reveal differences due to the different relaxation times of probed atoms in crystalline and amorphous environments.[13] The phase diagram in Fig. 2 shows a case in which the complex and solution fractions, and the salt content of each, were determined from NMR signals of polymer protons and fluorine anions contained in the salt.

It has already been stated that high ionic conductivity occurs only in the amorphous phase. Consequently the factors promoting an amorphous structure will now be considered. Generally, an amorphous parent polymer gives an amorphous complex. The following features are therefore beneficial.

(a) A random backbone, as in a copolymer or a sterically random polymer such as atactic poly(propylene oxide).
(b) Crosslinking the polymer while in the amorphous state.
(c) A broad molecular mass distribution, again impeding crystallization.

In some cases, amorphous complexes are formed from polymers which are normally crystalline. Rubidium and caesium salt complexes of PEO, for example, have been reported to be amorphous on account of the large size of the rubidium ion compared with the helix radius.[14] Metastable amorphous polymorphs are occasionally produced when prepared by evaporation of a solution in a liquid solvent.[15] Both cation and anion transport are required for a separation of phases of different composition, and therefore a low anion transport number could limit the rate of crystallization.

A very important issue as regards ionic conductivity is whether or not the anion is displaced from the cation by the polymeric ligand, and if so, by what distance. If the anion remains in the coordination sphere, the result is a solvated ion pair which, being a neutral species, does not conduct. Studies of ion pairing in liquid electrolyes have shown that ion pairs are dissociated efficiently if the solvent has a high Lewis basicity and the anion is large, or easily polarizable. In such a case the solvent can easily displace the anion from the ion pair. Therefore the suitable anions are those which produce strong (Brönsted) acids when protonated, e.g. I^-, ClO_4^-, $CF_3SO_3^-$, BF_4^-, $B(C_6H_5)_4^-$, etc.

In the context of coordination chemistry we can classify PEO as a member of the 'glyme' (*gly*col *me*thyl ether) series of complexing solvents:

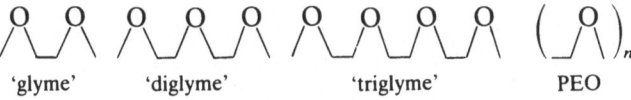

The above ligands are powerful in two respects. First, the Lewis base strength due to the oxygen is relatively high. Second, the glymes are multidentate, thereby showing the chelate effect (multiple coordination by a single ligand, avoiding the entropy penalty of several ligands ordered onto a single metal centre).

The cyclic crown ethers, also derived from the ethylene oxide unit, are amongst the strongest ligands known for complexing alkali metal ions, and are also selective for preferred ion sizes.[16] The ethylene oxide group also features as a predominant ligand in some acyclic ionophoric antibiotics (e.g. monensin[17]) which are known to wrap around the selected ion on complexation. Studies of some modified oligo(ethylene glycol)s and other medium-sized molecules of interest as phase transfer catalysts reveal that strong steric factors are involved in their reactions with ionic salts.[18]

In the light of the background chemistry above, it is not surprising to find that PEO and other polymers based on the O—C—C—O backbone are sterically favourable for complexation, as compared with other polyethers.[6] Poly(ethylene oxide) itself tends to form crystalline complexes with a helical structure wrapped around the guest cations.[19] The helix repeat distance is considerably reduced from that of the unsalted polymer, showing a strong attraction between the cations and the oxygen on the helix.

In the cases most relevant here, the complexes have amorphous structures, which are difficult to study with standard X-ray techniques. Spectroscopic techniques such as infrared (IR),[20] Raman,[20] or

TABLE 3
Examples of Ion Solvating Polymers

Solvating group	Polymer	Soluble salts	Reference
Ether	{ PEO { PPO	{ Almost all cations { Monovalent anions	2
Ester	^{a}PE succinate	$LiBF_4$	23
Ester	^{a}PE adipate	$LiCF_3SO_3$	24
Amine(—NH—)	^{a}PE imine	NaI	25

^{a} PE, polyethylene.

X-ray absorption fine structure (EXAFS),[21] for example, have been used to determine local atomic arrangements. On a larger scale, it may be speculated that helical segments are also formed in the amorphous complexes, but that the periodicity is randomly interrupted by reversals and phase modulation of the helical wave. Alternatively, a supposedly amorphous polymer may be microsegregated on a scale escaping detection either optically or by X-rays, into ion-rich and ion-poor regions in a manner similar to the phase segregation known to occur in some ionomers.[22]

The list of solvating polymers in Table 3 illustrates that ligands other than ether oxygens can provide the solvating function. The field is greatly expanded by considering copolymers, in which only one of the co-monomers need have the solvating capability discussed in this section. At the time of writing, a major application for polymeric electrolytes is anticipated in lithium solid state cells for high-energy-density batteries. This application, however, will require the electrolyte to be stable under severe reducing, oxidizing and basic conditions, and in some cases at relatively high temperatures (ca 100°C). Therefore some chemical functional groups, e.g. ketonic, amino, or alcoholic, may not be tolerated in the electrolyte. The same would apply to any potentially reactive anions. This restriction may well narrow the field of solvating polymers applicable in batteries to the polyethers, either alone or in conjunction with other non-reactive structural components.

3. THE KINETICS OF ION TRANSPORT

In contrast with crystals and glasses, organic polymers have not yet shown fast ion conduction in the real solid phase, i.e. the phase in which the host 'lattice', whether regular or disordered, is immobile. Conductivities high enough to be classed as 'fast' are generally attributed to amorphous phases above the glass transition temperature, T_g, where a liquid-like motion of the host is assumed to take place on a microscopic scale. The guest and host motions are considered as coupled—that is, host movement is an integral part of the ion transport mechanism. Instead of the solid state description of ion transport, in which ions of sufficient energy are modelled as squeezing through rigid potential energy barriers between fixed sites, we here take the liquid model, where ion motion is a result of motion of the potential barriers themselves. The starting point for the study is therefore an examination of the natural motion of polymer chain segments.

3.1. Free Volume Theory

According to the widely accepted free volume theory,[26-28] each polymer is characterized by a glass transition temperature, T_g, below which chain segments are essentially immobile. Above this temperature, the vibrational energy of a segment is sufficient to push against the hydrostatic pressure imposed by its neighbours and create a small amount of space surrounding its own volume in which the vibrational motion can occur. The extra space is called the free volume per segment, v_f, and varies randomly about a mean value, $\bar{v}_f$. Occasionally, v_f becomes as large as the chain segment itself (of volume v) and a translational movement, such as exchange with a neighbour, can take place. A statistical calculation shows that the probability, P, of v_f exceeding v is:

$$P = \exp(-v/\bar{v}_f) \tag{1}$$

This expression is analogous to the Arrhenius expression [$\exp(-E_A/kT)$] for the probability of particles surmounting an energy barrier, and is responsible for the main variations of visco-elastic and diffusional properties of polymers and liquids above T_g. The form of expression is universal although the interpretation of the parameters $\bar{v}_f$ and v can differ in detailed analyses. For example, $\bar{v}_f$ can be corrected for free volume overlap, and v replaced by a general 'activation volume', v^*, defined as the volume required for a conformational change to be possible. However, restricting ourselves to the simple theory, for the moment, we can identify the free volume fraction, v_f/v, with the macroscopic analogue:

$$f = \bar{v}_f/v = V_f/V \tag{2}$$

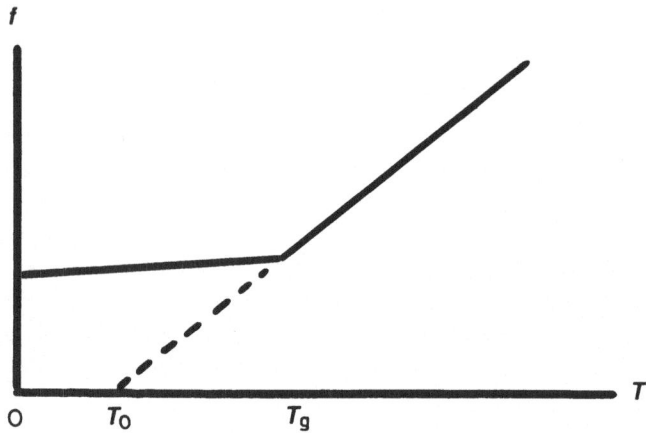

FIG. 4. Temperature variation of the free volume fraction, f.

where V_f is the total free volume in the system and V is the total volume of the system.

For the temperature range T_g to $T_g + 150°C$, say, which covers most applications, f varies as shown in Fig. 4. Its expansion coefficient, α, is constant, and its absolute value at a given temperature can be obtained by extrapolating from the hypothetical zero point of f at T_0. We can then write

$$f = \alpha(T - T_0) \quad \text{for} \quad T_g < T < T_g + 150°C \quad (3)$$

Whereas T_0 varies widely from one material to the next, α is almost constant at $0·0005/°C$. T_0 is generally estimated to be about 50°C below T_g. (Strictly, T_g is frequency-dependent, and quoted values depend on the method of measurement.) Therefore we can substitute in eqn (1) to obtain an equation for the temperature dependence of all transport properties.

$$P = \exp[-1/\alpha(T - T_0)] \quad (4)$$

or

$$P = \exp[-B/(T - T_0)] \quad \text{(where } B = v^*/\alpha v) \quad (5)$$

[The form of eqn (5) is more general, since it allows for the possibility of an activation volume other than v, such as v^*, to be incorporated in the parameter B.]

Corrected for small temperature dependences of pre-exponential factors, the diffusion coefficient, D, the conductivity, σ, and the reciprocal viscosity (fluidity), $1/\eta$ can all be expressed thus:

$$W(T) = W_0 \exp[-B/(T - T_0)] \qquad (6)$$

where $W(T) = DT^{-1/2}$ or $\sigma T^{1/2}$ or $1/\eta$, etc.

To eliminate pre-exponential factors, W_0 and the hypothetical T_0, polymer scientists have used the Williams–Landel–Ferry (WLF) equation,[27] eqn. (7), which follows from the above:

For any reference temperature, T_r,

$$\ln \frac{W(T)}{W(T_r)} = -B\left[\frac{1}{T - T_0} - \frac{1}{T_r - T_0}\right]$$

T_0 is eliminated by the transformations

$$C_1 = B/(T_r - T_0) \qquad C_2 = T_r - T_0$$

or

$$T_0 = T_r - C_2 \qquad B = C_1 \cdot C_2$$

so that:

$$\ln \frac{W(T)}{W(T_r)} = \frac{C_1(T - T_r)}{C_2 + (T - T_r)} \qquad (7)$$

The reference temperature, T_r, may be chosen anywhere between T_g and $T_g + 150°C$ according to experimental convenience. Thus a property can be specified over a wide temperature range, by tuning the parameters C_1 and C_2 to the best data fit. The values of the parameters will depend on the choice of reference temperature, but the calculated values of T_0 and B should be absolute characteristics of the material.

3.2. Application to Ion Conduction

The first important modification to the above theory required for application to ion transport concerns the value of B. Because the activation volume for ion movement is not necessarily the same as that for a polymer conformation change, the relevant value of B should be modified according to the ion size:[29]

$$\frac{B \text{ (conductivity)}}{B \text{ (visco-elasticity)}} = \frac{v^* \text{ (ion)}}{v^* \text{ (chain segment)}} \qquad (8)$$

The activation volume is, in theory, measurable by investigating the effect of pressure on ion mobility. However, initial attempts at this approach have been complicated by other effects, such as pressure dependence of crystallization in PEO.[30] Direct comparisons between the temperature dependences of visco-elastic and conductive properties of the PEO/NaBPh$_4$ samples led to the surprising conclusion that activation volumes for chain segment and ion mobility are equal.[29] This is plausible for cations, whose motion is probably dependent on a cooperative movement of the coordinated chain segments. Such simultaneous segmental motion was not, however, easily explained in anion transport, and the anion contribution was assumed to be negligible.

A second major complication is due to the possibility of limited salt dissociation, and its temperature dependence. Arrhenius terms added to eqn (1) to account for an activated jump and an activated dissociation were found to be appropriate in a treatment of undoped polymers of low conductivity.[31]

$$P = \exp(-v/\bar{v}f - E_A/kT - W/kT) \quad (9)$$

Strictly, the activated jump term, E_A/kT, should always be included, at least in principle, in order to give a theoretical basis for Ohm's law. However, recent WLF treatments of fast ion conducting polymers have shown straight lines without the consideration of activation terms, at least in dilute polymers.[30] Such results may be interpreted by a low activation energy for the ion jump process and complete salt dissociation. On the other hand, high dopant concentrations are expected to reveal a term due to salt dissociation, W/kT.

Another complication arises due to the strong dependence of T_0 on salt concentration. Large increases in glass transition temperatures on salt addition to liquid polyethers[2] have also been reported in cross-linked polymer networks.[30] The corresponding decrease in chain segmental mobility can be attributed to the ring structures resulting from the coordination of more than one ether oxygen onto the cation. Because of this effect, conductivities determined for one dopant concentration cannot be compared directly with those determined at another concentration without a knowledge of each T_0 value. Unfortunately, many studies of the concentrated doped polymers do not report a directly measured T_g or a wide enough range of conductivity versus temperature measurements for T_0 to be accurately deduced. However, in the case of PEO networks containing LiClO$_4$ at low concentrations, an excellent comparison has been made between conductivity and concentration at constant $T - T_g$, and this showed a constant ion mobility over the range studied.[32] A similar comparison with NaBPh$_4$ dopant showed similar mobilities in both PEO and PPO networks at constant $T - T_g$.[29]

The next complication is due to the mobility of the anion, which until recently was assumed to be negligible but now is known to be similar to that of the cation. Unless special experiments are performed (see below) the conductivity measured is the sum of that due to both cation and anion. A temperature dependence of the transference number, if known, could well be used to resolve two regions of the free volume plot, with one ion 'frozen out' at low temperatures due to a higher activation volume. Again, we suffer from a lack of data due to the fact that most of the methods used to determine transference numbers are only applicable at high temperature.

To resolve the above difficulties, an armoury of auxiliary techniques are being added to the simple mechanical and electrical studies. Notably, tracer diffusion[33] and NMR[34] have been used to investigate the most contentious issues of ion dissociation, and the decomposition of the conductivity into the product of concentration and mobility. At the present time conclusions of these issues are so contradictory that a discussion of results would be premature in this chapter, but welcome in a future review. Also omitted here are the configurational entropy theory,[35,36] which gives similar results to free volume theory, and the novel dynamic bond percolation (DBP) theory,[37] which represents a significant departure from the free volume approach.

3.3. Conclusions Regarding Kinetics

According to the evidence available to the author, the link between electrical and mechanical properties is strong enough to make some semi-quantitative statements and predictions.

First, decreasing T_0 is the only sure way of increasing ion mobility. As yet there is little hope that any changes in v^* can be made to bring about effects comparable with those due to changes in T_0. The proportional changes in ionic mobility due to a different T_0 will be in the first approximation the same as the proportional changes in appropriately defined mechanical properties of an amorphous polymer.

The balance between electrical mobility and mechanical strength in PEO/ionic salt electrolyte films has proved to be appropriate for battery use at 80°C, i.e. where the polymer is totally amorphous and at about 100°C above its glass transition temperature. Free volume theory predicts that the same balance, and similar absolute values of the properties, would exist at 20°C for an amorphous polymer chemically identical to PEO but having a glass transition temperature of −80°C. It also predicts that the variations in properties experienced due to an extended service temperature range, say 0–40°C, imposed on such a polymer, would be the same as in (non-crystalline) PEO between 60 and 100°C.

Free volume theory cannot show a way of increasing conductivity without affecting the mechanical properties simultaneously. For this we must increase salt dissociation. Poor salt dissociation is an important factor which explains why polysiloxanes which have glass transition temperatures (T_g) of about −80°C are poor ionic conductors unless specifically modified for that purpose. The simplest quantification of the salt dissociation factor is found from the conductivity of the non-polymeric chemical analogues. For example, it is unlikely that the dissociation factor of PEO will be dramatically improved upon by any manipulations of the chemistry, since a close non-polymeric analogue, 2M-$LiClO_4$ in dioxolane, has one of the best conductivities reported for aprotic organic electrolytes (about 1 S/m).

In conclusion to this section, it may be predicted that the conductivity of about 10^{-1} S/m in $(PEO)_8:LiClO_4$ at 100°C probably represents a maximum for an aprotic, self-supporting, homogeneous polymer, and that a similar situation could be achieved at room temperature with a polymer which had a transition temperature about −80°C.

4. ELECTRICAL MEASUREMENT TECHNIQUES

The methods used to study the ion conducting properties of polymers are essentially the same as those applicable to liquid electrolytes, except that the relatively low conductivities demand special geometries of measurement. Generally, a thin (about 100 μm) sample is pressed in between two metallic electrodes of about 1 cm^2 area. In this way, the conductance (SI unit the Siemen, S) measured is numerically two orders of magnitude higher than the conductivity, σ(S/cm) and is therefore well within range of normal impedance measurement equipment. With this geometry, however, a four-electrode measurement is difficult to arrange, and steps must be taken to resolve and eliminate the series impedance of the interfaces between the electrodes and the sample. Depending on the electrochemical properties of the electrodes, the impedance can be of either form shown in Fig. 5. In neither case will a simple d.c. resistance be valid, because the interfacial impedance will add an undetermined quantity to the sample impedance. Therefore, the frequency response analysis method is usually employed. The theory of this technique is beyond the scope of this chapter and may be found elsewhere.[38] Figure 5 is given to identify the features expected of ideal complex plane plots.

The ideal semi-circle plot is expected from a homogeneous sample, in which uniform current paths due to conductance and capacitance are parallel. The frequency at which the imaginary part is maximum is related to the conductivity and permittivity (ε) thus:

$$F(\max. Z'') = \sigma/2\pi\varepsilon \tag{10}$$

Such ideal plots are rarely obtained, although the conductivity value given by the arc diameter is usually interpreted as that of a single-phase material. Little attention has been given to the detailed shape of these arcs, which may originate from microscopic inhomogeneity[39] or non-Debye (i.e. frequency-dependent) capacitance.[40]

The low-frequency part of the impedance spectrum contains information regarding the electrode/electrolyte interface. With ideally blocking (inert) electrodes the spur should be vertical. Again this is rarely the case, and surface adsorption or even unexpected electrochemical reactions at the interface may be indicated. To minimize the latter, the amplitude of the potential should be kept small (say, 10 mV).

The use of non-blocking, i.e. reversible, electrodes assumes an electrochemical reaction at the interface, for example alternate plating or stripping of the (electrode-reversible) cation contained in the polymer. Ideally, a semi-circle in the medium-frequency range represents the interfacial charge-transfer resistance, which is related to the plating/stripping kinetics. The low-frequency impedance between selectively blocking electrodes can reveal the transference number of a blocked ion, e.g. a non-dischargeable anion. This is because the high-frequency conductance is the sum of all ion conductances in parallel, whereas as the frequency decreases substantially below the value $1/R_b C_{dl}$ (where R_b is bulk resistance and C_{dl} is double layer capacitance), non-blocked ions become the sole current carriers.

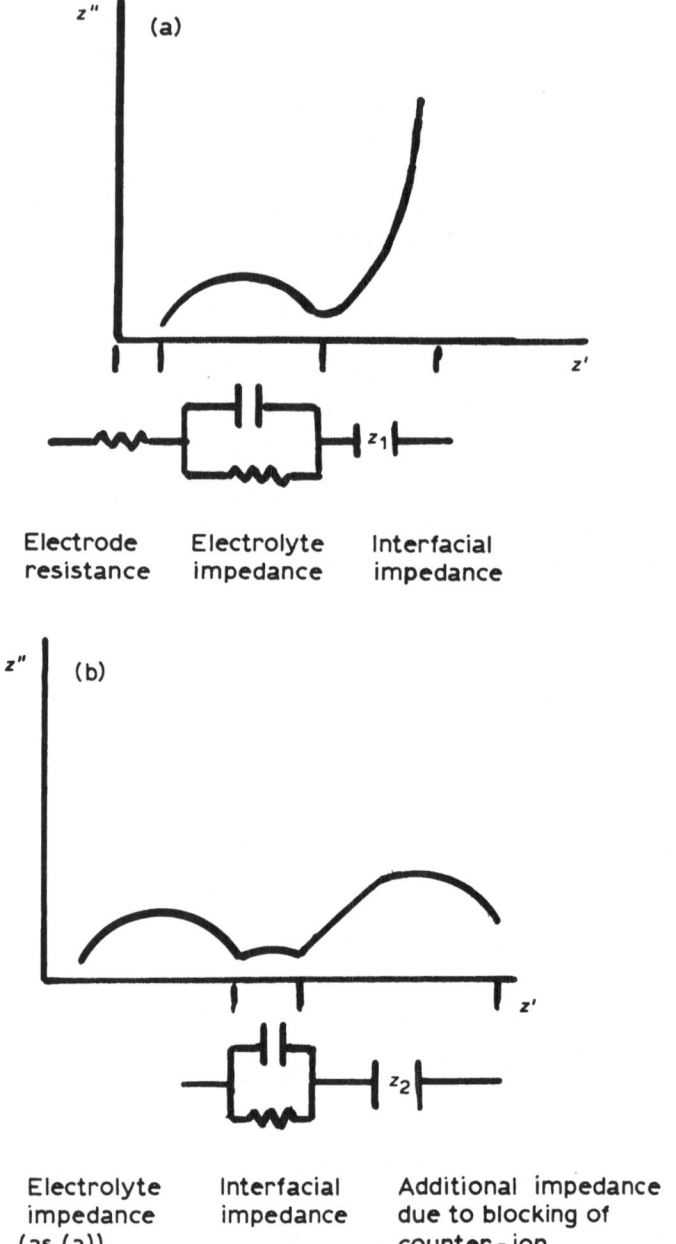

FIG. 5. Complex plane plots of electrical impedance of polymer electrolytes. (a) Blocking (e.g. Pt) electrodes; (b) non-blocking (e.g. Li) electrodes.

Constant current or other time domain experiments can provide all the above information and more. In fact, the a.c. technique is based on a linear current/potential relationship which is only valid strictly at small potential differences. Time domain experiments, however, are more subject to electrical noise, and in some cases more difficult to interpret than the frequency equivalent. For example, separation of the three characteristics shown clearly in the complex plane plot is not a simple matter in the time domain equivalent of Fig. 5.

Electrical methods are also useful in determining the chemical and electrochemical stability of the polymer at the electrode interface, information which is essential for the success of any device. Cyclic voltammetry[41] subjects the sample to alternately highly oxidizing and reducing potentials while monitoring the currents due to electrolysis. In this way, PEO has been shown to have an electrochemical stability window of up to 3·5 V. The high stability has been ascribed to the immobility of the structure over long distances, so that once the material nearest the electrode has completely reacted, no further reaction can occur. This is in contrast to the liquid electrolyte, in which all parts of the sample sooner or later reach the harsh conditions at the electrode interface by diffusion. Steady state d.c. methods have also been applied to quantify the stability, with similar results.

In the final reckoning, however, the only sure test of the polymer electrolyte performance is in the device itself. There is now a large volume of data concerning the cycling of cells with PEO as the electrolyte[42–44] but relatively little on other polymers.

5. CURRENT RESEARCH DIRECTIONS

Considerable efforts are now underway to develop polymers for the electrolyte in solid state lithium batteries. For a design based on electrode/electrolyte laminations 10–100 μm thick, conductivities of 10^{-2} S/m or better are required. Figure 6 shows that a number of electrolytes are available for use at about 80°C or above, but that only a few relatively untested compositions could suffice at room temperature. Therefore polymers with low T_g values are prime candidates for investigation. Polysiloxanes have extremely low T_g values due to the flexibility of the Si—O—Si linkage, but unfortunately these are poor solvents for ionic salts. Substituted polymers and copolymers of siloxanes with polyethers have been examined and shown to have

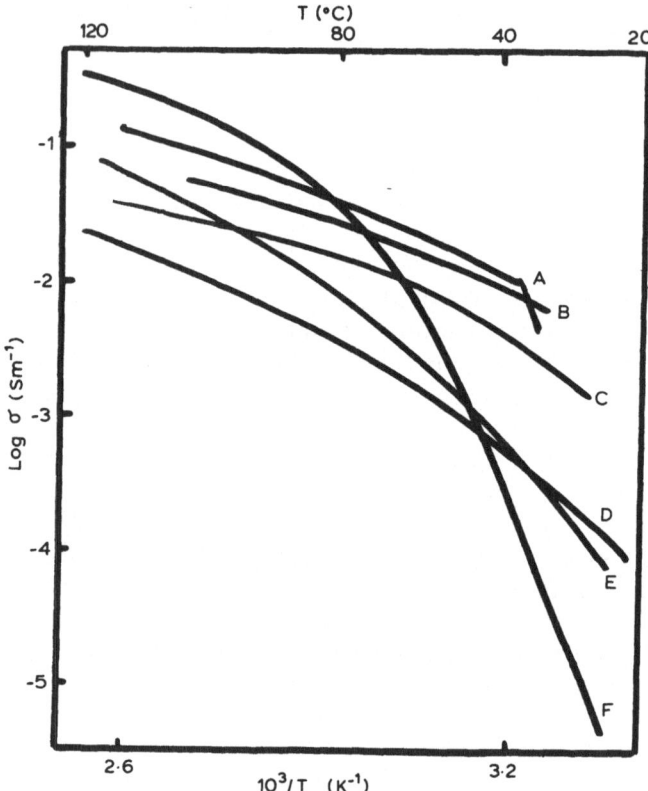

FIG. 6. Log (conductivity) versus reciprocal temperature for some polymeric electrolytes: A, plasticized PEO/LiCF$_3$SO$_3$ (12:1);[51] B, polyphosphazene/LiCF$_3$SO$_3$ (4:1);[47] C, Viton/PC/LiClO$_4$ gel polymer electrolyte;[2] D, PPO/LiCF$_3$SO$_3$ (9:1);[2] E, crosslinked PEO/LiClO$_4$ (8:1);[49] F, PEO/LiClO$_4$ (12:1)[2].

promising conductivity values.[45,46] The structurally similar, but more chemically tractable, polyphosphazenes have been reported to have the required conductivity of 10^{-2} S/m at ambient temperature.[47] However, in both cases other criteria must be satisfied before the material is declared suitable for the application.

In addition to a cation contribution to the conductivity of at least 10^{-2} S/m, the electrolyte in a lithium battery should have the following properties:

(1) Mechanical stability, at ambient temperature and, say, 50° above.
(2) Chemical stability to lithium and a powerfully oxidizing positive electrode.
(3) No mobile components which react with either electrode.

A specification of mechanical strength is required in order to describe how solid the material is. Amorphous polymers of low T_g are generally liquids up to molecular masses of about 100 000. At first sight, a high ionic mobility, requiring a low T_g value, seems incompatible with good mechanical stability. However, the use of fillers[48] and/or crosslinking agents[49] may be a solution to this problem. Another approach is to attach mobile, ion-solvating groups onto a robust backbone as a branched copolymer.[50] Conversely, a plasticizer may be added to a conducting polymer rigid enough to withstand the softening which accompanies an increase in conductivity.[51]

Apart from mechanical strength, a supposed advantage of a solid electrolyte is the immobility of any components which may otherwise react at the electrodes or cause concentration redistribution. The use of plasticizers, and filler-immobilized polymeric liquids to produce high conductivity, must therefore be treated with caution. Whether or not polymers with wholly tethered solvating groups will ever have conductivities approaching those of the best organic liquids will depend on the kinetics of transfer of ions from one coordinated arrangement to another.

Mobile anions continue to be a cause for concern, especially when salt redistribution may result in an insulating phase such as $P(EO)_{n<4}:LiCF_3SO_3$ at one electrode interface. Some interesting developments in this respect are found in the new class of *solvating ionomers*. An example is given by the structure I.[52] This is similar to

$$\underset{(I)}{\left[CH_2-\underset{\underset{\underset{O}{\overset{\|}{C}}}{\overset{|}{\underset{|}{C}}-CH_3}}{C} \right]_x \quad O-CH_2-CH_2SO_3^- \quad Li^+ \cdots (CH_2-CH_2-O)_y}$$

$P(EO)_8:LiCF_3SO_3$, except that now the anion has been bound to another polymer chain blended with the PEO. Therefore the lithium ion should be the only species mobile over long distances. The consequence will be a disappearance of the limiting current effect, with important implications for high-rate discharge. The obvious next step will be a polymer containing both ionic and solvating branches on the same backbone chain.

Perhaps one of the most exciting new developments expected in the field of conducting polymers is the concept of the polymeric insertion electrode. Inorganic insertion electrodes such as TiS_2 have the combined properties of electronic conductivity, redox capacity and ionic conductivity.[53] Most electron conducting polymers reported show only the first two, and rely on a liquid electrolyte for ion transport. However, a copolymer made from electron and ion conducting components could give a superior electrode material which could easily accommodate the strain of ion insertion, which is one of the fundamental problems associated with solid state devices made with hard solids.[54]

REFERENCES

1. A. E. Binks and A. Sharples, *J. Polymer Sci. A2*, 1968, **6,** 407.
2. M. B. Armand, *Solid State Batteries,* ed. A. Hooper and C. A. C. Sequeira, NATO ASI Series E 101, Martinus Nijhoff, Dordrecht, 1985, p 63.
3. J. Mocanin and E. F. Cuddihy, *J. Polymer Sci. C*, 1966, **14,** 313.
4. R. D. Lundberg, F. E. Bailey and R. W. Callard, *J. Polymer Sci. A1*, 1966, **4,** 1563.
5. B. E. Fenton, J. M. Parker and P. V. Wright, *Polymer,* 1973, **14,** 589.
6. M. E. Armand, J. M. Chabagno and M. J. Duclot, *Fast Ion Transport in Solids,* eds P. Vashishta, J. N. Mundy and G. K. Shenoy, North-Holland, Amsterdam, 1979, p. 131.
7. S. Atlung, *Solid State Batteries,* ed. A. Hooper and C. A. C. Sequeira, NATO ASI Series E 101, Martinus Nijhoff, Dordrecht, 1985, p. 129.
8. F. A. Cotton and G. Wilkinson, *Advanced Inorganic Chemistry,* 4th ed., Wiley, 1980, ch. 3.
9. M. J. Hannon and K. F. Wissbrun, *J. Polymer Sci., Polym. Phys. Ed.,* 1975, **13,** 113.
10. J. Weston and B. C. H. Steele, *Solid State Ionics,* 1982, **7,** 81.
11. A. Killis, J. F. LeNest, A. Gandini and H. Cheradame, *Solid State Ionics,* 1984, **14,** 231.
12. J. Runt and I. R. Harrison, *Polymers,* ed. R. A. Fava, Vol. 16 of *Methods of Experimental Physics,* ed. L. Marton, Academic Press, New York, 1980, p. 287.

13. M. Minier, C. Berthier and W. Gorecki, *J. Phys.*, 1984, **45**, 739.
14. D. F. Shriver, R. Dupon and M. Stainer, *J. Power Sources*, 1983, **9**, 383.
15. J. Weston and B. C. H. Steele, *Solid State Ionics*, 1982, **7**, 81.
16. R. M. Izatt, D. S. Bradshaw, S. A. Nielsen, J. D. Lamb, J. J. Christensen and D. Sen, *Chem. Rev.*, 1985, **85**, 271.
17. F. Vogtle and E. Weber, *Angew. Chem. Int. Ed. Engl.*, 1979, **18**, 753.
18. S. Yanagida, K. Takahashi and M. Okahara, *Bull. Chem. Soc. Japan.*, 1978, **51**, 3111.
19. J. M. Parker, P. V. Wright and C. C. Lee, *Polymer*, 1981, **22**, 1305.
20. B. L. Papke, M. A. Ratner and D. F. Shriver, *J. Electrochem. Soc.*, 1982, **129**, 1434.
21. C. R. A. Catlow, A. V. Chadwick, G. N. Greaves, L. M. Moroney and M. R. Worboys, *Solid State Ionics*, 1983, **9/10**, 1107.
22. S. Y. Yeo, *J. Electrochem. Soc.*, 1983, **130**, 533.
23. R. Dupon, B. L. Papke, M. A. Ratner and D. F. Shriver, *J. Electrochem. Soc.*, 1984, **131**, 586.
24. R. D. Armstrong and M. D. Clark, *Electrochem. Acta*, 1984, **29**, 1443.
25. C. K. Chiang, T. Davis, T. Takahashi and C. A. Harding, *Electrochem. Soc. Extended Abstracts*, 1985, **85**(1), 116.
26. T. G. Fox and P. J. Flory, *J. Polymer Sci.*, 1954, **14**, 315.
27. M. L. Williams, R. F. Landel and J. D. Ferry, *J. Amer. Chem. Soc.*, 1955, **77**, 3701.
28. M. H. Cohen and D. Turnbull, *J. Chem. Phys.*, 1959, **31**, 1164.
29. A. Killis, J. F. LeNest, H. Cheradame and A. Gandini, *Makromol. Chem.*, 1982, **183**, 2385.
30. J. J. Fontanella, M. C. Wintersgill, J. D. Calame, F. P. Pursel, D. R. Figueroa and C. J. Andeen, *Solid State Ionics*, 1983, **9/10**, 1139.
31. T. Miyamoto and K. Shibayama, *J. Appl. Phys.*, 1973, **44**, 5372.
32. A. Killis, J. F. LeNest, A. Gandini, H. Cheradame and J. P. Cohen-Addad, *Solid State Ionics*, 1984, **14**, 231.
33. A. V. Chadwick, J. H. Strange and M. R. Worboys, *Solid State Ionics*, 1983, **9/10**, 1155.
34. W. Gorecki, R. Andereau, C. Berthier, M. Armand, M. Mali, J. Roos and B. Brinkman, *Solid State Ionics*, 1986, **18/19**, 295.
35. G. Adam and J. H. Gibbs, *J. Chem. Phys.*, 1965, **43**, 139.
36. E. Williams and C. A. Angell, *J. Phys. Chem.*, 1977, **8**, 232.
37. S. D. Druger, M. A. Ratner and A. Nitzan, *Solid State Ionics*, 1983, **9/10**, 1115.
38. R. D. Armstrong and W. I. Archer, *Chem. Soc. Spec. Per. Rep., Electrochemistry*, 1980, **7**, 157.
39. A. Le Mehaute, in *Transport-Structure Relations in Fast Ion and Mixed Conductors*, eds F. W. Poulsen, N. Hessel Anderson, K. Clausen, S. Skaarup and O. Toft Sørensen, Riso National, Laboratory, Roskilde, 1985, p. 25.
40. A. K. Jonscher, *Phys. Stat. Sol. A*, 1975, **32**, 665.
41. M. Armand, M. J. Duclot and P. Rigaud, *Solid State Ionics*, 1981, **3/4**, 429.
42. M. Gauthier, D. Fauteux, G. Vassort, A. Belanger, M. Duval, P. Ricoux,

J. M. Chabagno, D. Muller, P. Rigaud, M. B. Armand and D. Deroo, *J. Electrochem. Soc.*, 1985, **132,** 1333.
43. A. Hooper, Solid State Batteries, ed. A. Hooper and C. A. C. Sequeira, NATO ASI Series E101, Martinus Nijhoff, Dordrecht, 1985, p. 399.
44. B. C. H. Steele, G. E. Lagos, P. C. Spurdens, C. Forsyth and A. D. Foord, *Solid State Ionics,* 1983, **9/10,** 391.
45. K. Nagaoka, H. Naruse, I. Shinohara and M. Watanabe, *J. Polymer Sci., Polym. Lett. Ed.*, 1984, **22,** 659.
46. A. Bouridah, F. Dalard, J. F. LeNest and H. Cheradame, *Solid State Ionics,* 1985, **15,** 233.
47. P. M. Blonski and D. F. Shriver, *J. Amer. Chem. Soc.*, 1984, **106,** 6854.
48. J. Weston and B. C. H. Steele, *Solid State Ionics,* 1981, **2,** 347.
49. J. R. MacCallum, M. J. Smith and C. A. Vincent, *Solid State Ionics,* 1984, **11,** 307.
50. F. M. Gray, J. R. MacCallum and C. A. Vincent, *Solid State Ionics,* 1986, **18/19,** 282.
51. I. E. Kelly, J. R. Owen and B. C. H. Steele, *J. Electroanal. Chem.*, 1984, **168,** 467.
52. D. J. Bannister, G. R. Davies and I. M. Ward, *Polymer,* 1984, **25,** 1291.
53. J. R. Owen, J. Drennan, G. E. Lagos, P. C. Spurdens and B. C. H. Steele, *Solid State Ionics,* 1981, **5,** 343.
54. B. C. H. Steele, Solid State Batteries, ed. A. Hooper and C. A. C. Sequeira, NATO ASI Series E101, Martinus Nijhoff, Dordrecht, 1985, p. 163.

CHAPTER 4

Organic Polymers as Electroactive Materials

ALAN G. MACDIARMID

University of Pennsylvania, Philadelphia, Pennsylvania, USA

and

MACRAE MAXFIELD

Corporate Research Center, Allied-Signal Inc., Morristown, New Jersey, USA

1. INTRODUCTION

Electroactive materials—materials which can undergo electrochemical oxidation or reduction reactions—provide the scientific basis for the storage of electrical energy in non-rechargeable (primary) or rechargeable (secondary) batteries. When electroactive materials are used as electrodes in batteries, they are often referred to as electrode-active. In primary batteries the electrochemical reaction which occurs during discharge cannot be efficiently reversed by the application of an external voltage. In secondary batteries the electrochemical discharge reaction can be reversed during the charge process with a high degree of efficiency by the application of an appropriate external potential. Until 1979 no serious attention had been given to the possible use of organic polymers as electrode-active materials in rechargeable batteries. It had been taken more or less for granted that the electroactive components of a useful rechargeable battery must consist of inorganic materials such as metals and metal oxides, etc.

In 1977[1] it was discovered that the organic polymer polyacetylene, $(CH)_x$ (Fig. 1), synthesized in a simple manner by the polymerization of gaseous acetylene, C_2H_2, could be chemically p-doped (partly

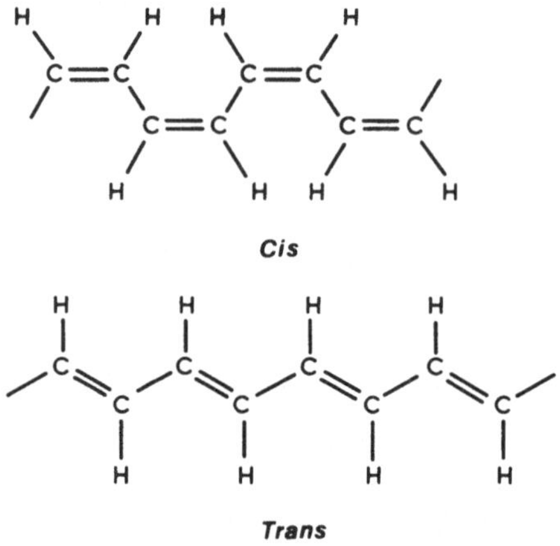

FIG. 1. *Cis* and *trans* isomers of $(CH)_x$.

oxidized) with a concomitant increase in its conductivity through the semi-conducting to the metallic regime. This introduced to condensed matter science new concepts of considerable theoretical and possible technological importance. It was found shortly thereafter[2] that the conductivity of polyacetylene could also be raised to the metallic regime by chemical *n*-doping (partial reduction). In 1979[3] it was discovered that *p*- or *n*-doping of $(CH)_x$ could be accomplished electrochemically and that these processes were electrochemically reversible. This led naturally to the conclusion that $(CH)_x$ and its various oxidized or reduced forms might act as good storage media for electricity, especially since it was found, perhaps surprisingly, that the stored charge could be delivered on demand at a relatively high rate.[4]

Since 1979 other conducting polymers such as poly(*p*-phenylene),[5] polypyrrole,[6] polythiophene,[7] polyaniline,[8] etc., have been discovered, and these have also been examined as potential electrode-active materials[9] for possible use in rechargeable batteries. These findings have stimulated much industrial and academic interest in the electro-chemistry of conducting polymers and their possible technological application in energy storage. This activity is related in no small part to the fact that the range of inorganic materials in batteries is limited, whereas the ability of the chemist to 'tailor-make' and modify chemical, electrochemical and physical properties of organic polymers presents an almost unlimited number of potential electrode-active materials exhibiting any desired, pre-selected properties.

This chapter will summarize the basic concepts related to the chemical and electrochemical doping of an organic polymer to the metallic regime and to the application of these concepts to the use of organic polymers as electroactive materials in rechargeable batteries. The coverage is intended to be exemplary rather than exhaustive. Furthermore, since by far the largest amount of published material has dealt with $(CH)_x$ electrodes, it is this material which will be discussed to the greatest extent. The basic concepts involving $(CH)_x$ electrodes also apply to other polymer electrodes such as poly(p-phenylene), polythiophene, polypyrrole and polyaniline. The chief advantages and problems in the use of such polymers for energy storage will be described and future trends in the field will be indicated.

2. p- AND n-DOPING OF POLYACETYLENE

When it was first discovered[10] that the conductivity of $(CH)_x$ could be increased by up to 12 orders of magnitude by reaction with small quantities of electron-accepting or electron-donating species, the phenomena were termed 'p-doping' and 'n-doping', respectively, by analogy with the doping of a classical semi-conductor, such as silicon. Selected examples of chemical p- and n-doping of $(CH)_x$ are given in Table 1. Phenomenologically this designation is correct, in that large

TABLE 1
Selected Dopants for $(CH)_x$

Substance	Conductivity (S/cm)
cis-$(CH)_x$	1.7×10^{-9}
trans-$(CH)_x$	4.4×10^{-5}
p-Doping (oxidation)	
I_2 vapor: $[(CH^{0.07+})(I_3^-)_{0.07}]_x$	5.5×10^2
AsF_5 vapor: $[(CH^{0.1+})(AsF_6^-)_{0.1}]_x$	1.2×10^3
$HClO_4$ (liquid or vapor):	
$[(CH\{OH\}_{0.08})^{0.12+}(ClO_4)_{0.12}^-]_x$	5×10^1
Electrochemical: $[(CH^{0.1+})(ClO_4^-)_{0.1}]_x$	1×10^3
n-Doping (reduction)	
Li naphthalide: $[Li_{0.2}^+(CH^{0.2-})]_x$	2×10^2
Na naphthalide: $[Na_{0.2}^+(CH^{0.2-})]_x$	2.5×10^1
Electrochemical: $[Li_{0.1}^+(CH^{0.1-})]_x$	10^1–10^2

increases in conductivity are observed when the material takes up very small quantities of certain chemical species. However, as a better understanding of the nature of the doping process in $(CH)_x$ developed, it has become apparent that the doping of conducting polymers is conceptually completely different from the doping of a classical semi-conductor. p-Doping of a conducting polymer refers to the partial oxidation of the polymer, for example

$$(CH)_x \rightarrow [(CH^{y+})]_x + (xy)e^- \qquad (1)$$

This may be accomplished either chemically or electrochemically.[11] In order to preserve electrical neutrality in the system a counter-anion, A^-, must also be provided:

$$[(CH^{y+})]_x + (xy)A^- \rightarrow [(CH^{y+})A_y^-]_x \qquad (2)$$

For example, when $(CH)_x$ is p-doped with iodine, the reaction can be regarded as occurring in the following steps:

$$(CH)_x + \frac{xy}{2}I_2 \rightarrow (CH^{y+})_x + (xy)I^- \qquad (3)$$

$$(xy)I^- + (xy)I_2 \rightarrow (xy)I_3^- \qquad (4)$$

$$(CH^{y+})_x + (xy)I_3^- \rightarrow [(CH^{y+})(I_3^-)_y]_x \qquad (5)$$

Analogously, n-doping refers to the partial reduction of the conducting polymer, for example

$$(CH)_x + (xy)e^- \rightarrow [(CH^{y-})]_x \qquad (6)$$

Again, in order to preserve electrical neutrality, a counter cation, M^+, must be provided:

$$[(CH^{y-})]_x + (xy)M^+ \rightarrow [M_y^+(CH^{y-})]_x \qquad (7)$$

In the n-doping of $(CH)_x$ by sodium naphthalide, the strongly reducing naphthalide radical anion, made by dissolving metallic sodium in a solution of naphthalene in tetrahydrofuran,[12] is used. The lowest energy π^* molecular orbital in $(CH)_x$ is apparently of lower energy than the π^* orbital of the naphthalide radical anion containing the unpaired electron, since the electron is spontaneously transferred from the π^* orbital of the naphthalene to the π^* orbital of the $(CH)_x$, when $(CH)_x$ is placed in the solution of sodium naphthalide. The naphthalide radical anion acts as the reducing agent:

$$(CH)_x + (xy)Nphth^{\cdot -} \rightarrow [(CH^{y-})]_x + (xy)Nphth \qquad (8)$$

while the Na⁺ ion acts as the 'dopant' counter-cation:

$$(CH^{y-})_x + (xy)Na^+ \rightarrow [Na_y^+(CH^{y-})]_x \tag{9}$$

3. BASIC CONCEPTS OF ELECTROCHEMISTRY

The non-electrochemist is reminded that some of the concepts discussed in Chapter 1 are necessary to the understanding of secondary battery systems that utilize electroactive polymers, in particular the following.

(i) All electrochemical reactions are a combination of two 'half-reactions' occurring at the electrodes. An oxidation reaction occurs at the anode and a reduction reaction occurs at the cathode. During discharge of the cell, the anode is negative and the cathode positive, but when a secondary battery is recharged, the negative electrode becomes the cathode and the positive electrode becomes the anode.

(ii) Each electrode reaction is characterized by a standard potential and the combination of anode and cathode potentials gives the cell reaction potential $E_{cell}^{\ominus}$, which must have a positive value if the cell reaction is to be spontaneous. $E^{\ominus}$ is close to the measured open circuit voltage, V_{oc}.

(iii) When a battery is discharged rapidly, it tends to become polarized due to kinetic factors which include local depletion of anions and/or cations. Of particular importance in the case of polymer electrodes is the current limiting effect of diffusion of dopant ions from the interior of a polymer electrode to the surface in contact with the electrolyte, and vice versa.[13]

(iv) It can be useful to study battery characteristics both galvanostatically (i.e. monitoring the charge or discharge voltage with time at a pre-selected constant applied current) and potentiostatically (i.e. monitoring current with respect to time under a constant pre-selected applied voltage).

(v) Useful performance parameters for a battery include the capacity in ampere hours (Ah), the energy density in terms of watt hours per kilogram (Wh/kg) or per litre (Wh/litre) and the power density in watts per kilogram (W/kg) or per litre (W/litre).

4. POLYACETYLENE AS AN ELECTRODE MATERIAL

The free-standing flexible, silvery films of cis-$(CH)_x$ synthesized by the usual Shirakawa method[14] are ~0·1 mm thick and have a bulk density of ~0·4 g/cm^3.[15] The density by flotation methods is ~1·2 g/cm^3; hence the films are approximately two-thirds void space and have a surface area of ~60–100 m^2/g.[16] Polyacetylene is insoluble in all solvents, begins to decompose in a vacuum at ~300°C and reacts slowly with air during a day or two.[15] The films consist of a mat of interconnecting fibrils whose typical diameter is ~200–800 Å. The thickness, bulk density and fibril diameter vary significantly according to variations in synthetic procedures. For example, thin ~2000 Å films consist of fibrils having a diameter of ~50–100 Å. Polyacetylene powder which exhibits the typical fibrillar morphology can also be produced.

Free-standing 0·1 mm cis-$(CH)_x$ films have been most commonly used in electrochemical and battery type studies.[13] The film is usually placed on a metal mesh gauze (current collector) of platinum, nickel or stainless steel, which is then folded tightly upon itself so as to encase the film. This assembly serves as the electrode. Glass filter paper or porous polypropylene film is used as a separator, i.e. 'spacer', between the anode and cathode materials. It sometimes is also used as a solid, porous medium to hold the electrolyte. Electrodes consisting of $(CH)_x$ films on which gold has been evaporated or sputtered on one side to serve as a current collector have also been used. Cis-$(CH)_x$ is usually preferred in constructing an electrode since it is more flexible than the (more thermodynamically stable) trans-isomer. If the trans-isomer should be required it can be prepared conveniently by heating the cis-isomer in a vacuum or in an inert atmosphere for ~1 h at ~120–150°C.[17] Since the cis-isomer is converted to the trans-isomer on electrochemical p- or n-doping,[18,19] the cis-isomer used in fabricating electrodes will have been converted to the trans-isomer after one or two electrochemical oxidation–reduction cycles.

Both p- and n-doped $(CH)_x$ are extraordinarily sensitive to even minute traces of oxygen (air) and/or water which, under certain experimental conditions, lead to destruction of the electroactivity of the material.[18] Minute quantities of unknown impurities in the electrolytes are also believed to degrade the material in certain cases. Hence best electrochemical results have been obtained when cells having the smallest possible volume employing the smallest amount of electrolyte are used. Most studies have been carried out employing cells which have been evacuated and sealed to exclude air.[13]

Although n-doped (reduced) $(CH)_x$ appears to react with air and/or water under all conditions,[19] it should be noted that p-doped (oxidized) $(CH)_x$ is relatively stable to water if the solution is acidic; indeed, battery-type studies have been carried out on p-doped $(CH)_x$ cathodes in aqueous electrolytes.[20] In general, increasing the effective acidity of either an aqueous or a non-aqueous electrolyte appears to be a good general method for increasing the chemical stability of p-doped $(CH)_x$, not only to water impurities, but possibly also to other unknown impurities by which the $(CH^{y+})_x$ polycarbonium ion is susceptible to nucleophilic attack.[21] It should also be noted that oxygen does not necessarily destroy the electroactivity of $(CH)_x$. As shown in Section 10, recent studies demonstrate that gaseous oxygen will actually p-dope $(CH)_x$ to the metallic regime in aqueous acid solution.[21]

Of extreme importance in determining the usefulness of a $(CH)_x$ electrode is the rate of diffusion of 'dopant' counter-ions to and from the bulk of the electrolyte through the void spaces in the film to the polymer fibrils. Of even greater importance is the rate of diffusion of the dopant ions from the surface of a fibril to its interior and from the interior to its surface. If any of these processes are slow, the rate of charging and discharging of a $(CH)_x$ electrode will be impaired. This results in polarization problems, and full utilization of the $(CH)_x$ for practical electrochemical energy storage purposes will not be realized.

In this respect it should be noted that *in situ* electrochemical visible/near-IR transmittance studies of ~ 2000 Å films of $(CH)_x$ (fibril diameter ~ 50 Å) on conducting glass show conclusively that the $(CH)_x$ can be readily oxidized or reduced throughout the whole fibril thickness.[22] One is not, therefore, dealing with electrical storage by capacitance effects.[23] Spectroscopic and electrochemical studies of these thin films and also of free-standing films give diffusion constants of $\sim 10^{-18}$ cm^2/s for diffusion of ClO_4^- and Li^+ ions into and out of the fibrils.[13] There is significant disagreement as to the validity of these values depending on how the raw experimental data are interpreted. Values as low as $\sim 10^{-12}$ cm^2/s have been reported.[24] The rate at which charge can be inserted into or removed from the interior of $(CH)_x$ fibrils will be limited by the rate of diffusion of the dopant counter-ion. The small values of the diffusion constants would suggest that the rate at which $(CH)_x$ electrodes could be charged or discharged would be small; however the experimentally determined power densities are unexpectedly large.[13] It has been suggested that when a potential is applied to a $(CH)_x$ electrode in an electrochemical cell, field-enhanced diffusion will occur, resulting in rates of diffusion of dopant ions into and out of $(CH)_x$ fibrils which are several orders of magnitude greater than those obtained above in the absence of an applied field.[25]

It should also be noted that in order for electrochemical oxidation or reduction of a material to occur at a rate sufficiently great for it to be used in batteries it must have an appropriate conductivity, even if small, in its lightly doped form. If its conductivity is very small, its rate of oxidation or reduction can be increased by mixing it, in the form of a powder, with ~5% of a conducting substance such as carbon black powder.

The above concepts, which have been studied primarily for $(CH)_x$, also presumably apply in appropriately modified forms to the electrochemistry of other conducting polymers. Considerably more work needs to be done in order to understand factors which control the rates of electrochemical oxidation and reduction of conducting polymers including engineering aspects, for example on the use of $(CH)_x$ film or powder, the method of current collection, and the type of current collectors needed.

5. THE POLYACETYLENE CATHODE

As pointed out in Section 3, the cathode of a battery is defined as the electrode at which reduction occurs. During discharge, this is the positive electrode; at initial charge, or under subsequent recharge, this same positive electrode becomes an anode. This section deals with the use of polyacetylene as a cathode during cell *discharge*. Thus, if $[(CH^{y+})A_y^-]_x$ is being reduced to a lower oxidation state, including neutral $(CH)_x$, then it is acting as the cathode. Many studies of this type have been carried out with p-doped $(CH)_x$. A large number of studies have also been performed in which neutral $(CH)_x$ is reduced to $[M_y^+(CH^{y-})]_x$ during discharge. Since such reactions also involve reduction of $(CH)_x$, the polymer is also functioning as a cathode in this type of reaction. Note that in the first type of reaction p-doped polyacetylene is involved; in the second type of reaction n-doped polyacetylene is involved. These two types of polyacetylene cathodes will be discussed below.

5.1. Use of p-Doped (Oxidized) Polyacetylene

When the positive and negative terminals of a d.c. power source are applied to a $(CH)_x$ and a Li electrode respectively in an electrolyte such as a solution of $LiClO_4$ in propylene carbonate (PC), electrochemical oxidation of polyacetylene proceeds readily. The electrode reactions are:

At the $(CH)_x$ (anode)

$$(CH)_x \rightarrow (CH^{y+})_x + (xy)e^- \qquad (10)$$

At the Li (cathode)

$$(xy)Li^+ + (xy)e^- \rightarrow (xy)Li \qquad (11)$$

resulting in the generalized overall charging reaction:

$$(CH)_x + (xy)Li^+ \rightarrow (CH^{y+})_x + (xy)Li \qquad (12)$$

For a specific anion, e.g. ClO_4^-, the resulting net reaction is:

$$(CH)_x + (xy)Li^+(ClO_4^-) \rightarrow [(CH^{y+})(ClO_4^-)_y]_x + (xy)Li \qquad (13)$$

If the power supply is removed and the two electrodes are connected by a wire, spontaneous electrochemical reactions (discharge reactions), which are the reverse of those given by eqns (10)–(13) occur as electrons pass through the wire from the Li to the $[(CH^{y+})(ClO_4^-)_y]_x$, regenerating $(CH)_x$ and $LiClO_4$. These types of reactions provide the basis for the use of $[(CH^{y+})A_y^-]_x$ as a possible cathode-active material in rechargeable batteries. Cells of this type employing electrolytes consisting of salts of (ClO_4^-), (AsF_6^-), (PF_6^-), (BF_4^-) and $(CF_3SO_3^-)$ dissolved in propylene carbonate or sulfolane have been studied.[26] Tetrahydrofuran (THF) is not suitable for extended use as a solvent since at potentials greater than ~3·2 V versus Li^+/Li it polymerizes. The most extensive investigations have been performed on cells employing solutions of $LiClO_4$ in propylene carbonate as the electrolyte and Li metal as the anode.

Such a cell may be charged in a series of constant potential steps until the final desired potential is achieved. Good results have also been obtained by charging at a constant applied current at the rate of 1% doping per hour (e.g. 0·142 mA for 1·2 cm² of $(CH)_x$ film; mass 6·9 mg).[13,27] The charging process can be interrupted periodically and the V_{oc} measured. The V_{oc} is found to fall rapidly during the first 5 min and then more slowly during the following 24 h, during which quasi-equilibrium conditions for the diffusion of (ClO_4^-) ions from the exterior to the interior of the $(CH)_x$ fibrils is attained.[13,27] A decrease of ~0.2 V was observed after 24 h, for example, when $(CH)_x$ was oxidized to 3.8 V versus Li^+/Li at a rate of 2% per hour. The V_{oc} (24 h) value of 3.6 V corresponds to ~3.8% oxidation as determined by the quantity of electricity involved in the charging process. The magnitude of the decrease on standing will depend on the rate of charging, the time elapsed between the termination of the charging step and the measurement of the open circuit voltage, etc.

If the open circuit voltage is measured 24 h after termination of the charging step and the $[(CH^{y+})(ClO_4^-)_y]_x$ is then reduced electrochemically back to a potential of 2.50 V versus Li^+/Li, characteristic of neutral $(CH)_x$, the number of coulombs liberated can be used to calculate the degree of oxidation corresponding to the open circuit voltage observed after the above 24 h period.

Although the potential of the electrode continues to fall slowly, at varying rates depending on the nature of the electrolyte, etc., due to self-discharge and related effects, most of the drop in potential during the first day is believed to be caused by diffusion effects within the $(CH)_x$ fibrils. This is consistent with the observation that the potential of the electrode *rose* by almost 1 V (from 2.50 to 3.40 V) during 20 min after completion of a constant current [0.55 mA/3.3 mg, ~1 cm² $(CH)_x$] discharge to 2.50 V.[13] This can be explained by the diffusion of ClO_4^- ions from the interior to the exterior of the fibrils during the 20 min stand period.

A typical 1 h charge cycle of a cell at a constant current of 0.189 mA/cm² from an initial uncharged open circuit voltage of 2.75 V to 3.78 V is given in Fig. 2.[27] From the quantity of electricity involved in this process, Q_{in} coulombs, the polyacetylene was oxidized to 2%, i.e. to a composition of $[(CH^{0.02+})(ClO_4^-)_{0.02}]_x$. It can be seen that the charging potential rose rapidly at first and then as the $(CH)_x$ became more highly conducting it increased only slowly. After a 16 h stand period following completion of the charge cycle, it was found that the open circuit voltage had dropped to the near-equilibrium potential of 3.55 V. When the cell was discharged at a constant current of 0.10 mA/cm² of film, controlled by means of a galvanostat, the discharge curve lay close to the V_{oc} (24 h) quasi-equilibrium curve during most of the discharge, indicating only slight polarization of the electrode. Some residual charge could still be removed from the film during the final constant potential discharge at 2.50 V.

Discharge characteristics of a $Li/LiClO_4(PC)/[(CH^{y+})(ClO_4^-)_y]$ cell

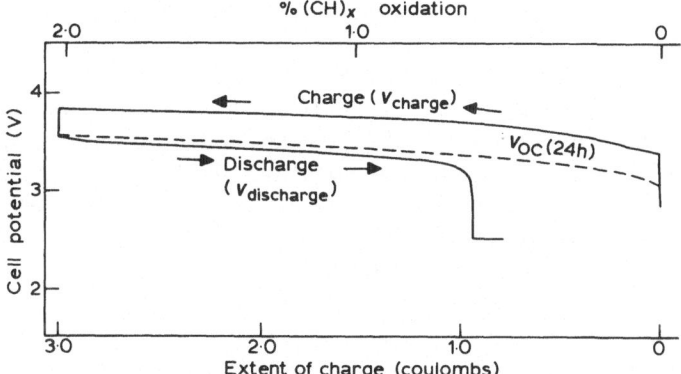

Fig. 2. Comparison of charge and discharge potentials and quasi-equilibrium cell potentials of a Li/LiClO$_4$(PC)/[(CH^{y+})(ClO$_4^-$)$_y$]$_x$ cell as a function of extent of charging (percentage oxidation of polyacetylene).

containing a cathode of 7·0% oxidized (CH)$_x$ are given in Fig. 3 for several different constant current discharges.[27] The discharge studies were performed immediately after a charge cycle was completed.

These curves are typical of those observed for Li/Li$^+$A$^-$/[(CH^{y+})A$_y^-$]$_x$ cells, but it should be stressed that they will depend, like those for any battery, sometimes significantly, on a variety of different parameters including the thickness of the film, the fibril diameter of

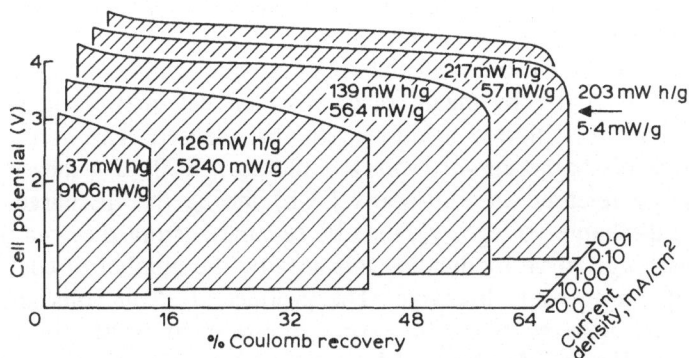

Fig. 3. Discharge characteristics of a Li/LiClO$_4$(PC)/[(CH^{y+})(ClO$_4^-$)$_y$]$_x$ cell containing a cathode of 7·0 mol % oxidized polyacetylene.

the $(CH)_x$, the nature of A^-, the nature of the organic solvent, the concentration of the electrolyte, the type of spacer, the method of attaching the current collector to the film, etc. In each case a sufficient quantity of electricity to oxidize the $(CH)_x$ to 7·0%, Q_{in}, was used in the charging (oxidizing) process. Constant applied discharge currents controlled by a galvanostat ranging from 0·01 to 20·0 mA/cm² of $(CH)_x$ film were employed in each of the five separate discharge studies. The discharge voltage, V_d, was recorded in each discharge until the $(CH)_x$ potential dropped to 2·50 V, characteristic of neutral $(CH)_x$.

The total quantity of electricity liberated (Q_{out}) on reaching this voltage was recorded. The percentage coulomb recovery at the end of a constant current discharge is therefore given by 100 (Q_{out}/Q_{in}). Residual charge was released during a final constant applied potential discharge at 2·50 V for ~16 h resulting in an increase in the total coulomb recovery, 100 ($Q_{out(total)}/Q_{in}$). For example, in the case of the 20 mA discharge, the coulombic recovery rose from 13·5% to 60%. Depending on discharge conditions, cell construction, etc., coulombic recoveries in the 90–95% range have been obtained in cells employing $(CH)_x$ cathodes oxidized to 6–10% levels.[26,28]

The cell potential, V_d, may be plotted against coulombs liberated during the discharge process, or against discharge time since the number of coulombs liberated at a constant current discharge is directly proportional to the discharge time. It may also be plotted against '% coulomb recovery' since the total coulombs liberated at any given instant of the discharge process are directly proportional to the percentage coulomb recovery at that stage in the discharge process. This latter method is used in Fig. 3.

Energy densities are calculated using the charge released on discharge to 2·50 V by means of curves identical to those in Fig. 3, except that V_d is plotted against coulombs released. The area under each such discharge curve, (volts × coulombs)/3600, is a measure of the energy released in watt-hours (W h) during the discharge. The average discharge voltage, $\bar{V}_d$, may be calculated from the area under each discharge curve by dividing this area by the number of coulombs released during each discharge. The product of $\bar{V}_d$ in volts and the current in amperes gives the average power in watts during the discharge. Energy densities and average power densities are based only on the mass of the $[(CH^{0.07+})(ClO_4^-)_{0.07}]_x$ employed and on the mass of the Li metal consumed in the discharge reaction. The energy density and power densities in Fig. 3 are given as milliwatt-hours/gram (mW h/g) and milliwatts/gram (mW/g) respectively. These values are, of course, numerically equal to the corresponding values in watt-hours/kilogram (W h/kg) and watts/kilogram (W/kg), respectively.

As is the case in the discharge of all batteries, greater discharge rates result in a smaller coulombic recovery, i.e. a smaller effective capacity at the time the potential of the battery has dropped to a value characteristic of its discharged state. The resulting energy density, $Q_{out} \times \bar{V}_d$, is also smaller because of the smaller quantity of electricity involved and because the average discharge voltage, $\bar{V}_d$, is also smaller at higher discharge rates. Under ideal circumstances the stored coulombs and stored energy are still present in the cathode material at the end of a rapid discharge, and indeed can be recovered in a subsequent slower discharge. Their incomplete release during rapid discharge is due to kinetic factors—'polarization' effects—related to a number of factors including concentration gradients in the electrolyte and in the electrode material. In the case of $[(CH^{y+})A_y^-]_x$ electrodes, the rate of removal or addition of charge is believed to be due primarily to the limiting rate of diffusion of the dopant ion within the polyacetylene fibrils.[13,24,26]

The constant current discharge characteristics given in Fig. 3 are very satisfactory in that they show good plateau regions, i.e. the discharge voltage stays relatively constant until it drops rapidly when the battery approaches its discharged state. This is what is sought in a battery.

The energy densities compare favorably with, for example, those of a lead/acid battery which has a theoretical energy density (energy density at an infinitely slow rate of discharge) of 186 W h/kg[29] based only on the masses of electrolyte materials involved in the discharge reactions. The observation (see Fig. 3) that an organic polymer can release its stored charge on demand at such a high rate, i.e. that it exhibits a good power density, is surprising. However, it is very difficult to predict what power density a completely packaged polyacetylene cell of this type would have since it depends greatly on engineering aspects involved in the fabrication of the battery.

A solid state battery employing oxidized polyacetylene as the cathode-active material has also been reported.[30] The cell, $Ag/RbAg_4I_5/[(CH^{0.07+})(I^-)_{0.07}]_x$, operates at room temperature with a V_{oc} of 0·65 V, an energy density of ~10 W h/kg, a current density up to 50 mA/cm² of $(CH)_x$ and a long lifetime, with a test cell functioning for at least two years. Although the voltage and energy density of such a cell are low, the uses suggested for it include powering electronic devices and liquid crystal displays.

In order for a battery to have commercial usefulness it must exhibit good retention of charge when standing in its charged state. It must also be capable of undergoing repeated charging and discharging over many hundreds of cycles. As far as can be gathered from published reports at the present time, the $[(CH^{y+})A_y^-]_x$ cathode shows less than ideal characteristics in both these respects with a variety of anions in a variety of electrolytes, at least when $y > \sim 0.04$.[26] It tends to lose its charge on standing, and on repeated cycling it is chemically destroyed, resulting in a serious loss in capacity.[31] At lower doping levels these characteristics appear to be greatly improved.[27,28] However, very recent studies[32] have shown that when the p-doped polyacetylenes, $[(CH^{y+})A_y^-]_x$ [where $y = 0.06$–0.08 and $A^- = (BF_4^-)$, (PF_6^-) and $(CF_3SO_3^-)$], are held in a vacuum at temperatures between 60 and 80°C for several hours they decompose spontaneously with the formation of BF_3, PF_5, CF_3SO_3H, etc. Hence, if spontaneous decomposition occurs even in the absence of an electrolyte at temperatures at which batteries might be expected to be exposed, it seems clear that p-doped polyacetylene can in all probability not be used as a cathode in a functional commercial battery. It should be noted, however that n-doped polyacetylene is stable on heating, $[K_y^+(CH^{y-})]_x$ for example being stable up to $\sim 220°C$.[32]

5.2. Use of n-Doped (Reduced) Polyacetylene

As shown earlier, the electrochemical oxidation of $(CH)_x$ in a cell consisting of a $(CH)_x$ electrode and a Li electrode in an electrolyte such as $LiClO_4$ in propylene carbonate is a thermodynamically non-spontaneous process. However, the reduction potentials of the Li^+/Li couple and $(CH)_x$ couples show that the electrochemical reduction of $(CH)_x$ in such a cell is a thermodynamically spontaneous process.[33] Thus if the electrodes are connected by a wire, a current flows spontaneously and the $(CH)_x$ is reduced. This may be regarded as the discharge reaction of a battery cell where $(CH)_x$ is acting as the cathode. The electrode reactions are:

At the $(CH)_x$ (cathode)

$$(CH)_x + (xy)e^- \rightarrow (CH^{y-})_x \qquad (14)$$

At the Li (anode)

$$(xy)\text{Li} \rightarrow (xy)\text{Li}^+ + (xy)e^- \qquad (15)$$

resulting in the overall reduction (discharge) reaction:

$$(\text{CH})_x + (xy)\text{Li} \rightarrow [\text{Li}_y^+(\text{CH}^{y-})]_x \qquad (16)$$

Other alkali metals may be used in place of lithium. The charging reactions are the reverse of those given by eqns (14)–(16). If the study is performed in a cell using an electrolyte of $\text{Bu}_4\text{N}^+ \text{ClO}_4^-$ in THF in which the $(\text{CH})_x$ and Li are separated by a glass frit to prevent diffusion of Li^+ ions from the Li to the $(\text{CH})_x$, then n-doped $(\text{CH})_x$ containing Bu_4N^+ dopant cations can be obtained, i.e.

$$(\text{CH})_x + (xy)\text{Li} + (\text{Bu}_4\text{N}^+)(\text{ClO}_4^-) \rightarrow [(\text{Bu}_4\text{N}^+)_y(\text{CH}^{y-})]_x + (xy)\text{LiClO}_4 \qquad (17)$$

The coulombic recovery is much better than for p-doped polyacetylene. For example, after a 48 h stand period following reduction to a given level, the quantity of electricity involved in oxidizing the $[\text{Li}_y^+(\text{CH}^{y-})]_x$ (where $y \leq 0.1$) back to neutral $(\text{CH})_x$ were equal, within experimental error, to the quantity of electricity used in the reduction process.

In view of the above observations, it is therefore not surprising to find that both $(\text{CH})_x$ and $(\text{CH}^{y-})_x$ (where $y \leq 0.1$) show excellent stability in a $\text{LiClO}_4(\text{THF})$ electrolyte over a period of four months (at which time the study was terminated), as shown in Fig. 4.[34] As can be seen, the open circuit voltage, V_{oc}, between a $(\text{CH})_x$ electrode and a $[\text{Li}_{0.07}^+(\text{CH}^{0.07-})]_x$ electrode remained constant during this period, indicating absence of any detectable self-discharge or degradation. As can be seen from Fig. 4 the cell also exhibited stable, short circuit current characteristics throughout the period, and good discharge characteristics were obtained at the end of the period. An analogous study of $(\text{CH})_x$ and $[\text{Na}_{0.07}^+(\text{CH}^{0.07-})]_x$ electrodes in a $\text{NaPF}_6(\text{THF})$ electrolyte showed similar stability.[35]

Constant current discharge studies were performed on Li/$[\text{Li}_y^+(\text{CH}^{y-})]_x$ and Na/$[\text{Na}_y^+(\text{CH}^{y-})]_x$ cells from $y = 0$ to $y = 0.10$ as described for the Li/$[(\text{CH}^{0.07+})(\text{ClO}_4)_{0.07}^-]_x$ cells discussed previously.[35] The corresponding discharge curves are given in Figs 5 and 6

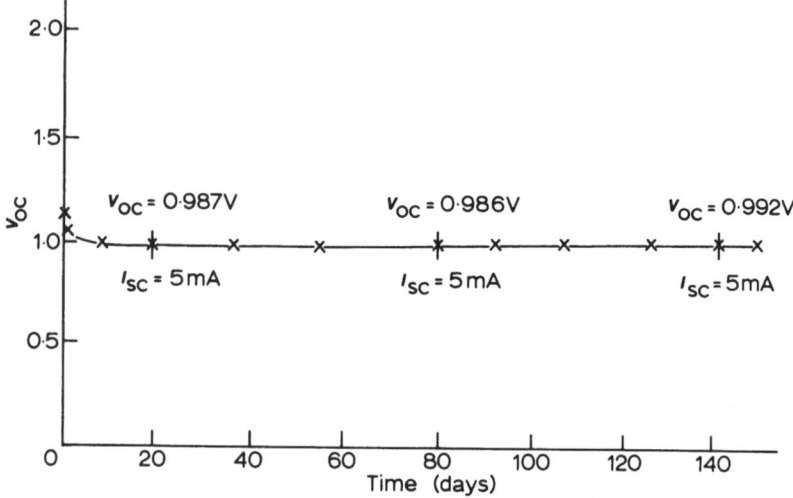

Fig. 4. Open circuit voltage, V_{oc}, and short circuit current, I_{sc}, versus stand time, demonstrating the stability in potential and current of a $[Li^+_{0.07}(CH^{0.07-})]_x/LiClO_4(THF)/(CH)_x$ cell.

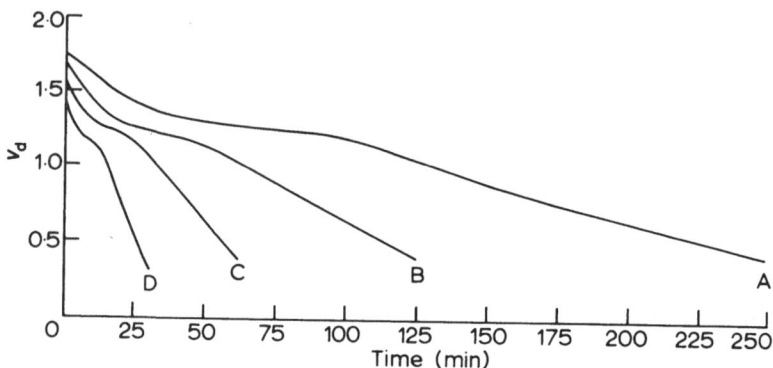

Fig. 5. Constant current discharge characteristics of polyacetylene reduced to 10·0 mol %, i.e. $[Li^+_{0.10}(CH^{0.10-})]_x$, at 0·16 mA (curve A), 0·32 mA (curve B), 0·64 mA (curve C) and 1·28 mA (curve D) in a $Li/LiClO_4(THF)/(CH)_x$ cell employing 3·2 mg (0·9 cm²) of $(CH)_x$.

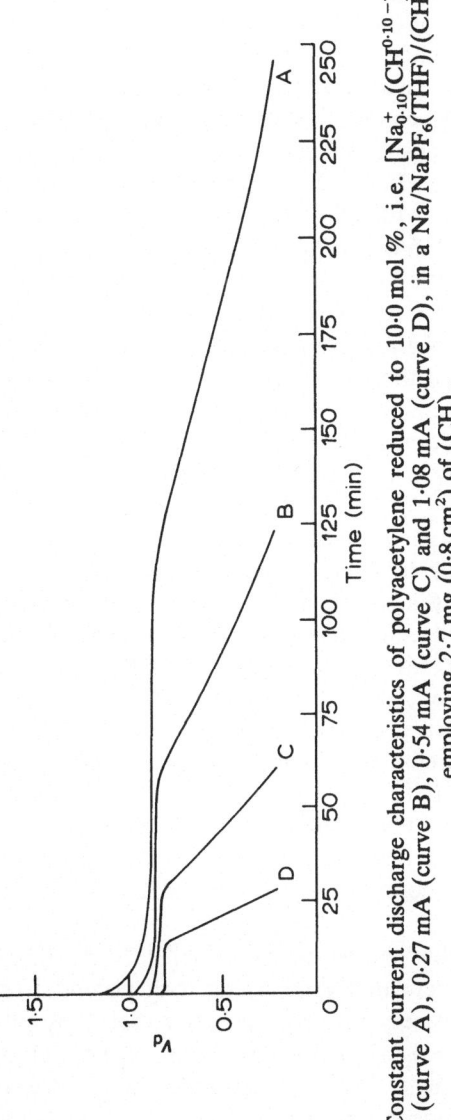

FIG. 6. Constant current discharge characteristics of polyacetylene reduced to 10·0 mol %, i.e. $[Na_{0.10}^{+}(CH^{0.10-})]_x$, at 0·135 mA (curve A), 0·27 mA (curve B), 0·54 mA (curve C) and 1·08 mA (curve D), in a $Na/NaPF_6(THF)/(CH)_x$ cell employing 2·7 mg (0·8 cm^2) of $(CH)_x$.

respectively. The different discharge voltages result from the fact that the standard reduction potential of the Na^+/Na couple is 0·35 V more positive than that of the Li^+/Li couple.

The discharge curves of the $Li/[Li_y^+(CH^{y-})]_x$ cells are associated with the following characteristics after a final 16 h discharge at 2·50 V (average discharge voltage; total coulombic recovery; energy density; average power density): Curve A, 0·98 V, 101·4%, 201·5 W h/kg, 48·9 W/kg; Curve B, 0·96 V, 100·7%, 197·6 W h/kg, 95·9 W/kg; Curve C, 0·92 V, 101·8%, 188·9 W h/kg, 184·2 W/kg; Curve D, 0·81 V, 100·1%, 166·6 W h/kg, 322·8 W/kg. The capacity for each of the above discharge reactions is ~205 A h/kg. The energy densities and power densities are based only on the mass of $(CH)_x$ employed and on the mass of Li or Na consumed in the discharge reaction. The discharge curves of the $Na/[Na_y^+(CH^{y-})]_x$ cells are associated with the following characteristics (average discharge voltage; total coulombic recovery; energy density; average power density): Curve A, 0·66 V, 101·0%, 115·4 W h/kg, 28·1 W/kg; Curve B, 0·65 V, 100·4%, 113·2 W h/kg, 55·1 W/kg; Curve C, 0·63 V, 101·1%, 109·7 W h/kg, 106·76 W/kg; Curve D, 0·61 V, 101·5%, 105·7 W h/kg, 205·6 W/kg. The capacity for each of the above discharge reactions is ~174 A h/kg.

It can be seen that the general shapes of the discharge curves and the change in cell characteristics with increasing discharge rates are similar to those for the $Li/[(CH^{0·07+})(ClO_4^-)_{0·07}]_x$ cells described earlier. As expected, in view of the more negative reduction potentials and smaller relative atomic mass of Li compared with Na, the $Li/[Li_y^+(CH^{y-})]_x$ cells have higher energy and power densities per kg than the $Na/[Na_y^+(CH^{y-})]_x$ cells. The Na cells, however, show a flatter and larger plateau effect. The plateau effects in the Li- and Na-doped polyacetylene have been shown to be related to a 'staging' effect[36] analogous to that observed in the intercalation of graphite with alkali metals.

The recycling characteristics of the $Li/[Li_y^+(CH^{y-})]_x$ cell, which have been studied on repeated discharging to a potential characteristic of 10·0% doped polyacetylene, are far better than those observed for the $Li/[(CH^{0·07+})(ClO_4^-)_{0·07}]_x$ cells. However, after 100 deep discharge cycles using a $LiClO_4$(THF) electrolyte, some degradation of the polyacetylene has been observed and the capacity is reduced to ~60% of its original value.[37]

It is believed that at n-doping levels greater than ~8–10%, the $(CH^{y-})_x$ species is a sufficiently strong reducing agent to reduce the

$(ClO_4)^-$ ion in the $LiClO_4$ electrolyte in the $Li/[Li_y^+(CH^{y-})]_x$ cells and at doping levels >14% the $(CH^{y-})_x$ can reduce the (PF_6^-) ion in the $NaPF_6$ electrolyte of $Na/[Na_y^+(CH^{y-})]_x$ cells.[38]. This seriously limits the capacity of the cells. However, two different approaches have been announced[38] very recently which appear to overcome the problem and permit the attainment of stable higher n-doping levels.

The first approach[38] involves the discovery of an electrolyte which is inert to highly n-doped polyacetylene. By changing the electrolyte salt from $LiClO_4$ to an organoborate salt, $Li^+BR_4^-$ (where R = phenyl, n-butyl) or $Na^+(BEt_4^-)$ or $K^+[B(n-Bu)_4^-]$ the maximum value of y can be extended to 0·18 for $[Li_y^+(CH^{y-})]_x$ and to 0·16 for $[Na_y^+(CH^{y-})]_x$ and $[K_y^+(CH^{y-})]_x$ in solvents such as THF and 2-methyl-THF. These doping levels can be attained at an applied potential of ~ +0·20 V versus Li^+/Li in the case of the lithium salts. Coulombic recoveries in excess of 98% can be obtained on reoxidizing to $(CH)_x$ after a stand period of two weeks or more. The voltage of the Li battery changes from ~0·5 to 0·15 V as y changes from 0·09 to 0·18. The voltage of the $[Na_y^+(CH^{y-})]_x$ electrode changes from 0·7 to 0 V as y changes from 0·08 to 0·15. The theoretical capacity of the $[Li_{0.18}(CH^{0.18-})]_x$ electrode has the promisingly high value of 338 A h/kg.

The second approach[39] involves the chemical attachment to the surface of the n-doped polyacetylene fibrils of a very thin layer of a different polymer which separates the $[M_y^+(CH^{y-})]_x$ in the interior of the polymer fibrils from any chemically reactive species in the electrolyte. The thin layer of the polymer which has been attached to the polyacetylene is, however, an ionic conductor so that alkali metal ions can pass through it, into or out of the $[M_y^+(CH^{y-})]_x$. This method appears to be effective in using n-doped polyacetylene in an electrolyte containing a solvent such as propylene carbonate in which n-doped polyacetylene is otherwise not very stable. However, the rates of charging and discharging at a given potential are somewhat reduced because of the smaller ionic conductivity of the attached polymer layers compared with the liquid electrolyte. The attachment of the surface coating is the result of a reaction between reduced polyacetylene and either an epoxide such as ethylene oxide or a cyclic sulfonate ester such as 1,4-butane sultone.[39] The products are a poly(ethylene oxide)-type polymer and an alkyl sulfonate salt, respectively, bonded to the polyacetylene surface. In the former case, the coating is best applied by exposing alkali-metal-doped polyacetylene to the monomer in the form of a gas or dissolved in an inert solvent,

$$M^+\left[H-\underset{\|}{C}-\right]^- + \overset{O}{\underset{CH_2-CH_2}{\diagup\diagdown}} \longrightarrow M^+\left[H-\underset{|}{C}-CH_2-CH_2-O\right]^- \quad (18)$$

$[M_y^+(CH^{y-})]_x$ surface-coated $[M_y^+(CH^{y-})]_x$

In the latter case, the sultone may be added to the propylene carbonate or THF electrolyte of an electrochemical cell containing a polyacetylene electrode. The sultone reacts at the polymer surface as soon as the electrode is reduced:

$$M^+\left[H-\underset{\|}{C}-\right]^- + \begin{array}{c} O \quad\;\; O \\ \diagup \;\; \diagdown\!\!\!\diagup \\ H_2C \quad\;\; S\!\!=\!\!O \\ | \quad\quad | \\ H_2C \quad CH_2 \\ \diagdown\; C \;\diagup \\ H_2 \end{array} \longrightarrow M^+\left[H-\underset{|}{C}-(CH_2)_4-SO_3\right]^- \quad (19)$$

$[M_y^+(CH^{y-})]_x$ surface-coated $[M_y^+(CH^{y-})]_x$

Neither perchlorate anion, propylene carbonate, nor excess sultone reacts with the polymer, regardless of its state of charge after the surface barrier has been formed.

Cycling of a $[Li_y^+(CH^{y-})]_x/LiClO_4$(propylene carbonate)/Li cell between 1·5 and 0·2 V resulted in y changing from 0·08 to 0·13.[39] The composition of the $[Li_y^+(CH^{y-})]_x$ changed from $[Li_{0.08}^+(CH^{0.08-})]_x$ to $[Li_{0.13}^+(CH^{0.13-})]_x$ during the discharge process; hence it was acting as a cathode. After 81 cycles at 1·0 mA/cm^2 (78°C) the coulombic efficiency was still >99% and the capacity remained unchanged. Electrodes coated with the sulfonate polymer which is a better ionic conductor than the poly(ethylene oxide) polymer, were cycled (18°C) between $[Li_{0.05}^+(CH^{0.05-})]_x$ and $[Li_{0.15}^+(CH^{0.15-})]_x$. Coulombic efficiencies and capacity retention were similar to those with the poly(ethylene oxide) polymer.

6. THE POLYACETYLENE ANODE

Polyacetylene, in its n-doped (reduced) form, $[M_y^+(CH^{y-})]_x$, can act as an anode in a battery when, during the discharge reaction of a cell, it becomes oxidized, i.e. the negative charge residing on the (CH) unit is reduced in magnitude. Although, in principle, the $(CH^{y-})_x$ polycarbanion could be oxidized past neutral $(CH)_x$ to the regime where it becomes p-doped, e.g. to give $(CH^{y+})_x$, in practice oxidation is terminated when it is oxidized to neutral $(CH)_x$. This is because solvents which are chemically compatible with the $(CH^{y-})_x$ polycarbanion are usually incompatible to varying extents with the $(CH^{y+})_x$ polycarbonium ion.

A simple example of a cell in which polyacetylene is acting as an anode is that depicted in Fig. 4. The discharge reactions for such a cell are:

Anode (oxidation)

$$(CH^{y-})_x \rightarrow (CH^{(y/2)-})_x + (xy/2)e^- \qquad (20)$$

Cathode (reduction)

$$(CH)_x + (xy/2)e^- \rightarrow (CH^{(y/2)-})_x \qquad (21)$$

giving the overall discharge reaction:

$$(CH^{y-})_x + (CH)_x \rightarrow 2(CH^{(y/2)-})_x \qquad (22)$$

If the counter-cation, e.g. Li^+, is included, the net discharge reaction becomes:

$$[Li_y^+(CH^{y-})]_x + (CH)_x \rightarrow 2[Li_{y/2}^+(CH^{(y/2)-})]_x \qquad (23)$$

Hence, if the anode and the cathode in an electrolyte such as $LiClO_4$ in THF are connected by an external wire, electrons flow from the anode to the cathode through the wire. The charge reaction, accomplished by applying a negative potential to the original anode and a positive potential to the original cathode results in the above reactions occurring in the reverse direction.

During recent years considerable research has been carried out on rechargeable batteries using a Li anode and a TiS_2 cathode.[40] During discharge the TiS_2 becomes 'intercalated' or 'doped', using conducting polymer terminology, with Li^+ according to the reactions:

Anode (oxidation)

$$xLi \rightarrow xLi^+ + xe^- \qquad (24)$$

Cathode (reduction)

$$(TiS_2)_x + e^- \rightarrow (TiS_2^-)_x \qquad (25)$$

giving the net discharge reaction:

$$x\text{Li} + (\text{TiS}_2)_x \rightarrow [\text{Li}^+(\text{TiS}_2^-)]_x \qquad (26)$$

Such batteries show many desirable features, but possess disadvantages associated with lithium anodes: specifically, slow reaction of the lithium electrode with the organic electrolyte and inadequate cycle life due to the growth of lithium dendrites.

Since the $[\text{Li}_y^+(\text{CH}^{y-})]_x$ anode shows many desirable features and avoids many of the problems associated with metal anodes, studies have been carried out on a TiS_2 cell in which the Li anode has been replaced by a $[\text{Li}_y^+(\text{CH}^{y-})]_x$ anode using a $\text{LiClO}_4(\text{THF})$ electrolyte.[41] The net discharge reaction is, therefore:

$$[\text{Li}_y^+(\text{CH}^{y-})]_x + (\text{TiS}_2)_x \rightarrow [\text{Li}_y^+(\text{TiS}_2^{y-})]_x + (\text{CH})_x \qquad (27)$$

If 10% reduced $(\text{CH})_x$ is employed, eqn (27) becomes:

$$10[\text{Li}_{0.10}^+(\text{CH}^{0.10-})]_x + (\text{TiS}_2)_x \rightarrow [\text{Li}^+(\text{TiS}_2^-)]_x + 10(\text{CH})_x \qquad (28)$$

It is important to note that the potential of $[\text{Li}^+(\text{TiS}_2^-)]_x$ versus Li^+/Li, which is ~ 1.8 V, is essentially identical to the potential of neutral $(\text{CH})_x$ versus Li^+/Li (1·81 V). Hence full utilization of the reducing and oxidizing power of both the $(\text{CH}^{0.10-})_x$ and TiS_2 electrodes is possible.

A $[\text{Li}_{0.10}^+(\text{CH}^{0.10-})]_x/(\text{TiS}_2)_x$ cell has an open circuit voltage of 1·65 V. The cell possesses good stability, and the electrochemical reaction given by eqn (28) has been found to be completely reversible.[41] It has a theoretical energy density of 111 W h/kg. A cell containing a $[\text{Na}_{0.10}^+(\text{CH}^{0.10-})]_x$ anode, a TiS_2 cathode and a $\text{NaPF}_6(\text{THF})$ electrolyte has a V_{oc} of 1·80 V. It also shows good stability and reversibility and has a theoretical energy density of 79 W h/kg.

The higher charge levels afforded by the borate salts discussed in Section 5.2 should make possible the construction of cells having a greater capacity and higher average cell voltage. For example, a $[\text{Li}_{0.18}^+(\text{CH}^{0.18-})]_x/(\text{TiS}_2)_y$ cell could in principle be cycled between useful cell voltages of 2·20 and 1·40 V involving a compositional change in the $[\text{Li}_y^+(\text{CH}^{y-})]_x$ between $y = 0.18$ and 0·09. The use of an n-doped polyacetylene anode in cells of this type and those containing other medium- and high-potential cathodes appears to warrant considerably more study.

7. BATTERIES EMPLOYING POLYACETYLENE ANODES AND CATHODES

If appropriate energy densities and recyclability could be attained, a most attractive rechargeable battery would be one in which both anode and cathode were fabricated from $(CH)_x$. An example of a cell using a polyacetylene anode and a polyacetylene cathode has already been described in Section 6. However, since on complete discharge only half the charge in the $(CH^{y-})_x$ electrode has been utilized the energy density is too low for any practical use. The cell does, however, serve as an illustration of the concept of a stable rechargeable battery consisting of two polymer electrodes.

A far more interesting cell involves the use of a $(CH^{y-})_x$ anode and a $(CH^{y+})_x$ cathode. A cell potential of $\sim 2 \cdot 5$ V is obtained for a value of $y \sim 0 \cdot 07$. In a typical example, a cell consisting of two pieces of $(CH)_x$ film (each ~ 2 mg; $\sim 0 \cdot 5$ cm$^2 \times 0 \cdot 01$ cm) in a 1M solution of $(n\text{-Bu}_4\text{N}^+)(\text{PF}_6^-)$ in THF was charged to p- and n-dope the electrodes to 7%.[11] The discharge reactions resulting when the electrodes are connected by a wire are:

Anode (oxidation)

$$(CH^{0 \cdot 07-})_x \rightarrow (CH)_x + (0 \cdot 07x)e^- \quad (29)$$

Cathode (reduction)

$$(CH^{0 \cdot 07+})_x + (0 \cdot 07x)e^- \rightarrow (CH)_x \quad (30)$$

giving the overall discharge reaction:

$$(CH^{0 \cdot 07-})_x + (CH^{0 \cdot 07+})_x \rightarrow 2(CH)_x \quad (31)$$

If the appropriate dopant ions are incorporated this becomes:

$$[(n\text{-Bu}_4\text{N}^+)_{0 \cdot 07}(CH^{0 \cdot 07-})]_x + [(CH^{0 \cdot 07+})(\text{PF}_6^-)_{0 \cdot 07}]_x \rightarrow$$
$$2(CH)_x + (0 \cdot 07x)(n\text{-Bu}_4\text{N}^+)(\text{PF}_6^-) \quad (32)$$

The open circuit potential of such a cell was $\sim 2 \cdot 5$ V and the short circuit current was ~ 80 mA per cm^2 of film.[11] The magnitude of this short circuit current is most promising, particularly in view of the large size of the $(n\text{-Bu}_4\text{N}^+)$ dopant cation which might have been expected to decrease the rate of discharge because of its expected smaller rate of diffusion as compared, for example, with the Li$^+$ ion.

A novel solid state battery employing a $[\text{Na}_y^+(CH^{y-})]_x$ anode and a

$[(CH^{y+})(I_3^-)_y]_x$ cathode separated by a solid electrolyte of NaI in poly(ethylene oxide) has also been described.[42] The electrochemical reactions involved are analogous to those given by eqns (29) to (32), the (n-Bu$_4$N$^+$) and (PF$_6^-$) ions being replaced by Na$^+$ and (I$_3^-$) ions respectively. Depending on experimental conditions, open circuit voltages ranging from 2·8 to 3·5 V and short circuit currents ranging from 1 to 12 mA have been obtained.

It should be noted that those solvents in which $(CH^{y+})_x$ is most stable (propylene carbonate, sulfolane, etc.) are those in which $(CH^{y-})_x$ is least stable, and vice versa. However, in principle, it is not necessary to have the same conducting polymer for both anode and cathode. The judicious choice of different conducting polymers for anode and cathode or the protection of a conducting polymer electrode with a film of ionically conducting polymer as described in Section 5.2 may result in a cell in which each material in its appropriate oxidation state is stable and behaves well in the same electrolyte.

8. BATTERIES USING OTHER CONDUCTING POLYMERS

Although most published work to date on the use of organic polymers as electroactive materials in rechargeable batteries has been concerned with polyacetylene it is appropriate to mention briefly other organic polymers which show electrochemical activity and whose possible use in rechargeable batteries has been suggested or might be considered.

Shortly after it was found that (CH)$_x$ could be electrochemically oxidized and reduced and that it showed potential as an electrode-active material in rechargeable batteries, it was announced that poly(p-phenylene) (PPP), I, could also be oxidized and reduced electrochemically to the metallic conducting regime and that it could serve as an electrode-active material.[43] Since then it has also been demonstrated that polypyrrole,[6] II, and polythiophene,[44] III, can be electrochemically oxidized to the metallic regime and electrochemically reduced back to the parent polymer. It has also been found that polyquinolines such as IV can be electrochemically reduced to the

(IV; A = H, Cl, OCH$_3$, etc.)

metallic regime and electrochemically oxidized back to the parent polymer and that they can serve as an anode in a rechargeable battery. Although each of these materials shows certain desirable characteristic features which make it worthy of consideration as a practical anode and/or cathode material, it is apparent from published data that their capacities are all inferior to those of polyacetylene. This stems from the small molar mass, 13, of the monomer (CH) unit of polyacetylene which can be oxidized to approximately $(CH^{0.10+})_x$,[26] and reduced to approximately $(CH^{0.18-})_x$.[38] This immediately places the practical usefulness of these other polymers as electrode materials in question. However, poor capacity effects can, in principle, be compensated by higher average discharge voltages, greater stability and higher utilization of electrode material at higher discharge rates. All of these factors can affect significantly the practical energy density of batteries.

Poly(p-phenylene) is a potentially inexpensive material synthesized as a powder which can be fabricated into an electrode having mechanical stability and electrical conductivity without the aid of conductive additives.[43] Although its capacity is approximately only half that of $(CH)_x$, most of its capacity in its n-doped, reduced form, $[M_y^+(C_6H_4^{y-})]_x$, can be delivered during discharge at a lower, and within a more narrow, voltage range than is possible in the case of $[M_y^+(CH^{y-})]_x$. This makes it particularly attractive as an anode material as a possible substitute for lithium metal. A serious problem with lithium anodes in rechargeable batteries is that on charging, when lithium is plated out on the lithium electrode, dendrites tend to be formed, particularly at high charging rates. These dendrites can grow to a considerable length and in so doing can short-circuit a cell internally. As was the case with $(CH)_x$, alkali metal borate salts permit poly(p-phenylene)[43] to be reduced to a greater extent than when an electrolyte containing $LiClO_4$ is used. The resulting polymer is stable on standing in the electrolyte. By applying a potential as low as 0·15 V versus Li^+/Li, to a poly(p-phenylene) electrode in a $Li^+(BPh_4^-)$/THF electrolyte, the compound $[Li_{0.44}^+(C_6H_4^{0.44-})]_x$ is formed. This material is stable in the electrolyte. Furthermore, on cycling a poly(p-phenylene) cathode in a cell comprised of a sodium anode and a $Na^+(BEt_4^-)$/THF electrolyte, between 0·05 and 0·75 V the value of y in $[Na_y^+(C_6H_4^{y-})]_x$ was observed to change from 0·41 at 0·05 V to zero at 0·75 V. Hence the total capacity of the poly(p-phenylene) electrode was utilized over a 0·7 V range. In general, more of the capacity of n-doped poly(p-phenylene) can be utilized in a potential range suitable for its use as an anode than in the case of n-doped polyacetylene. This is because the voltage plateaus associated with different stages of alkali metal intercalation are less pronounced[36] in poly(p-phenylene). This is especially true for the sodium complex whose composition can be varied between $y = 0.08$ and 0·32 simply by changing the applied potential from 0·50 to 0·30 V. Thus 75% of the capacity of the electrode can be delivered over a 0·2 V range during discharge.[36] Poly(p-phenylene) can also be electrochemically oxidized to $[(C_6H_4^{0.16+})A_{0.16}^-]_x$,[43] and hence it can be used as a cathode in a cell with, for example, a lithium anode. However, even at a lower level of oxidation such as $[(C_6H_4^{0.1+})]_x$, it has a reduction potential versus Li^+/Li of 4·4 V. At such a potential, it will be expected to react chemically with most commonly employed electrolytes.[43]

In the case of polythiophene the maximum reported oxidation level obtained electrochemically is $[(C_4H_2S^{0.2+})]_x$.[44] At this level of oxidation its reduction potential versus Li^+/Li is ~4 V. In an electrolyte such as $LiClO_4$ in propylene carbonate it is reported to exhibit good electrochemical reversibility on cycling between its oxidized and neutral forms.

Polypyrrole also appears to have good potential as a cathode since it can be electrochemically oxidized and reduced back to the neutral form with little irreversible degradation.[6] Oxidation commences at ~3 V versus Li^+/Li. Again, its capacity is less than that of polyacetylene since its maximum level of oxidation is approximately $[(C_4H_3N^{0.33+})]_x$.

The polyquinolines, whose maximum level of reduction per repeat unit is ~0·7, although exhibiting a smaller capacity than the $(CH^{0.18-})_x$ anode, show very high rates of charging and discharging and good recyclability.[45] The reduction potential of the neutral polymer varies significantly according to the nature of the substituent species.

Considerable interest has been shown very recently in 'polyaniline', known for approximately 100 years, forms of which have been called 'aniline black', 'nigraniline', 'emeraldine', etc. Whereas essentially all the studies described in this review have involved *non*-aqueous electrolytes, the electrochemistry of certain forms of polyaniline have been investigated in both non-aqueous[46–49] and aqueous electrolytes,[49–52] in the latter electrolyte both as a cathode or as an anode material. As is shown below, the energy density of certain types of cells using polyaniline electrodes seems most attractive.

The emeraldine base and salt forms of polyaniline can be synthesized either by chemical or electrochemical oxidation of the cheap industrial chemical, aniline ($C_6H_5NH_2$), in aqueous or non-aqueous media without the necessity of excluding air.[46–52] Preliminary observations have been reported on the use of the emeraldine base form of polyaniline, V, as a cathode material in an aqueous $ZnCl_2/HCl$

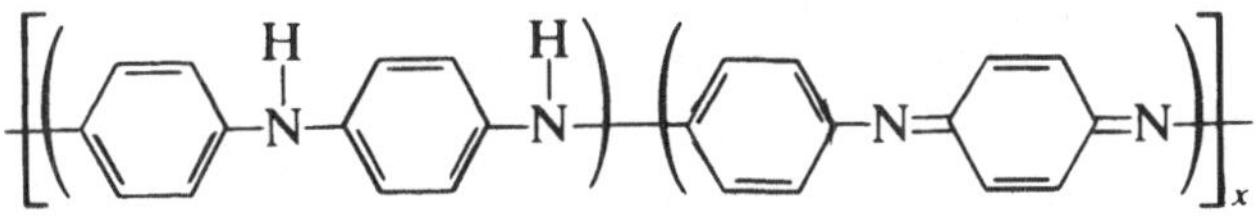

(V)

solution (pH ~ 4) together with an amalgamated zinc anode.[51] The proposed discharge reactions are:

$$xZn \rightarrow xZn^{2+} + (2x)e^- \quad (33)$$

at the anode, and

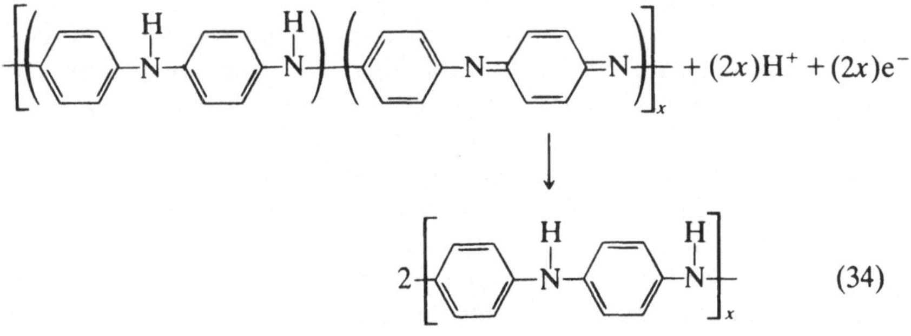

$$[\ldots]_x + (2x)H^+ + (2x)e^- \downarrow 2[\ldots]_x \quad (34)$$

at the cathode, giving the overall discharge reaction:

$$\left[\left(\underset{}{\bigcirc}-\underset{H}{N}-\underset{}{\bigcirc}-\underset{H}{N}\right)\left(\underset{}{\bigcirc}-N=\underset{}{\bigcirc}=N\right)\right]_x + x\text{Zn} + (2x)\text{HCl}$$

$$\downarrow$$

$$2\left[\underset{}{\bigcirc}-\underset{H}{N}-\underset{}{\bigcirc}-\underset{H}{N}\right]_x + x\text{ZnCl}_2 \quad (35)$$

Both the polyaniline and the zinc can be repeatedly electrochemically oxidized and reduced in the pH range between ~2 and ~4 with coulombic efficiencies of almost 100%. When the polyaniline is oxidized and reduced between ~1·4 and ~0·8 V (versus Zn^{2+}/Zn) in a battery charge/discharge mode the capacity is only very slightly reduced after 200 cycles. A semi-theoretical energy density of 120 W h/kg is obtained for an average experimental discharge voltage of 1·1 V using only the mass of the polyaniline employed and the mass of the zinc consumed in the discharge reaction. This is ~62% of the energy density of a nickel–cadmium battery (192 W h/kg).

The utilization of the capacity of the polyaniline based on the discharge reaction given by eqns (34) and (35) is 100%.[53] Such cells also show excellent retention of charge on standing. The capacity of the oxidized form can be compared with the capacity of a similar mass of 7% oxidized polyacetylene. Thus 13 g (1 mol) of $(\text{CH}^{0·07+})_x$ utilize 0·07 mol of electrons on reduction to $(\text{CH})_x$. Thirteen grams of emeraldine base also utilize 0·07 mol of electrons on reduction according to eqn (34). In view of the fact that oxidized polyacetylene does not show good recyclability or retention of charge on standing, it appears that the emeraldine form of polyaniline, insofar as its capacity and recyclability are concerned, shows promise as a cathode material.

Very preliminary studies[51] show that the emeraldine form of polyaniline can act as an anode in dilute aqueous acid with a PbO_2 cathode, the *idealized* discharge reaction being shown in eqn (36).
Such a cell shows good recyclability and a semi-theoretical energy density of ~90 W h/kg based on an experimental average discharge voltage of ~1 V.

Detailed electrochemical studies of polyaniline have been carried out in non-aqueous electrolytes and its use as a cathode in a

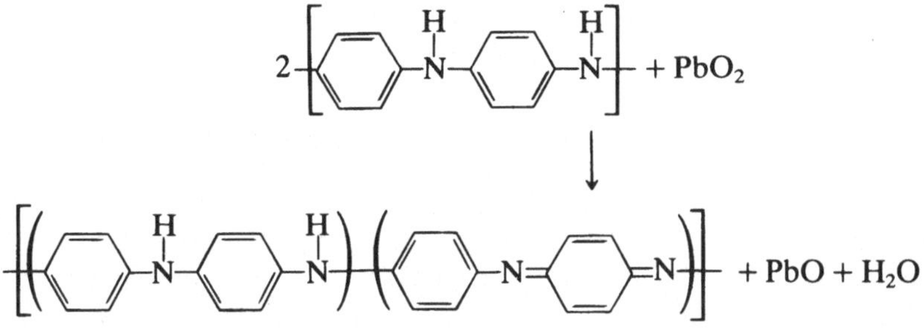

$$2\left[\underset{\text{H}}{\bigcirc}-\underset{|}{\text{N}}-\bigcirc-\underset{|}{\text{N}}\right] + PbO_2$$

$$\downarrow$$

$$\left[\left(\bigcirc-\underset{\text{H}}{\text{N}}-\bigcirc-\underset{\text{H}}{\text{N}}\right)\left(\bigcirc-\text{N}=\bigcirc=\text{N}\right)\right] + PbO + H_2O$$

(36)

$LiClO_4$/propylene carbonate electrolyte with a Li anode has been described.[46,47,49] It is stated to have an energy density of 352 W h/kg and to show excellent rechargeability.[49] The energy density is based upon the mass of polyaniline placed in the cell.

More recent studies have shown that the emeraldine base form of polyaniline acts as a most promising cathode in a $LiClO_4$/propylene carbonate electrolyte showing very good recyclability[54] and essentially 100% coulombic efficiency. During the first charge it is converted to the emeraldine hydroperchlorate salt. The utilization of the polyaniline is 100%, based on the discharge reaction:

$$\left[\left(\bigcirc-\text{N}=\bigcirc=\text{N}\right)\left(\bigcirc-\underset{(ClO_4^-)}{\overset{H}{\text{N}^+}}=\bigcirc=\underset{(ClO_4^-)}{\overset{H}{\text{N}^+}}\right)\right]_x + (2x)\text{Li}$$

$$\downarrow$$

$$\left[\left(\bigcirc-\text{N}=\bigcirc=\text{N}\right)\left(\bigcirc-\underset{\text{H}}{\text{N}}-\bigcirc-\underset{\text{H}}{\text{N}}\right)\right]_x + (2x)LiClO_4$$

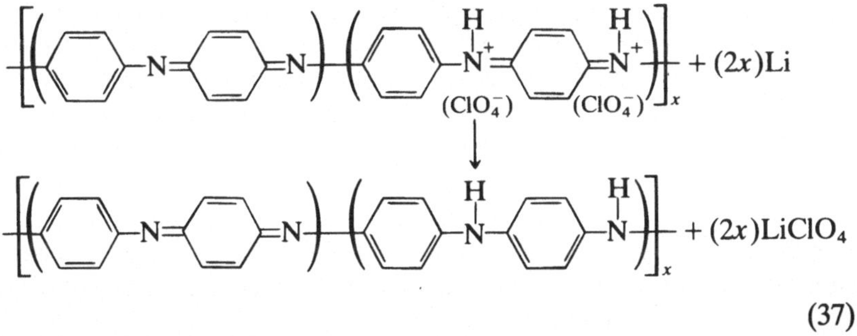

(37)

The charge reaction is the reverse of the above. The capacity based only on the mass of emeraldine base placed in the cell is 148 A h/kg. It is 93 A h/kg based on the discharge reaction given by eqn (37) where the mass of the $LiClO_4$ is also included. The corresponding energy density calculated for an experimentally determined average discharge voltage of 3·43 V is 319 W h/kg. The excellent energy density and recyclability of this polymer, together with its low cost, suggest that it holds particular promise for practical use in rechargeable batteries. In view of the number of chemically modified derivatives of polyaniline which are known and which it should be possible to synthesize, it seems that considerable fine-tuning of cell potential with corresponding increase in energy density might also well be possible.

Without doubt, new organic polymers suitable for electrical energy storage will be discovered during the next few years. As more information becomes available it will be possible to determine what particular property of a polymer is necessary to ensure high doping levels, good retention of charge on standing and good recyclability.

9. PRESENT STATUS OF BATTERIES EMPLOYING POLYMER ELECTRODES

The electrochemistry of organic polymers involving oxidation or reduction of the polymer backbone and its application to electrical storage is still a very new field of endeavor (the first preliminary paper on this subject appeared in 1979).[3] Since the commercial development of a successful battery often takes more than ten years from the first laboratory discovery of the phenomenon on which it is based, it would be most surprising if a battery employing an electroactive polymer were already in commercial production. Not only are new chemistry and electrochemistry involved, but also new engineering aspects concerning, for example, something as apparently simple as the method of attachment of an organic polymer to a current collector. A disadvantage of all polymer electrodes is that they will, in general, be expected to exhibit smaller volumetric energy densities than conventional batteries, since organic materials are usually less dense than metals and inorganic substances. To what extent this is a serious disadvantage will depend on the use to which the cell in which they are incorporated might be put.

Electrodes using polymers as electrode-active materials differ from conventional materials in that the polymer does not dissolve and is not redeposited during charge/discharge cycles, although some swelling of the polymer does occur during oxidation or reduction. This absence of dissolution and redeposition is expected to promote longer life for polymer electrodes due to the absence of mechanical changes in electrode dimensions, etc., such as those observed with many conventional electrodes. Also, in most batteries, complete deep discharge leads to rapid deterioration. Hence, only partial utilization of the energy density of a battery can be achieved if it has to be only partly discharged in order to prolong its effective operating life. The energy density of a polymer electrode can, however, in many cases be completely utilized in repeated deep discharges at good discharge rates without adverse effects. For practical use this may in many instances compensate for the smaller gravimetric and/or volumetric energy density of polymer batteries compared with some conventional rechargeable batteries. If the charge/discharge cycles are held between potentials so that the polymer always retains sufficient conductivity to permit appropriately rapid charging and discharging, both the effective gravimetric and volumetric energy densities in a completely packaged battery may be larger than otherwise expected. This is because the necessity of introducing a conductive filler such as carbon black will be reduced. It may be noted that the flexibility of polymeric films may be significant in their use in certain types of thin batteries.

The number of different types of batteries possible with inorganic materials is limited, whereas the number of potential electrode-active materials involving organic substances is essentially 'unlimited' in view of the enormous number of different types of organic materials—monomers, oligomers, polymers (either conducting or non-conducting)—which are potentially synthesizable. It would appear that the use of organic substances as electrode-active materials in batteries for certain fields of application is still in its infancy.

10. CONCLUSIONS

The discovery that the backbone of an organic polymer having an appropriate type of bonding can be reversibly oxidized and reduced electrochemically with, in certain cases, little irreversible degradation opens up a new potential field for electrical energy storage. It shows that certain types of organic polymers can act as a storage medium for electricity and that the stored charge can be delivered on demand at a rate which could be useful in practical cells. In view of the newness of the field, and the potentially enormous number of synthesizable new polymers, it seems improbable that the best polymers have necessarily yet been found.

The question whether non-aqueous or aqueous electrolytes will be preferable for a specific purpose cannot be answered until much more information has become available. The use of aqueous electrolytes will, in general, be expected to result in cells having a smaller potential and, hence, a smaller energy density than batteries employing non-aqueous electrolytes unless the polymer can be more highly oxidized or reduced. Real long-term advances in the field can only be accomplished by detailed chemical and electrochemical studies of different types of polymers and of the principal and side reactions which occur during their storage and recycling under a wide variety of experimental conditions.

ACKNOWLEDGEMENT

Work in this manuscript originating from the University of Pennsylvania was supported by the Department of Energy, contract no. DE-AC02-81-ER1083, the Office of Naval Research, the University of Pennsylvania Materials Research Laboratory through NSF–MRL grant no. DMR-82-16718 and by NSF grant no. DMR-80-22870.

REFERENCES

1. H. Shirakawa, E. J. Louis, A. G. MacDiarmid, C. K. Chiang and A. J. Heeger, *J. Chem. Soc. Chem. Commun.*, 1977, 578; C. K. Chiang, C. R. Fincher, Jr, Y. W. Park, A. J. Heeger, H. Shirakawa, E. J. Louis, S. C. Gau and A. G. MacDiarmid, *Phys. Rev. Lett.*, 1977, **39**, 1098.
2. C. K. Chiang, M. A. Druy, S. C. Gau, A. J. Heeger, E. J. Louis, A. G. MacDiarmid, Y. W. Park and H. Shirakawa, *J. Amer. Chem. Soc.*, 1978, **100**, 1013.
3. P. J. Nigrey, A. G. MacDiarmid and A. J. Heeger, *J. Chem. Soc. Chem. Commun.*, 1979, 594.
4. P. J. Nigrey, D. MacInnes, Jr., D. P. Nairns, A. G. MacDiarmid and A. J. Heeger, *J. Electrochem. Soc.*, 1981, **128**, 1651; K. Kaneto, M. Maxfield, D. P. Nairns and A. G. MacDiarmid, *J. Chem. Soc., Faraday Trans. I*, 1982, **78**, 3417.
5. L. W. Shacklette, R. R. Chance, D. M. Ivory, G. G. Miller and R. H. Baughman, *Syn. Metals*, 1979/80, **1**, 307.
6. A. F. Diaz, K. K. Kanazawa and G. P. Gardini, *J. Chem. Soc. Chem. Commun.*, 1979, 535; A. F. Diaz and K. K. Kanazawa, *Extended Linear Chain Compounds*, ed. J. S. Miller, Plenum Press, New York, 1982, pp. 417–27.
7. T. Yamamoto, K. Sanechika and A. Yamamoto, *J. Polym. Sci., Polym. Lett. Ed.*, 1980, **18**, 9.

8. A. G. MacDiarmid, J. C. Chiang, M. Halpern, W. S. Huang, S. L. Mu, N. L. D. Somasiri, W. Wu and S. Yaniger, *Mol. Cryst. Liq. Cryst.*, 1985, **21**, 173 and references therein.
9. L. W. Shacklette, R. L. Elsenbaumer and R. H. Baughman, *J. Phys. (C3)*, 1983 **44**, 559; L. W. Shacklette, R. L. Elsenbaumer, R. R. Chance, J. M. Sowa, D. M. Ivory, G. G. Miller and R. H. Baughman, *J. Chem. Soc. Chem. Commun.*, 1982, 361; G. Tourillon and F. Garnier, *J. Electroanal. Chem.*, 1982, **135**, 173; K. Kaneto, K. Yoshino and Y. Inuishi, *Japanese J. Appl. Phys.*, 1983, **22**, L567; A. G. MacDiarmid, S. L. Mu, N. L. D. Somasiri and W. Wu, *Mol. Cryst. Liq. Cryst.*, 1985, **121**, 187.
10. A. G. MacDiarmid and A. J. Heeger, *Syn. Metals*, 1979/80, **1**, 101.
11. D. MacInnes, Jr, M. A. Druy, P. J. Nigrey, D. P. Nairns, A. G. MacDiarmid and A. J. Heeger, *J. Chem. Soc. Chem. Commun.*, 1981, 317.
12. C. K. Chiang, S. C. Gau, C. R. Fincher, Jr, Y. W. Park, A. G. MacDiarmid and A. J. Heeger, *Appl. Phys. Lett.*, 1978, **33**, 18.
13. K. Kaneto, M. Maxfield, D. P. Nairns, A. G. MacDiarmid and A. J. Heeger, *J. Chem. Soc., Faraday Trans. I*, 1982, **78**, 3417.
14. H. Shirakawa and S. Ikeda, *Polym. J.*, 1971, **2**, 231; H. Shirakawa, T. Ito and S. Ikeda, *Polym. J.*, 1973, **4**, 460; T. Ito, H. Shirakawa and S. Ikeda, *J. Polym. Sci., Polym. Chem. Ed.*, 1974, **12**, 11.
15. T. Ito, H. Shirakawa and S. Ikeda, *J. Polym. Sci., Polym. Chem. Ed.*, 1975, **13**, 1943.
16. F. E. Karasz, J. C. W. Chien, R. Galkiewicz, G. E. Wnek, A. J. Heeger and A. G. MacDiarmid, *Nature (London)*, 1979, **282**, 286.
17. J. C. W. Chien, F. E. Karasz and G. E. Wnek, *Nature (London)*, 1980, **285**, 390; M. Rolland, P. Bernier, S. Lefrant and M. Aldissi, *Polymer*, 1980, **21**, 1111.
18. J. M. Pochan, H. W. Gibson and F. C. Bailey, *J. Polym. Sci., Polym. Lett. Ed.*, 1980, **18**, 447; J. C. W. Chien, *J. Polym. Sci., Polym. Lett. Ed.*, 1981, **19**, 249; J. M. Pochan, D. F. Pochan, M. Rommelmann and H. W. Gibson, *Macromolecules*, 1981, **14**, 110; J. M. Pochan, H. W. Gibson and J. Harbour, *Polymer*, 1982, **23**, 439; H. W. Gibson and J. M. Pochan, *Macromolecules*, 1982, **15**, 242, A. G. MacDiarmid, Y. W. Park, A. J. Heeger and M. A. Druy, *J. Chem. Phys.*, 1980, **73**, 946.
19. T.-C. Chung, A. Feldblum, A. J. Heeger and A. G. MacDiarmid, *J. Chem. Phys.*, 1981, **74**, 5504.
20. W. Wu, R. J. Mammone and A. G. MacDiarmid, *Syn. Metals*, 1985, **10**, 235.
21. R. J. Mammone and A. G. MacDiarmid, *J. Chem. Soc., Faraday Trans. I*, 1985, **81**, 105; *Syn. Metals*, 1984, **9**, 143.
22. A. Feldblum, A. J. Heeger, T.-C. Chung and A. G. MacDiarmid, *J. Chem. Phys.*, 1982, **77**, 5114.
23. P. J. Nigrey, D. MacInnes, Jr, D. P. Nairns, A. G. MacDiarmid and A. J. Heeger, *J. Electrochem. Soc.*, 1982, **129**, 1270.
24. F. G. Will, *Electrochem. Soc. Extended Abstracts*, 1983, **83** (1), 838.

25. J. H. Kaufman, J. W. Kaufer, A. J. Heeger, R. B. Kaner and A. G. MacDiarmid, *Phys. Rev. B,* 1982, **26,** 2327.
26. G. C. Farrington, B. Scrosati, D. Frydrych and J. DeNuzzio, *J. Electrochem. Soc.,* 1984, **131,** 7; S. L. Mu, S. J. Porter, W. Wu and A. G. MacDiarmid, *J. Electrochem. Soc.,* 1984, **131,** 7.
27. M. Maxfield, S. L. Mu and A. G. MacDiarmid, *J. Electrochem. Soc.,* 1985, **132,** 838.
28. M. Maxfield, J. F. Wolf, P. R. Jow and L. W. Shacklette, *186th Amer. Chem. Soc. Natl. Mtg,* INDE-16, 1983.
29. J. O'M. Bockris, B. E. Conroy, E. Yeager and R. E. White (eds) *Comprehensive Treatise of Electrochemistry,* Plenum Press, New York, 1981, p. 308.
30. F. Beniere, D. Boils, H. Canepa, J. Franco, A. LeCorve and J. P. Louboutin, *J. Phys. (C3),* 1983, **44,** 567.
31. M. R. Winkle, R. T. Gray and M. J. Hurwitz, *J. Electrochem. Soc.,* 1983, **130,** 242C; K. Abe, F. Goto, T. Yoshida and H. Morimoto, *Electrochem. Soc. Extended Abstracts,* 1983, **83**(1), 836.
32. R. L. Elsenbaumer, G. G. Miller, Y. P. Khanna, E. McCarthy and R. H. Baughman, *Abstr. Electrochem. Soc. Meeting, Toronto, Ontario, Canada, May 12–17, 1985*; *J. Electrochem. Soc.,* 1985, **99C,** 132; R. H. Baughman, P. Delanoy, N. S. Murthy, G. G. Miller, H. Eckhardt and L. W. Shacklette, *Mol. Cryst. Liq. Cryst.,* 1984, **106,** 415; P. Delannoy, G. G. Miller, N. S. Murthy, C. E. Forbes, H. Eckhardt, R. L. Elsenbaumer and R. H. Baughman, *APS National Meeting, March, 1985,* DE-7.
33. A. G. MacDiarmid, R. J. Mammone, R. B. Kaner and S. J. Porter, *Phil. Trans. R. Soc. Lond.,* 1985, **A314,** 3; A. G. MacDiarmid, R. J. Mammone, J. R. Krawczyk, III and S. J. Porter, *Mol. Cryst. Liq. Cryst.,* 1984, **105,** 89.
34. R. B. Kaner, A. G. MacDiarmid and R. J. Mammone, in *Polymers in Electronics,* ed. T. Davidson, American Chemical Society, Washington, DC, 1985, pp. 575–84.
35. R. B. Kaner, S. J. Porter and A. G. MacDiarmid, *J. Chem. Soc., Faraday Trans. I,* 1986, **82,** 2323.
36. R. H. Baughman, N. S. Murthy and G. G. Miller, *J. Chem. Phys.,* 1983, **79,** 515; R. H. Baughman, N. S. Murthy, G. G. Miller, L. W. Shacklette and R. M. Metzger, *J. Phys. (Paris) (C3),* 1983, **44,** 53; R. H. Baughman, L. W. Shacklette, N. S. Murthy, G. G. Miller and R. L. Elsenbaumer, *Mol. Cryst. Liq. Cryst.,* 1985, **118,** 253.
37. R. B. Kaner and A. G. MacDiarmid, *Syn. Metals,* 1986, **14,** 3.
38. L. W. Shacklette, J. E. Toth, N. S. Murthy and R. H. Baughman, *J. Electrochem. Soc.,* 1985, **132,** 1529.
39. M. Maxfield, J. F. Wolf, G. G. Miller, J. E. Frommer and L. W. Shacklette, *J. Electrochem. Soc.,* 1986, **133,** 117.
40. M. S. Whittingham, *J. Electrochem. Soc.,* 1976, **123,** 315; M. S. Whittingham, *Prog. Solid State Chem.,* 1978, **12,** 41; S. Atlung, K. West and T. Jacobsen, *J. Electrochem. Soc.,* 1979, **126,** 1311.

41. J. Caja, R. B. Kaner and A. G. MacDiarmid, *J. Electrochem. Soc.*, 1984, **131,** 2744.
42. C. K. Chiang, *Polymer*, 1981, **22,** 1454.
43. L. W. Shacklette, R. L. Elsenbaumer and R. H. Baughman, *J. Phys (C3)*, 1983, **44,** 559; L. W. Shacklette, R. L. Elsenbaumer, R. R. Chance, J. M. Sowa, D. M. Ivory, G. G. Miller and R. H. Baughman, *J. Chem. Soc., Chem. Commun.*, 1982, 361.
44. T.-C. Chung, J. H. Kaufman, A. J. Heeger and F. Wudl, *Phys. Rev. B*, 1984, **30,** 702; K. Kaneto, K. Yoshino and Y. Inuishi, *Japanese J. Appl. Phys.*, 1983, **22,** L567, L412.
45. Y. S. Papir, V. P. Kurkov and S. P. Current, *Electrochem. Soc. Extended Abstracts*, **83**(1), 820; A. H. Schroeder, Y. S. Papir and V. P. Kurkov, *Electrochem. Soc. Extended Abstracts*, **83**(1), 822.
46. J. B. Travers, J. Chroboczek, F. Devreux, F. Genoud, M. Nectschein, A. Syed, E. M. Genies and C. Tsintavis, *Mol. Cryst. Liq. Cryst.*, 1985, **121,** 195.
47. E. M. Genies, A. A. Syed and C. Tsintavis, *Mol. Cryst. Liq. Cryst.*, 1985, **121,** 181.
48. A. Kitani, M. Kaya and K. Sasaki, Abstracts "24th Battery Symposium in Japan," Nov. 10–11, 1983, Osaka, Japan.
49. A. Kitani, J. Izumi, J. Yano, Y. Hiromoto and K. Sasaki, *Bull. Chem. Soc. Japan*, 1984, **57,** 2254; M. Kata, A. Kitani and K. Sasaki, *Denki Kagaku*, 1984, **52,** 847.
50. R. de Surville, M. Josefowicz, L. T. Yu, J. Perichon and R. Buvet, *Electrochim. Acta*, 1968, **13,** 1451; M. Josefowicz, L. J. Yu, J. Perichon and R. Buvet, *J. Polym. Sci., Polym. Chem. Ed.*, 1969, **22,** 1189 and references therein.
51. A. G. MacDiarmid, S. L. Mu, N. L. D. Somasiri and W. Wu, *Mol. Cryst. Liq. Cryst.*, 1985, **121,** 187.
52. A. G. MacDiarmid, J. C. Chiang, M. Halpern, W. S. Huang, S. L. Mu, N. L. D. Somasiri, W. Wu and S. Yaniger, *Mol. Cryst. Liq. Cryst.*, 1985, **121,** 173.
53. N. L. D. Somasiri and A. G. MacDiarmid, 1984, unpublished observations.
54. L. S. Yang, W. S. Huang and A. G. MacDiarmid, *Syn. Metals*, in press; L. S. Yang and A. G. MacDiarmid, *J. Electrochem. Soc.*, in press.

CHAPTER 5

Polymer Modified Electrodes: Preparation and Characterisation

A. ROBERT HILLMAN

School of Chemistry, University of Bristol, UK

ABBREVIATIONS

acac	acetylacetonate
AES	Auger electron spectroscopy
amphen	5-amino-1,10-phenanthroline
bipy	2,2'-bipyridine
bpz	2,2'-bipyrazine
CPE	carbon paste electrode
Cp	cyclopentadienyl anion
GC	glassy carbon
MV	methylviologen
NPV	normal pulse voltammetry
NQ	1,4-naphthoquinone
ϕ	benzene nucleus
Pc	phthalocyanine
phen	phenanthroline
PLL	poly(L-lysine)
PLV	poly(L-valine)
PMS	phenazine methosulphate
porph	porphyrin
PTS	*p*-toluenesulphonate anion
PVP	poly(4-vinylpyridine)
PVF	poly vinylferrocene
(R)RDE	(rotating) ring disc electrode
SEM	scanning electron microscopy

SERS	surface enhanced Raman spectroscopy
TAPP	tetra-aminophenyl porphyrin
TBAP	tetrabutylammonium perchlorate
TCNQ	tetracyanoquinodimethane
TEAP	tetraethylammonium perchlorate
TEAT	tetraethylammonium tetrafluoroborate
TPP	tetraphenylporphyrin
tpy	2,2',2''-terpyridine
TTF	tetrathiafulvalene
VDQ	vinyldiquat
vibipy	4-vinyl-4'-methyl-2,2'-bipyridine
viphen	4-vinyl-1,10-phenanthroline
vipy	4-vinylpyridine
XPS	X-ray photoelectron spectroscopy

1. INTRODUCTION

A modified electrode is created by deliberately coating a clean electrode surface with a thin film of a chosen material, with the objective of altering its properties in some desirable fashion. Work in this area began in 1975 when Murray's group covalently attached silane derivatives to a variety of materials. Following this a flood of publications appeared, describing the simultaneous or subsequent immobilisation of redox groups to metals and semi-conductors by condensation-type reactions. These systems, the forerunners of polymer modified electrodes, have been reviewed elsewhere[1-3] and will not be dwelt upon here. Whilst the synthetic elegance of these monolayer derivatisation schemes is obvious and many stable interfaces were created, it became clear that most applications would benefit from polymeric coatings and polymer modified electrodes were born.

To see why polymers are advantageous it is instructive to consider the objectives and requirements of surface modification. Usually, the objective is to immobilise redox centres on the surface, endowing the latter with the specific molecular identity and reactivity of the former, frequently to mediate charge transfer to target solution species, as illustrated in Fig. 1. In 'useful' systems it may be hoped to satisfy the subtle and complex chemical/environmental/steric requirements through the chosen mediator in its polymer matrix, although many systems studied to date involve model outer-sphere electron transfer couples.

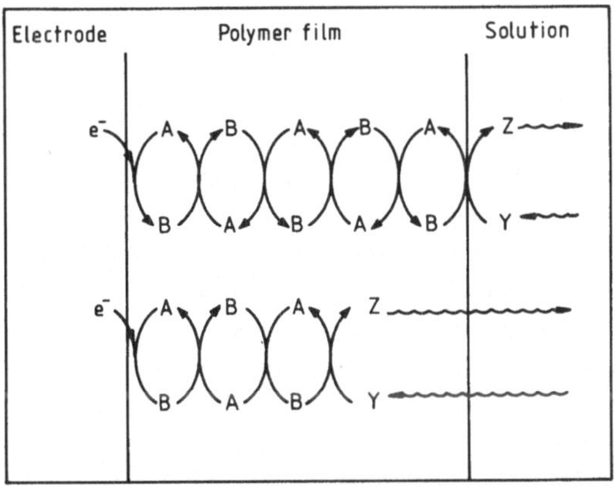

FIG. 1. Schematic representation of the mediated reduction of the solution species Y by the reduced partner B of the surface attached A/B couple. Reaction at the outer interface and, following partition, within the film are depicted, together with regeneration of B.

Bare or monolayer-derivatised electrodes only permit reaction in an essentially two-dimensional zone next to the electrode, whilst homogeneous reactions occur in a three-dimensional zone, permitting a far greater reaction flux. By using a thicker modifying film, one may hope to compete successfully[4] with homogeneous catalysis, since, in principle, one now has a three-dimensional reaction zone here also. For this situation to obtain, charge transport and solution species diffusion through the polymer must be rapid and partition of solution species favourable, important points to which we shall return in both this and the following chapter. In such a catalytic application, we are thus hoping to take advantage of cheap electrode materials (as the underlying support does not 'see' the target species), facile separation (clearly advantageous for the product, but also important for expensive or toxic mediators), and continuous mediator regeneration by the necessarily nearby electrode, plausibly in a flow-through system. It must be remembered, however, that a relatively small number of redox centres are employed, so they must remain firmly attached to the electrode and undergo clean reaction. For example, a current requirement of 1 A cm^{-2} from a (typical) film having $10^{-7} \text{ mol cm}^{-2}$ of mediator requires a turnover rate of 10^2 s^{-1}, implying many turnovers for a useful lifetime. This calculation also makes it plain why monolayer modifications ($\sim 10^{-10}$ mol cm^{-2}) will not give useful currents for realistic turnover rates. For other applications, such as preconcentration for analytical purposes, corrosion protection by charge transfer prevention, and optical displays by immobilised redox-state-dependent chromophores, the advantages of polymeric (i.e. thicker) films are obvious.

Many modified electrode systems have been devised and exploited in such diverse areas as photoelectrochemistry, bioelectrochemistry, conducting polymers, sensors and catalysis by problem-oriented scientists who have no particular allegiance to polymeric systems (or even an electrochemical approach in some cases). It will therefore be necessary on occasions to refer to monolayer systems inasmuch as they provide an alternative strategy which the polymer chemist must supersede. The strategy of presentation here is roughly chronological: early work centres on film formation/deposition/elaboration procedures, to be followed by electrochemical characterisation and structural determination by spectroscopic and other physical measurements. More complex issues of reactivity towards solution species have recently been put on a sound theoretical footing for simple and fairly realistic conditions of operation, opening the way to useful applications: these two aspects are dealt with in the following chapter.

2. PREPARATION

This section sets out to classify, in a broad way, the preparative procedures used for polymer modified electrodes. In principle this might be done via underlying substrate, polymer type or immobilised functionality. Since the substrate is envisioned as being less intimately concerned with the ultimate reactivity and properties than is a bare electrode, this would be a poor choice. The scheme chosen here is to classify by the immobilisation procedure, as this gives a better overall view of how one might attempt to attach a chosen entity to an electrode surface and helps to highlight the relative merits and limitations of the different strategies. The principal division is whether preformed polymers are used or polymerisation and coating are performed simultaneously. In either event, further elaboration of the film is possible, to introduce (extra) redox groups or to modify film properties, for example to stabilise it by cross-linking. The surfaces produced are presented in tabular form, according to this scheme, together with some broad generalisations as an aid to the aspiring electrode modifier.

2.1. Pre-formed Polymers

This was originally the most popular approach, largely due to the commercial availability of a wide range of known polymers and a vast literature covering the design, synthesis and characterisation of new ones. Furthermore, one can prepare, purify, characterise and, if necessary, fractionate the derivatising species before use. Frequently, there is no specific electrode/polymer bond and one relies on an unspecified adsorption/insolubility combination, with the result of considerable generality with respect to substrate electrode. Coating methods are usually very simple, the most common ones being droplet evaporation ('painting'), spin coating (whereby a more even film is produced by the droplet evaporation process on a rapidly spinning electrode) and a straightforward dip or dip/dry procedure.

2.1.1. Pre-functionalised polymers

Table 1 collects together details of modified electrodes produced by deposition of polymers in the final form in which they are required. The most obvious point is the preponderance of systems based on the vinyl function, largely redox functionalised polyethylenes and polystyrenes, although limited use of poly(acrylic acid)-based systems has been made. This is largely a result of the stability of the spine and the availability of such polymers together, perhaps, with the partially sighted view of polymer chemistry possessed by the specialist electrochemist. The majority of the remaining systems are the result of condensation (co-)polymerisations of mono- or di-functionalised monomers based on ester or amide formation. Variable extent of redox centre incorporation has been achieved by two approaches: firstly, variable degree of functionalisation of homopolymers and, secondly, copolymerisation of fully functionalised monomers with non-electroactive monomers. In both cases, the changes may not be straightforward and, particularly in the latter case, the unfunctionalised elements may result in different degrees of swelling, solubility (and thus stability of attachment), partition coefficients for solution species and 'clustering', as well as the more obvious increased separation of redox centres. Even this last example is not simple for, whilst increased functionalisation may facilitate site–site electron exchange by virtue of their proximity, it may also, by crosslinking

TABLE 1
Polymer Modified Electrodes Prepared from Completely Formed Polymers

Polymer	Substrate	Coating method	References
PVF (26 K)	Pt	Droplet evaporation.	5–7
(26 K)	Pt	Spin coating.	8
(26 K and 'somewhat lower')	Pt	Adsorption.	7, 9
(16 K, 26 K)	Pt	Electroprecipitation.	6, 10–14
(26 K and 'somewhat lower')	Pt	Photodeposition.	7, 9
PVF- co-acrylonitrile	Pt	Electroprecipitation.	13
PVP—M			
M = Ru(bipy)(tpy)$^{2+}$	Pt, GC	Dip/dry.	15
M = Ru(bipy)$_2$X, where			
X = PVP, H$_2$O, OH$^-$, CH$_3$CN, Cl$^-$, CN$^-$, NO$_2^-$, N$_3^-$	Pt, GC	Dip/dry.	16, 17
X = PVP	SnO$_2$		
	n-TiO$_2$	Dip/dry.	18
	Pt	Droplet evaporation.	19
X = Cl$^-$, pyr	GC, Pt	Dip or droplet evaporation.	20, 21
X = Cl$^-$	Au, Ir, Rh		
	n-MoSe$_2$,	Droplet evaporation.	22
	n-WSe$_2$		
M = ZnTPP	n-SnO$_2$	Droplet evaporation for PVP/ZnTPP solution mixture.	23
M = Ru(bipy)(PVP)$_3$	Ag	Dip.	24, 25

![structure1]	Pt/I	Dip/dry.	26
![structure2]	GC	Dip/dry in PS/CoPc solution.	27
![structure3]	Pt Pt	Dip. Dip/dry.	28, 29 30, 31

(continued)

TABLE 1—contd.

Polymer	Substrate	Coating method	References
$\{[CH_2-CH]_x[CH_2-CH]_y\}_n$ — pyridine–Ru(bipy)$_2$–pyridine / phenyl; $\{[CH_2-CH]_u[CH_2-CH]_v\}_m$ — phenyl-CH$_2$–N$^+$-bipyridinium–N$^+$–CH$_3$ / phenyl; $x = 0.033$, $y = (1-x)$, $u = 0.34$, $v = (1-u)$	Graphite	Successive droplet evaporation of solutions of the two polymers to give the two bilayer combinations.	32

Structure	Substrate	Method	Ref.
Ru(bipy)$_2$ structure; $x = 0.905$, $y = 0.047$, $z = 0.048$	Si-doped ⟨100⟩ n-GaAs	Droplet evaporation.	33
CH$_2$X / CH$_2$O·CO·TTF; $z = 0.15, 0.30, 0.87$; $y = 1 - z$; X = Cl, SCN	Pt	Spin or dip coating.	9, 34–36
CH$_2$Cl / CH$_2$–R	Pt	Droplet evaporation of suspension, spin coating.	8, 36–39

(continued)

TABLE 1—contd.

Polymer	Substrate	Coating method	References
[structure: R=O–φ–, pyrazoline ring with N–N–φ–R'] R' = H, OCH₃ $y \sim 0.51, 0.72, 0.85$ (R' = OCH₃), $0.11, 0.65$ (R' = H) $x = 0.99 - y$ (1% crosslinked)		(R' = H, crosslinked by cation radical based dimerisation.)	
[structure with morpholine-φ– substituent, N–N–φ–OCH₃] $y = 0.47, 0.65$	Pt	Spin coating.	8, 36
R = O₂C–(FeCp₂) $y = 0.60$	Pt	Spin coating.	8, 36

Structure	Electrode	Method	Ref.
$R = o\text{-}\phi\text{-TTF}$, $y = 0.70$	Pt	Spin coating.	8, 36, 38, 40
$R = -N-N=C\phi_2$, $y = 0.65$			
(carbazole structure), $y \simeq 0.96\text{-}0.97$	Pt	Spin coating. (Crosslinked by dimerisation.)	39
$R = -\overset{+}{\underset{}{N}}\!\!-\!\!\phi\!\!-\!\!CO-NH-R'$ $R' = C_2H_4$–(phenol with OH, OH)–$CH(R'')-(CH_2)_2CO_2H$ $R'' = H, CO_2H$	GC	Droplet evaporation.	41–45
$R = -\overset{+}{\underset{}{N}}\!\!-\!\!\phi\!\!-\!\!N^+\!\!-\!\!R'$ $R' = CH_3$	C/Pt	Carbon powder/polymer slurry dried on Pt.	46
$R' = H$	GC	Dip/dry.	47

(*continued*)

TABLE 1—*cont*.

Polymer	Substrate	Coating method	References
[structure: styrene-co-(methyl viologen benzyl) copolymer; $x = 0.34$, $y = 0.66$]	Graphite	Droplet evaporation.	48
[structure: styrene-co-(benzoquinone) copolymer; $x = 0, 0.75$; $y = 1.0, 0.25$]	Pt	Dip/dry or droplet evaporation.	49

Structure	Substrate	Method	Ref.
polymer with -C₆H₄-CH₂-⁺NR₃	Graphite	Dip/dry.	50
R = C₂H₅, C₆H₁₃			
polymer with -CO₂H	Graphite	Dip/dry.	50
R'-C(CH₂)ₙ-C(=O)-R, R' = H	Graphite	Dip/dry (also amide-bonded to surface—CO₂H).	51
R = L-(-NH-CH(CH(CH₃)₂)-C(=O)-OCH₃)			
R = 37% [-NH(CH₂)₂-C₆H₃(OH)₂], 63% OH	GC	Dip/dry.	52

(*continued*)

TABLE 1—contd.

Polymer	Substrate	Coating method	References
$R = -\underset{\underset{CH_3}{\|}}{N}-Th$ (0·83%) (also co-acrylic acid and co-vinylpyridine) $R' = CH_3$	Pt	Droplet evaporation.	53
$R = -NH$—⟨phenazine-S structure⟩—$N(CH_3)_2$	GC	Dip/dry.	47
$R = 41\%$ —$NH(CH_2)_2$—⟨catechol⟩—OH, 59% OH	GC	Dip/dry.	44, 45, 52, 54
$R = CH_3$, $R = CH_2-CH_2-CH_2-P\!\!\begin{array}{c}R''\\R'''\end{array}$ $R'' = R''' = \phi$, $CH_2CH_2P\phi_2$ $R'' = \phi$, $R''' = CH_2CH_2P\phi_2$ As complex: $Mo(N_2)_2(P)_4$	n-Si	Dip/dry.	55
	n-Si, C	Drop evaporation.	56

R = (o)—NH(Fe^{III}TPP), ‑(O—C$_2$H$_4$—$^+$N⟨py⟩)$_4$(Fe^{III}porphyrin)	GC	Dip or spin coating.	57
[poly(vinyl anthraquinone) structure]	Pt	Dip/dry.	30
[poly(N-vinylcarbazole) structure]	Pt	Droplet evaporation (electrochemical polymerisation also).	58
[poly(vinyl ether)—O—R structure]; R = RhB, merocyanine	n-SnO$_2$	Droplet evaporation.	59

(*continued*)

TABLE 1—contd.

Polymer	Substrate	Coating method	References
$\left(N\underset{R}{\overset{H}{\diagdown}}\right)_x\left(N\diagdown\right)_y$ $x = y = 0.5$	GC	Dip/dry.	47
$R = -\overset{\underset{\displaystyle\parallel}{O}}{C}-\phi-CH_2-\overset{+}{N}\diagdown\phi\diagdown N$ $y = 1 - x$ $x = 0.05, 0.10, 0.23, 0.38, 0.44, 0.53, 0.95$	GC, Hg	Dip/dry.	60
$R = -\overset{\underset{\displaystyle\parallel}{O}}{C}-\phi-N=N-\phi$ $x = 0.47, 0.65, 0.75$ $y = 1 - x$	GC/Hg	Dip/dry.	44, 45, 61–63
$R = -\overset{\underset{\displaystyle\parallel}{O}}{C}-2(AQ)$ $x = 0.13$ $y = 1 - x$	n-SnO$_2$	Drop evaporation.	59, 64
$R = -\overset{\underset{\displaystyle\parallel}{O}}{\underset{\underset{\displaystyle\parallel}{S}}{C}}-CH_2-N\diagdown$ (thiazolidine-benzothiazole structure with CH_2CH_3) $y = 1 - x$	n-SnO$_2$	Drop evaporation.	59

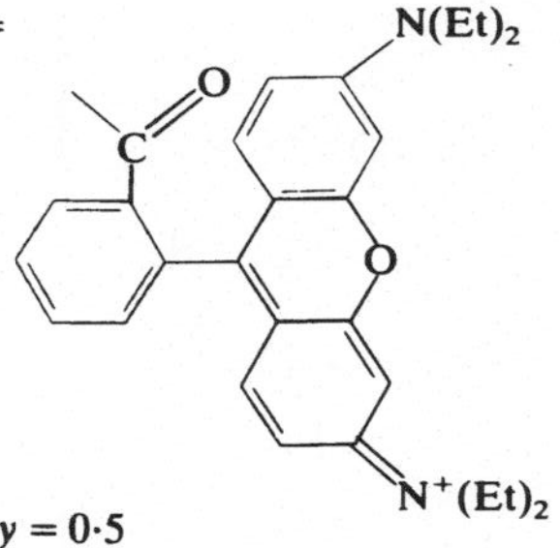	GC, Hg	Dip/dry.	65
	SnO$_2$	Dip/dry.	66
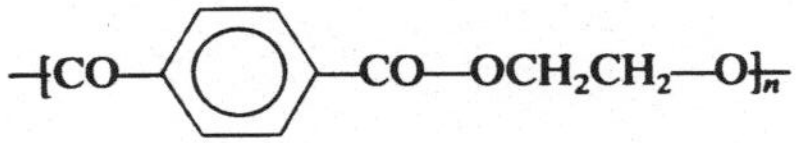	Pt n-Si	Dip/dry. Dip/dry.	55, 67, 68
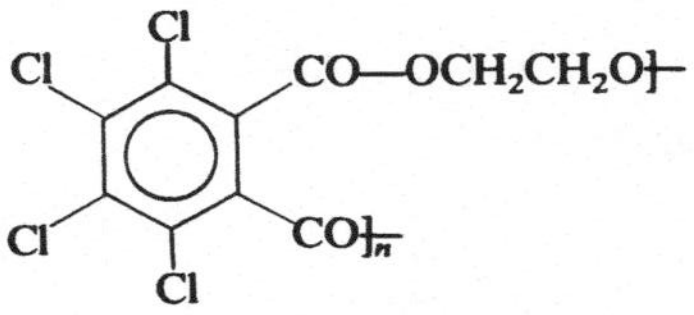	Pt	Dip/dry.	68

(continued)

R = (rhodamine structure with N(Et)$_2$ and N$^+$(Et)$_2$ groups)

$x = y = 0.5$

R = —C(=O)—Cp′—Co—Cp

R = H —[CO—C$_6$H$_4$—CO—OCH$_2$CH$_2$—O]$_n$—

—[CO—C$_6$Cl$_4$(CO)—OCH$_2$CH$_2$O]$_n$—

TABLE 1—contd.

Polymer	Substrate	Coating method	References
$2\left[-(CF_2CF_2)_v-(CFCF_2)_w\!\!-\!\!O\!-\!C_3F_6\!-\!O\!-\!C_2F_4\!-\!SO_3^-\right]$ with $H_3C\!-\!N^+\!\!-\!\!\bigcirc\!\!-\!\!\bigcirc\!\!-\!\!N^+\!\!-\!\!CH_3 \;\; X^-\;X^-$ and $-(N^+\!\!-\!\!CH_2\!-\!\bigcirc\!\!-\!\!\bigcirc\!\!-\!\!N^+\!\!-\!\!CH_2\!-\!\bigcirc\!-\!CH_2)_n\;\;X^-\;X^-$	Graphite	Drop evaporation of mixed solutions.	69
p-Xylene species: $X^- = \frac{1}{n}\!\left(\!-\!\bigcirc\!-\!SO_3^-\right)_n$	SnO$_2$ Graphite	Dip or spin coating. Polymer/binder /graphite paste coated on graphite.	70 71
$\left[-(CF_2CF_2)_v-(CFCF_2)_w-O-C_3F_6-O-C_2F_4-SO_3^-\right]\;\frac{1}{w}$	Graphite	Droplet evaporation of mixed solutions.	69

Cl⁻ *m*-Xylene species: $X = \frac{1}{n}$ [structure with SO₃⁻]	Graphite	Drop evaporation.	72
o-Xylene species: $X = \frac{1}{n}$ [structure with SO₃⁻]	Graphite	Droplet evaporation of mixed solutions.	48, 69
[−NH−C₃H₆−N⁺ ... N⁺−C₃H₆−NH− pyrimidine polymer structure with Br⁻ counterions]	SnO₂	Dip or spin coating.	70
	GC	Adsorption.	73

(continued)

TABLE 1—contd.

Polymer	Substrate	Coating method	References
[structure: polymer with NC, CN, (CH₂)₄, O groups]	Pt	Spin coating (some heat treatment: 2–3 min, 130–140°C).	74–78
[structure: polymer with biphenyl NH, C=O groups]	Hg, GC	Dip/dry.	65
[structure: metallophthalocyanine polymer with R₂ substituents and M center]	GC, Graphite	Dip/dry.	79

$2 \leq n \leq 5$		(Polymers not fully characterised.)	
$M = Fe(0.9, 3.0, 4.6, 4.7$ wt%); $Ni(3.7$ wt%); $Co(5.9$ wt%); remainder $2[H]$			
$3 < n < 8$	Active C	(Polymers uncharacterised.)	80
$M = 2[H]$, Fe; $R = 8[H]$, $8[OH]$, $8[OCH_3]$			
n unknown			
$M = Fe$	C	C/FePc/PTFE pasted.	81
$M = Fe$			82
	Graphite	Dip/dry.	51, 83, 84
	Pt, PbO_2	Dip.	85–87
	Pt/polypyrrole	Dip.	85–87
	PbO_2/polypyrrole	Dip.	86
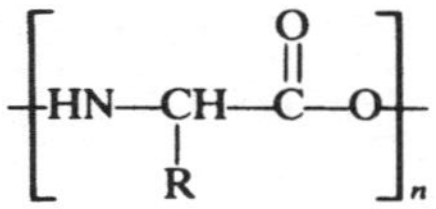	Pt, Pt/polypyrrole, Pt- or graphite-bonded polypyrrole	Dip.	85

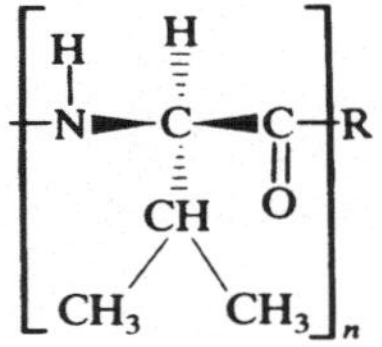

structure: cyclic unit $[-N(H)(H)-C(H)-C(=O)-R]_n$ with $CH(CH_3)_2$ side chain, $n \sim 20$, $R = OH$

$[-HN-CH(R)-C(=O)-O-]_n$

$R = -CH_2CH(CH_3)_2$

(*continued*)

TABLE 1—contd.

Polymer	Substrate	Coating method	References
R=—CH$_2$CH$_2$C(=O)O—CH$_2$—C$_6$H$_5$	Pt, Pt or graphite/ polypyrrole, Pt- or graphite-bonded polypyrrole	Dip.	87
[—HN—CH(CH$_3$)—C(=O)—O—]$_x$[—NH—CH((CH$_2$)$_4$NH$_2$)—C(=O)—O—]$_y$, $x = 0.40$, $y = 0.60$	Pt	Dip/dry (120°C in air, 20 min).	67
Mo complex with N$_2$, Pϕ_2, Pϕ_2CH$_3$ ligands, tethered via —C(=O)—N—H to polymer	Si, GC	Droplet evaporation	56, 88
Mo complex with N$_2$, Pϕ_2, Pϕ ligands, tethered via —NH—CO— to polymer	Si, GC	Drop evaporation.	56, 88

interactions, hinder necessary ion/solvent motion accompanying this process. These effects will be dealt with in Section 3 in more detail, but their importance is emphasised here because, even for these pre-functionalised pre-formed polymers, degrees of functionalisation are not always, and sequence lengths almost never, quoted.

A few of the polymers are designed not to function via charge transfer mediation but through the provision of specific environmental factors, such as the stereochemical ones provided by the poly(L-valine)[51,83–87] and poly(lysine-co-alanine)[67] films. It is noteworthy that in favourable cases not only primary thermodynamic properties, such as the formal potential E', but also more subtle kinetic features, such as stereospecificity, are retained on surface immobilisation. Limitations and modifications of this statement, largely the result of the very high effective surface concentration, are discussed in Section 3.

It is encouraging to see that some advantage has been taken of alternative synthetic approaches. This is well illustrated by the viologen redox function, which has been immobilised by covalent attachment (to polystyrene), by electrostatic binding (to Nafion: see Table 3) and by using it as an integral part of the polymer spine in poly(o-, m- and p-xylylviologen) {in conjunction with simple (halide) and polymeric [poly(styrene sulphonate)] counter-ions} and a cyanuric chloride copolymer. Spinal incorporation of the redox group has also been used for the tetracyanoquinodimethane (TCNQ) ester and cobaltocene polymers in Table 1.

Prior coordination of metal centres to ligand functionalised polymers is not very common—they are usually incorporated after film deposition in the case of pre-formed polymers. Simple entrapment (e.g. of cobalt phthalocyanine in polystyrene) is comparatively rare, presumably because a stronger redox centre/polymer interaction is required to retain the former at the surface.

Whilst certain materials (Pt, GC, graphite) appear more frequently, this approach is relatively general: indeed, many workers use substrates interchangeably to suit the experiment, e.g. Pt sheets or gauzes for preparative-scale electrolyses and doped In_2O_3 or SnO_2 on glass for visible transmission measurements. Instances where this assumption will be invalid include systems involving covalent polymer/electrode bond formation and phenomena where the substrate properties are important, such as modified semi-conductor photoelectrochemistry.

Largely unaddressed questions are media limitation (via film solubility) and polymer characterisation. In the former instance, it may be that many observations are unreported, as failed experiments. The lack of polymer characterisation—relatively few molecular weights, let alone distributions, are reported—is surprising, for prior characterisation is a primary advantage of this approach.

2.1.2. Post-coating functionalisation of pre-formed polymers

The most obvious method of attaching metal redox centres to a polymeric ligand is by displacement of a labile ligand (frequently H_2O) from the former. Examples of this type are collected in Table 2. All employ N- or P- donors. Although thermal equilibration with a solution of the appropriate ion is usual, photo-assisted substitution has been used.[21]

Much more popular are the ion exchange systems (Table 3) where one replaces simple ions (K^+, Na^+, halides, etc.) with the redox ions of interest. The extent, stability and rate of incorporation are dependent on electrostatic interactions between the polymer and the oppositely charged redox ion and, in certain cases (notably Nafion, discussed in Section 3), on hydrophobic interactions, particularly for organic ligands/low charge density polymer systems such as $Ru(bipy)_3^{2+}$/Nafion. Because of the interplay of electrostatic and hydrophobic interactions, generalisations are difficult, but electrostatic binding favours highly charged ions such as $Fe(CN)_6^{3-/4-}$, $Mo(CN)_8^{3-/4-}$, $Ru(NH_3)_6^{2+/3+}$, even in solutions devoid of them. The degree of functionalisation, i.e. loading of exchanged ion, is a more complex issue here: absolute determinations of partition coefficients are relatively few (see Section 3), and in any case the system may be under kinetic control, rather than in equilibrium with the solution, as a result of slow polymer solvation/structural changes.

By far the most popular anion exchanging system is poly(4-vinylpyridine)(PVP), either in protonated (PVP·H^+) or quaternised (PVP·R^+) form. There are advantages to both forms: the former allows *in situ* control of exchange capacity through pH, but has a limited pH range for binding (pH ≤ 5), whilst the latter gives access to the entire pH range. Most of the cation exchangers are sulphonates [Nafion, poly(styrene sulphonate) (PSS$^-$) and poly(vinyl sulphonate)], as they give access to a wider pH range than carboxylates [a carboxylate version of Nafion and poly(acrylic acid)].

It is pleasing to see that the power and range of the synthetic variables is beginning to be used more fully. For example, use has

TABLE 2
Pre-formed Polymers Coordinatively Functionalised after Coating

Final polymer	Substrate	Comments	References
![structure with pyridine-M] $n \simeq 7 \times 10^3$ (M.wt $\simeq 7.4 \times 10^5$)			
M = Ru^{III}(EDTA)	Graphite	H$_2$O ligand exchanged.	89–91
M = Ru(NH$_3$)$_5^{2+}$	Graphite	H$_2$O ligand exchanged.	91
M = Cu^{2+}	Graphite	Cl$^-$ ligand exchanged.	91
(M.wt $\simeq 10^5$)			
M $\simeq$ 0.2 Ru(bipy)$_2$Y	GC	Photosubstitution of Cl$^-$ ligand.	21
Y = H$_2$O			
Y = ClO$_4^-$			
Y = CH$_3$CN			
(M.wt = 7.4 × 10^5)			
M ≤ 0.33 Fe(CN)$_5^{3-}$	Graphite	H$_2$O ligand exchange when PVP/Fe(CN)$_5$(OH$_2$)$^{3-}$ mixed and solvent evaporated.	92, 93
—[CH$_2$—CH]$_n$— | CN—M n unknown	Graphite		
M = Ru^{III}(EDTA)		H$_2$O ligand exchanged.	91
M = Ru(NH$_3$)$_5^{2+}$		H$_2$O ligand exchanged.	91

(continued)

TABLE 2—cont.

Final polymer	Substrate	Comments	References
$\left[\begin{array}{c}\text{—(CH}_2\text{—CH)}_n\text{—}\\ |\\ \text{P}\phi_2\end{array}\right]_3$ Ni^0	C	Ni added to polymer films as Ni^{2+}.	94
(2% crosslinked by divinyl/benzene) —(CH$_2$—CH)$_n$— (Pϕ_2)$_x$Ni^0(Pϕ_3)$_y$	Carbon	Ni^{2+}/Pϕ_3 solution reduced in presence of polymer film (procedure unspecified).	94
[structure with Fe(phenanthroline) complex bearing two $-\text{SO}_3^-$(P$^+$)$_n$ sulfonate groups]$_3$	Pt, SnO$_2$	(P$^+$)$_n$·($^-$O$_3$S—R) spin coated on to electrode, dried and immersed in Fe^{II} solution.	95

128

P^+ =

─(CH$_2$─CH)$_n$─ (90–100% quaternisation),
         |
         [pyridinium-N$^+$-CH$_3$]

─(CH$_2$─CH─N$^+$H)─
              |
              CH$_3$

(approximately 50% quaternisation),

─(CH$_2$─C)$_n$─
         | \
     C$_2$H$_4$─N$^+$(CH$_3$)$_3$
         |
         CO$_2$H

TABLE 3
Pre-formed Polymers Modified by Ion Exchange

Final polymer	Substrate	Species ion exchanged into film/comments	References
![PVP structure] $H \cdot \frac{1}{x}[M^{x-}]$		PVP applied to electrode as neutral species, then protonated at pH ≤ 5.	
$M = Fe(CN)_6^{3-/4-}$	Pyrolytic graphite (PG)	$Fe(CN)_6^{3-/4-}$	93, 96–99
$M = Mo(CN)_8^{4-}$	PG	$Mo(CN)_8^{4-}$	99–101
$M = W(CN)_8^{4-}$	G	$W(CN)_8^{4-}$	97
$M = IrCl_6^{2-/3-}$	PG	$IrCl_6^{2-}$; 5% L-lysine copolymer used in ref. 103.	98, 99, 102, 103
$M = Cr_2O_7^{2-}$	Pt	$Cr_2O_7^{2-}$; thin films obtained by washing.	104, 105
$M = $ Rose Bengal	SnO_2	Rose Bengal	145
$M = $ (croconate dye structure)	SnO_2	At pH 3, where PVP is protonated, but dyes are dianions.	106, 107

[structure: polymer with pyridinium groups bound to tetracyanoquinodimethane-type anions and squarate-type dianions] $z = 0.94, y = 0.062$		
[structure: $-O-Si(C_3H_6-O-CO-)-O-$ silane polymer with $\frac{1}{x}[M^{x-}]$] $M = Fe(CN)_6^{3-}$ $M = IrCl_6^{3-}$	Pt	$\left.\begin{array}{l}Fe(CN)_6^{3-}\\ IrCl_6^{3-}\end{array}\right\}$ Individually and together. Silane both lightly cross-links and bonds to 'PtO'.	108, 109
[structure: polymer of vinylpyridinium units with $R \cdot \frac{1}{x}[M^{x-}]$] $R = CH_3, M = Fe(CN)_6^{3-}$ $0.19 \leq z \leq 0.88, y = 1 - z$	Graphite		110

(continued)

TABLE 3—contd.

Final polymer	Substrate	Species ion exchanged into film/comments	References
$R = CH_2-\langle\bigcirc\rangle$, $M = Fe(CN)_6^{3-}$ $0.10 \leq z \leq 0.92$, $y = 1-z$ $M = IrCl_6^{3-}$	Graphite	At pH 8, only PVP·R$^+$ binds; at pH 1, PVP·H$^+$ also binds.	99, 110
	Pt/I	Degree of quaternisation unknown.	26
[structure with pyridine groups]$_u$ [structure]$_w$$_n$ R·Fe(CN)$_6^{3-/4-}$ $R = C_mH_{2m+1}$, $m = 8, 10, 12$; $v > 0.95$; $u = 1-v$ $R = CH_3$, $v \sim 0.5$–0.6	GC	PVP·R$^+$ spin coated, then electrode immersed in Fe(CN)$_6^{3-/4-}$ solutions. Some o-xylene crosslinking	111 111
[pyridine-Ru structure]$_n$ (H$_2$O)Ru(bipy)$_2^{2+}$ · $\frac{1}{2}$[Fe(CN)$_6^{4-}$]		With time, H$_2$O ligand replaced by CN$^-$, to give bridged Ru(CN)Fe species.	112

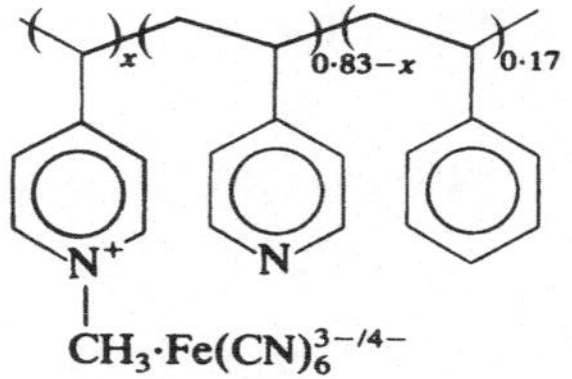

[structure: polymer with pyridinium, x=0.17 phenyl, 0.83−x pyridine units]

$CH_3 \cdot Fe(CN)_6^{3-/4-}$

$x = 0.37$
$x = 0.83$

GC, Pt

$Fe(CN)_6^{3-/4-}$

113

Some *o*-xylene crosslinking. Synchrotron radiation crosslinked.

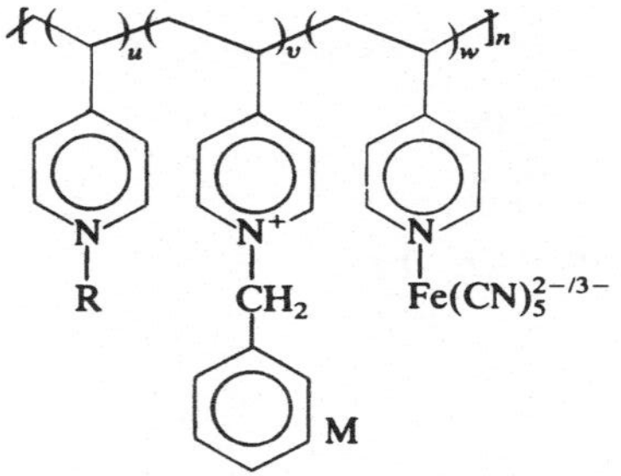

[structure: copolymer with pyridine-Fe(CN)5, pyridinium-CH2-Ph-M, and pyridine-R units]

$Fe(CN)_5^{2-/3-}$

$u = 0.6, v = 0.2, w = 0.2$
$R = $ nothing, H^+
$M = Ru(NH_3)_6^{3+}$, $IrCl_6^{3-}$, $Mo(CN)_8^{4-}$, $Fe(o\text{-phen})_3^{2+}$

$(H_2O)Fe(CN)_5^{2-}$ coordinated to PVP during coating.

Graphite

99

Variation of net polymer charge by pH and potential.

(*continued*)

TABLE 3—contd.

Final polymer	Substrate	Species ion exchanged into film/comments	References
$\text{+(NH-CH-CO)NH(CH}_2\text{)}_k\text{-NH+}$ $\quad\|$ $(CH_2)_4$ $\|$ $^+NH_3$			
$\text{+(CO-CH-NH)}_m\text{CO+(CH}_2\text{)}_k\text{-CO+}_n \cdot M^{x-}$ $\qquad\|$ $(CH_2)_4$ $\|$ $^+NH_3$			
$k = 6\text{–}10,\ m = 20\text{–}50,\ n \simeq 100$	Graphite	Heating at 80°C increases stability by crosslinking. $pH \leqslant 11$ protonates amines.	114–116
$M = Mo(CN)_8^{4-}$ $M = W(CN)_8^{4-}$ $M = Fe(CN)_6^{3-}$ $M = Fe(EDTA)^{-/2-}$			
$M = Co(C_2O_4)_3^{3-}$		Polymer structure given as in published 'correction'.	
$\text{+(CF}_2\text{CF}_2\text{)}_x\text{(CFCF}_2\text{)}_y\text{+}_n$ $\qquad\qquad\|$ $\qquad\qquad O$ $\qquad\qquad\|$ $C_3F_6\text{-O-}C_2F_4\text{-SO}_3^-$ M^{z+}		'Nafion'	

M = Ru(bipy)$_3^{2+}$	Graphite		117, 118
	GC		119–121, 132
	Pt		122
M = Co(bipy)$_3^{2+}$	Graphite		123
	GC		120
M = Fe(bipy)$_3^{2+}$	GC		120
M = Os(bipy)$_3^{2+}$	GC		121, 124
M = Fe(tpy)$_2^{2+}$	Graphite/GC		119
M = Ru(NH$_3$)$_6^{3+}$	GC		120, 132
M = Ru(NH$_3$)$_6^{2+}$, Co(tpy)$_2^{3+}$ mixtures	Graphite		125
M = CpFeCp—CH$_2$N$^+$(CH$_3$)$_3$,	Graphite		126
M = Co(tpy)$_2^{2+}$ mixtures	Graphite		126
M = Fe(o-phen)$_3^{2+}$	GC		120
M = Fe(Cp)(Cp—CH$_2$N$^+$(CH$_3$)$_3$)	GC, Au		120, 121, 127, 132
M = MV^{2+}	GC		132, 120, 119
	Pt		119
M = Ru(bipy)$_3^{2+}$, MV^{2+} mixtures	Graphite		48, 119, 128
M = Co(TPP)	Graphite		32
	Graphite	Film successively soaked in TPP and Co solutions; Ru(NH$_3$)$_6^{3+}$ used as 'e$^-$-shuttle'.	129
M = TTF$^+$	Pt	Potential cycling leads to (TTF)Br$_{0-7}$ formation.	130, 131
M = FeCp$_2^+$	Au		127

(continued)

TABLE 3—contd.

Final polymer	Substrate	Species ion exchanged into film/comments	References
	Graphite	Viologen oligomer.	133

Structure	Electrode	Notes	Ref.
$v = 0.033$; $u = 1 - v$; ($n \approx 20$–25); $M = ClCH_2-\phi-(N^+-CH_2-\phi-CH_2)_8Cl$	Graphite	Ru oligomer.	133
(M.wt = 7×10^4, except^a 3×10^6)	Graphite	p-Xylylviologen oligomer.	128
$SO_3^- \cdot \frac{1}{z} M^{z+}$; $M = Ru(bipy)_3^{2+}$	Pt	Spin coated.	134, 135
$M = Co(bipy)_3^{2+}$	Graphite	Drop evaporation.	110
$M = Co(phen)_3^{2+}$	Graphite		110
$M = Ru(NH_3)_6^{3+}$	Graphite		110
$M = [Ru(bipy)_2(OH_2)_2]^{2+}$	Graphite		110
$M = [Ru^{III}(bipy)_2(OH_2)]_2^{4+}O$	CPE	Film-coated C ion	136
$M = Cu^{2+}$	CPE	exchanged and pasted.	136
$-C_3H_6-Si(OCH_3)_{3-x}-(O)_x-$; $SO_3^-\,\frac{1}{z}M^{z+}$	CPE		137

(*continued*)

TABLE 3—contd.

Final polymer	Substrate	Species ion exchanged into film/comments	References
$M = Cr(bipy)_3^{2+}$, $Ru(NH_3)_6^{3+}$, MV^{2+} [structure: copolymer with 0.90 $-C_6H_4SO_2^-$, 0.05 $-C_6H_4SO_3H$, 0.05 $-C_6H_4SO_2-C_6H_4-$ units]	Pt	Crosslinked by HCl(g) or heating to 50°C *in vacuo*.	138
$Ru(bipy)_3^{2+}$ [poly(vinyl sulfone-phenyl) structure], $n \simeq 35$	Pt, GC	Ion exchange at pH 9.	139

$-[CH_2-CH]_n-$
       $\|$
       CO_2^-
       M
(M.wt = 2.5×10^5)
$M = Ru(NH_3)_6^{3+}$, $Ru(bipy)_3^{2+}$, $Co(bipy)_3^{2+}$, $Co(phen)_3^{2+}$
	Graphite	pH > 5	110, 140

$-[CH_2-CH]_n-$
       $\|$
       SO_3^-
       M
$M = Ru(bipy)_3^{2+}$, $Co(bipy)_3^{2+}$, $Ru(NH_3)_6^{3+}$, $Co(phen)_3^{2+}$
	Graphite	Cations lost 'fairly rapidly' in pure supporting electrolyte.	110

		Cations lost 'fairly rapidly' in pure supporting electrolyte.	

| | Graphite | Dual functionality. | 141 |

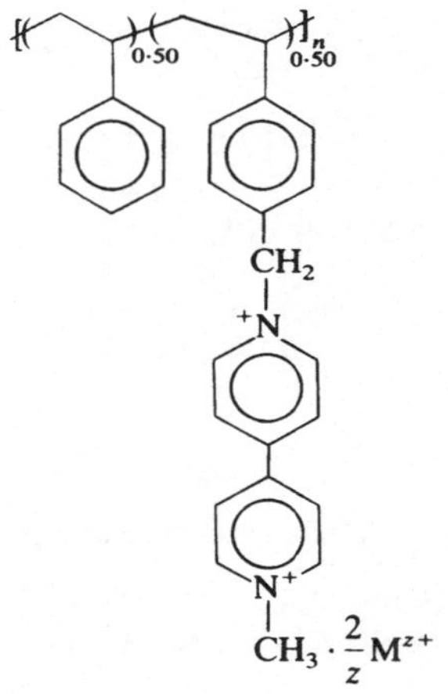

(M. wt ≈ 4×10^4)
M = $Fe(CN)_6^{3-/4-}$, $IrCl_6^{2-/3-}$, $Mo(CN)_8^{3-/4-}$

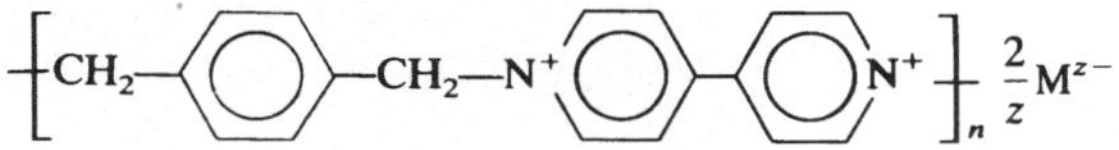

(M.wt > 10^4)

M = $Mo(CN)_8^{4-}$

| | Graphite | Drop evaporation or electrodeposition of radical cation. Dual functionality. | 141 |

TABLE 3—contd.

Final polymer	Substrate	Species ion exchanged into film/comments	References
(M. wt = 1.1×10^4) M = $(Pt^0)ClO_4^-$	n-Si	Droplet evaporation or electrochemical preparation (as radical cation polymer) followed by ion exchange of $PtCl_6^{2-}$ and photoreduction to Pt^0.	142
$\mathrm{-[(C_2F_4)_{0.79}(C_2F_3)_{0.21}]_n-}$ $\quad\quad\quad\mid$ $\quad\quad\quad O$ $\quad\quad\quad\mid$ $\mathrm{C_3F_6-CO_2^- \cdot \frac{1}{x} M^{x+}}$ M = $Ru(NH_3)_6^{3+}$, $Ru(NH_3)_5(\text{iso-nic})^{2+}$, $Os(bipy)_3^{2+}$, $Fe(Cp)[Cp'-CH_2-\overset{+}{N}(CH_3)_3]$	Graphite	Carboxylate analogue of Nafion.	125
Montmorillonite clay, $X \cdot \frac{1}{x} M^{x+}$ X = colloidal Pt^0		Colloidal Pt improves film uniformity and prevents cracking.	143
M = $Ru(bipy)_3^{2+}$	SnO_2		
M = $Fe(bipy)_3^{2+}$	SnO_2, GC, Pt		
M = $Ru(NH_3)_6^{3+}$	SnO_2		
M = $Fe(Cp)(Cp'-CH_2-\overset{+}{N}(-CH_3)_3)$	SnO_2		
M = MV^{2+}	SnO_2		
$X = -[CH_2-CH]_n-$ $\quad\quad\quad\quad\mid$ $\quad\quad\quad\quad OH$ M = Δ-$Ru(phen)_3^{2+}$	SnO_2	Stereoselective functionalisation.	144

been made of pendant[133] or spinally integrated[14] redox groups in the ion exchanging polymer to provide potential-dependent ion exchange capacity. In work from the same laboratory[99] a more complex system, employing several strategies simultaneously in a PVP system, has been developed as follows. Equal fractions (20%) of the pyridine groups were quaternised and bonded to —Fe(CN)$_5^{2-/3-}$ centres, to provide a polymer of net negative charge. The remaining (60%) pyridine sites were then available for protonation. Through pH (via the pyridine sites) and potential (via the —Fe(CN)$_5^{2-/3-}$ sites) it was then possible to produce a polymer of net negative, zero or positive charge. By studying anion and cation uptake, the authors were able to demonstrate preferential (partial or total) intra-film charge compensation. We may expect further exploitation of these versatile multi-functionalised systems in future.

Ion exchange polymers may be crosslinked, either deliberately via silane components[108,138] or heat treatment[114-116] or inadvertently through multiple electrostatic interactions of highly charged ions. The silane may also increase stability by the formation of covalent bonds to the underlying substrate.[109] Whilst some crosslinking may be desirable from a stability viewpoint, an excessive amount may inhibit spinal flexibility and/or solvent/ion motion to the point where electroactivity on reasonable timescales is significantly diminished.

Ion-exchanged systems may be further modified electrochemically, for example by photoreduction of $PtCl_6^{2-}$ to colloidal Pt^0 [142] or the formation of mixed-valent $(TTF)Br_{0.7}$.[130-131] The latter example is particularly interesting, for a Nafion/TTF$^+$, $(TTF)Br_{0.7}$ film embodying both ionic and electronic conduction modes was constructed. (Further examples of this exchange/reduction combination are found in Table 5 for polymers produced *in situ*.)

Ion exchange of polymers into Nafion has also been demonstrated, although evidence for homogeneity or otherwise, whether under kinetic or thermodynamic control, is absent. Bifunctionalisation of Nafion with cobalt tetraphenylporphyrin (CoTPP) and $Ru(NH_3)_6^{3+}$ has been carried out;[129] here the $Ru(NH_3)_6^{3+}$ is added as an 'electron shuttle' to ferry charge between the electrode and the porphyrin centres, as the latter are unable to do this alone at a rate sufficient to be catalytically useful.

Miscellaneous examples of pre-formed polymers modified on the electrode surface other than by coordination or ion-exchange processes are assembled in Table 4. These are principally crosslinking (by

TABLE 4
Surface Attached Polymers Modified by Non-coordinative/Ion-exchange Strategies

Final polymer	Substrate	Treatment and result	References
[structure: ferrocene-containing polymer with CO–O–C₃H₆Si(–O–)₃ group] $x = 0.88, 0.72, 0.63, 0.59, 0.38, 0.36$ $y = 1 - x$	Pt	Polymer film exposed to H_2O/HCl vapour to induce crosslinking.	146
[structure: copolymer with quaternary ammonium $H_2C-{}^+N(CH_3)_2-CH_2-$ groups and DCIP (2,6-dichloroindophenol) moiety]	Pt	Polymer film/DCIP occlusion exposed to ^{60}Co γ-radiation to crosslink, bond to electrode and bond DCIP.	147

142

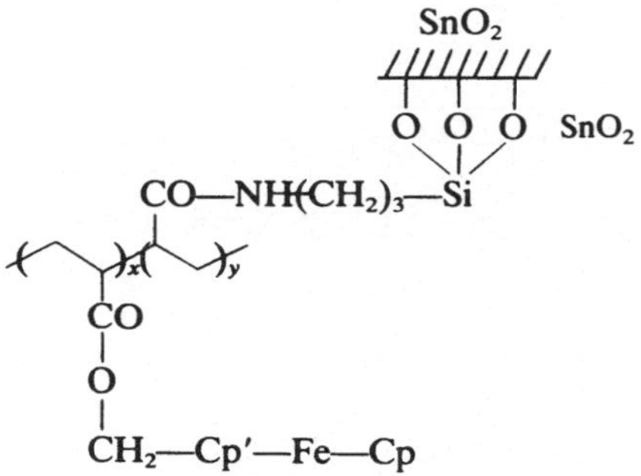	SnO_2	Silylated surface exposed to acid chloride polymer. Condensation of ferrocene redox centres onto unreacted acid chloride groups via amide formation.	148
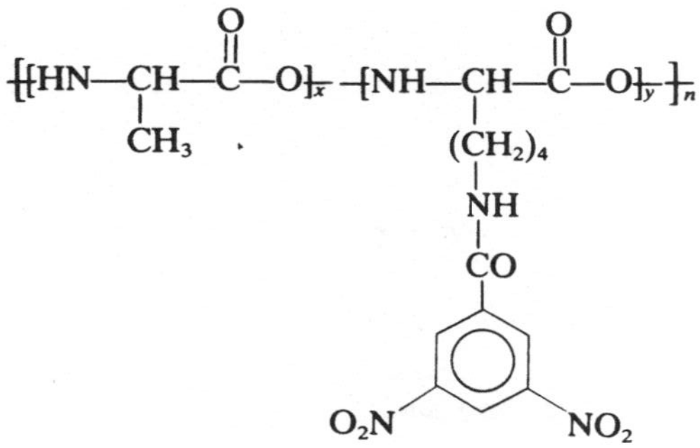 $x = 0{\cdot}40,\ y = 0{\cdot}60$	Pt	Dip-coated lysine/copolymer film exposed to dinitrobenzoyl chloride (amide condensation).	67

(continued)

TABLE 4—contd.

Final Polymer	Substrate	Treatment and result	References
[structure: polymer with fluorenone and anthraquinone groups linked via N-containing backbone, subscripts 0.5n and 0.5]	GC	Dip/dry or droplet evaporation, followed by cathodic reduction to give fluorenone crosslinking.	44
[structure: poly(styrene) with R substituent, subscript n]	Pt, GC	Mixtures and bilayers formed by sulphonamide formation reactions between ~SO_2Cl and appropriate amines.	139

R = 90%—SO_2R', 5%—SO_3H, 5%—SO_2
$n \simeq 35$
R' = —NH—ϕ—FeCp$_2$, —NH—ϕ—N(CH$_3$)$_2$, —NH-phenRu(bipy)$_2^{2+}$

silanes, γ-irradiation and cathodic coupling) and condensation reactions. In the case of the ferrocene/silane copolymer,[146] the authors are to be commended for an excellent polymer characterisation, including sequence lengths.

Of these systems, the most popular are the ion-exchange polymers. This is despite the fact that, unlike in the completely pre-formed pre-functionalised systems in Table 1, one does not know and cannot control homogeneity of redox site distribution, so that isolated pockets of sites may be produced which are unable to participate in film electrochemistry. The reasons for their success would seem to lie in their ease of preparation, general applicability, stability and versatility with regard to redox centre incorporation: a single polymer may give access to immobilisation of a whole series of redox systems, individually or collectively.

2.2. Polymers Formed/Coated Simultaneously

Simultaneous coating/polymerisation methods fall into three principal categories: condensation of silanes, electrochemically initiated polymerisation and a miscellaneous group of plasma initiated or *in situ* polymerisations. In all cases, the objective is to maintain (or augment in a controlled way) monomer reactivity during polymerisation on the surface: the extent to which this is achieved for each approach is considered in turn. Whilst this objective makes characterisation very important, measurements of such fundamental quantities as molecular weight may in fact be very difficult because of polymer/surface and crosslinking bond formation by reactive entities produced during polymerisation. Furthermore, many important properties may not be intrinsic to the monomer at all, but result from polymerisation—the polyaromatics of Table 6 are a prime example. For these systems, then, the structural (spectroscopic) characterisation described in Section 4 is of paramount importance.

2.2.1. Polymeric silanes

The systems summarised in Table 5 are largely an outgrowth of the early monolayer modified electrode studies, in which it was observed that relaxation of the stringent dry conditions usually employed for the bonding of hydrolytically unstable silanes to oxide surfaces resulted in coverages far in excess of a monolayer. The monomers are di- or tri-alkoxy or halogen substituted silanes, and possess one or two silane functions, so that suitable conditions may result in extensive crosslinking. The generality of the method seems well established, although the

TABLE 5
Silane-Based Polymer Modified Electrodes

Silane monomer	Substrate	Comments	References
R—N$^+$—〈phenyl〉—〈phenyl〉—N$^+$—R R = R' = —(CH$_2$)$_3$—Si(OCH$_3$)$_3$	Pt, W, p-Si, Indium tin oxide (ITO)	Coating/polymerisation by immersion in silane solution containing trace H$_2$O, unless otherwise stated. PdCl$_4^{2-}$ ion exchanged into film and reduced to Pd^0.	149, 150, 158
	Au, Pt, SnO$_2$	Deposition/polymerisation by holding in (or cycling into) viologen cation radical potential region.	151–153
	Pt, p-Si	Deposition as above or by immersion procedure. Ion exchange of Fe(CN)$_6^{4-}$ Ru(CN)$_6^{4-}$, IrCl$_6^{2-}$ and PtCl$_6^{2-}$.	154, 155
	p-Si	PtCl$_6^{2-}$ ion exchanged and reduced to Pt^0.	156, 157
	n-Si	Fe(CN)$_6^{3-}$, PtCl$_4^{2-}$, PtCl$_6^{2-}$ ion exchanged in and latter two reduced to Pt^0.	159

Structure	Substrate	Method	Ref.
R = R' = —CH₂—⟨C₆H₄⟩—Si(OCH₃)₃	Pt, SnO₂, n-Si		160
R = R' = —CH₂—⟨C₆H₄⟩—(CH₂)₂—Si(OCH₃)₃	Pt, GC, SnO₂	Spin coating with air/H₂O blown over droplet, or droplet evaporation followed by thermal (90°C) crosslinking.	161
R = —CH₂—⟨C₆H₄⟩—(CH₃)Si(OCH₃)₃, R' = —CH₃	Pt	Droplet evaporation/thermal (50°C) polymerisation.	162
	Pt, GC, SnO₂	Droplet evaporation or spin coating in air/H₂O vapour, then heating at 50°C.	161
[ferrocene]SiCl₂ (X = Cl)		Polymerisation/coating by simple exposure to silane solutions with trace H₂O.	
X = Cl	Au, Pt, Ni	Including single crystals, polished and 'textured' surfaces.	6, 163–166, 183
	n-Si		164, 165, 167–171
	n-Ge		172
	n-GaAs		173
X = CH₃	n-Si	Polished and 'textured' single crystals.	171

(continued)

TABLE 5—contd.

Silane monomer	Substrate	Comments	References
Ph-SiCl₃ (ferrocenyl)	n-Ge	Polymerisation/coating by simple exposure to silane solution with trace H₂O.	166, 172
	Au, Pt, n-Si		163, 166, 174
Fc-C₅H₄-Si(OCH₃)₃	Au, Pt		163, 165, 166
(CH₃O)₃Si-(substituted ferrocene)	n-Ge		165, 166
(Me₄Cp)Fe(Cp-CH₂-SiCl₂-Cp)	Pt, ⟨111⟩n-Si	Immersion in silane solution (in the dark).	175
Cl₂Si(ferrocenyl)₂	Pt, n-Si		176

Compound	Substrate	Method	Ref.
$H_2N(CH_2)_2NH(CH_2)_3Si(OCH_3)_3$ ('en' silane)	Pt	Immersion in silane solution with trace H_2O, then amide formation with ferrocene carboxylic acid species[a] in presence of coupling agent.	177
$H_2N(CH_2)_2NH(CH_2)_3Si(OCH_3)_3$	Pt	Spin coating under $N_2/H_2O/HCl$ vapour, protonation and ion exchange of $Fe(CN)_6^{4-}$.	178
[structure: $NH-(CH_2)_2-\overset{CH_3}{\underset{CH_3}{N^+}}-(CH_2)_3-Si(OCH_3)_3$ on chloronaphthoquinone]	Pt, W, p-WS$_2$	Extended immersion in silane solution. $Fe(CN)_6^{3-/4-}$ may be ion exchanged into film.	179

[a] [ferrocene with CH_2CO_2H substituent on phenyl ring]

(*continued*)

TABLE 5—contd.

Silane monomer	Substrate	Comments	References
![structure: O=C(Cl)(CH$_2$)$_3$-Si(CH$_3$)Cl$_2$]	Pt, GC	Drop evaporation or spin coating followed by amide formation with aminophenylferrocene or tetrakis(p-aminophenyl)porphyrin (metallated in CoCl$_2$ or CuCl$_2$ solutions).	180
RuL$_2$L′ L = bipy, L′ = [4,4′-bipyridine with C$_2$H$_4$SiCl$_3$ substituent]	Pt, SnO$_2$	Extended immersion in silane solution at 25–70°C.	181, 182
L = phen, L′ = [phenanthroline with C$_2$H$_4$SiCl$_3$ substituent]	Pt		182

$Cl_2Si[(CH_2)_3CN]_2$	Ni	Immersion in silane solution followed by $-Fe(CN)_5^{3-}$ coordination by displacement of H_2O ligand.	183
NR¹R² — C₆H₄ — NR³R⁴ $R^1 = R^2 = CH_3$, $R^3 = C_2H_5$, $R^4 = (CH_2)_4-Si(OCH_3)_3$ $R^1 = R^2 = R^3 = R^4 = -(CH_2)_3-Si(OCH_3)_3$	Au, Pt, SnO₂, n-Si		184
pyridyl-CH₂CH₂-Si(OCH₃)₃	Pt	Protonation (pH ≤ 5) allows ion exchange of $M(CN)_y^{x-}$ [M = Fe, Ru, Mo, Co; $y = 6$ or 8; $x = 3$ or 4], $(NQ)SO_3^-$, Indigo Carmine, I^-.	185

majority of the electrodes derivatised are Pt, Au, n-Si (all with some surface oxide present) and SnO_2, and most of the modifiers are viologen or ferrocene redox centre based. Control over the final product, notably with respect to film thickness and crosslinking, tends to be established by each author (through variation of reaction time, solvent, silane concentration, water content of reaction medium and temperature) in an *ad hoc* way: the phrase 'in our experience' is a recurrent one. Control of crosslinking seems more generally established through choice of monomer, two silane functions and three hydrolytically replaceable substituents promoting it. For example, Willman[161] has observed that their singly silylated viologen derivative gave well behaved polymer films, but that the polymer from the doubly functionalised analogue was so extensively crosslinked that only a small fraction of redox sites in the resultant three-dimensional network were electroactive on the cyclic voltammetric time scale. The question of molecular weight is essentially unaddressed, firstly because of variable crosslinking effects and secondly because many (most?) of the chains are likely to be covalently bound to the surface, like their monomeric forebears.

Whilst the formation of substrate —O—Si bonds is not universal (see, for example, the attachment of Calabrese's naphthoquinone silane to p-WS_2),[179] the stability of this linkage along with that of the polysiloxane group enhances film longevity and has contributed greatly to the popularity of these systems, despite the inability to control and measure molecular masses in a predictive fashion. Furthermore, whilst a number of the examples in Table 5 use redox functionalised monomers, subsequent introduction of redox groups to the polysiloxane film is possible by coordinative,[183] covalent bond formation,[177,180] ion exchange,[154,155,179] and protonation and ion exchange[178,185] strategies, and the ion exchange/electrochemical reduction[149,150,156–159] approach is open. In short, the advantages of silane chemistry may be coupled to the full range of incorporation procedures.

More recently, an increasingly common method of coating has been electrochemical polymerisation (see Table 6). In this approach, one uses the electrode as an initiator, generating radical intermediates in the vicinity of the electrode, which, it is hoped, will react with incoming monomer to produce polymer which will be deposited on the surface. With the exception of benzophenone mediation to styrene,[206] initiation is intended (and generally believed) to be direct. The rate of initiation is controlled by the electrode potential and measured by the current, giving simple and precise control. It should, however, be borne in mind that some systems involve more than simple initiation: stoichiometric oxidation may also occur, the charge for which must also be accounted for. (Prominent amongst these are the polyaromatic systems, such as those from pyrrole and thiophene monomers.) Whilst, from the current/time relationship it is in principle possible to deduce the film thickness and growth law, the number of examples of such *in situ* monitoring of the film deposition process is comparatively rare.[271,276,303] The basic reason for this is that one cannot assume that all radical intermediates produce polymer or that all polymer produced ends up on the electrode.

The systems listed fall essentially into four main categories: vinyl monomers, phenols, anilines and aromatic nuclei (principally pyrroles and thiophenes). In the case of the vinyl systems we may make several observations. Firstly, initiation is generally anionic and relies on the vinyl group being conjugated to a more readily reducible entity, such as a bipyridyl or viologen unit. Exceptions, though still involving the conjugation requirement, are *N*-vinylcarbazole,[207–210] the protoporphyrins[211,212] and the redox inactive vinylpyridines[213] and styrenes.[213–215] Mengoli's acrylic-acid-based system[216] fails to meet the conjugation criterion and initiation is probably indirect, relying on the presence of persulphate. Secondly, most monomers incorporate a redox centre, frequently a coordinated transition metal (Fe, Ru, Os), to which several vinyl moieties are attached. Thirdly, monovinyl species are more difficult to polymerise, but can readily be copolymerised with di- and tri-vinyl monomers, e.g. $[Ru(vibipy)(bipy)_2]^{2+}$ with $[Ru(vibipy)_3]^{2+}$.[188] The resulting polymers are likely to be branched, rather than linear, and indeed Murray[188] has suggested that they are more realistically viewed as clusters than polymers.

The phenols and anilines are anodically polymerised. Dubois'[221–228] group have studied the electronic and steric effects of a wide range of substituents on phenol polymerisation. A number of systems involve phenol (e.g. hydroxyphenazine[232,233]) or aniline (e.g. thionine,[250–255] 5-aminophenanthroline[249]) species as part of a fused aromatic redox system. In these instances, behaviour as a substituted phenol/aniline has been deduced intuitively,[249] by use of substituent 'blocking' effects[232] or by direct structural methods.[255]

The aromatic nuclei present a rather different picture: they do not 'bring' a redox site with them, but form one on polymerisation—the polymer spine itself. In the formation of these systems, the choice of

TABLE 6
Electrochemically Polymerised Systems

Monomer	Substrate	Comments	References
$M(vibipy)_3^{2+}$			
$M = Ru$	Pt, Au, GC, SnO_2, TiO_2	Includes 'bilayer' and 'sandwich' assemblies.	5, 161, 186–195
$M = Fe$	Pt, GC, SnO_2, TiO_2	Includes 'bilayer' and 'sandwich' assemblies.	5, 186–188, 193, 194, 196
$[Ru^{II}(vibipy)_2(Cl^-)_2]$	Pt, Au, GC, SnO_2, TiO_2	Also copolymer with $Ru(bipy)_2(vipy)_2^{2+}$. Some Cl^- exchange with solvent to give $[Ru(vibipy)_2(CH_3CN)_x(Cl)_{2-x}^-]^+$ ($x = 1, 2$).	188
$[Ru(vibipy)(bipy)_2]^{2+}$	Pt, Au, GC, SnO_2, TiO_2	Also copolymer with divinyl and trivinyl substituted monomer	186, 188, 197
$[M(vipy)_2(bipy)_2]^{2+}$			
$M = Ru$	Pt, Au, GC, SnO_2, TiO_2	Includes 'bilayer', 'sandwich' and metal particulate assemblies, and behaviour in molten salt media	186, 188, 193, 198–201
$M = Os$		Includes 'sandwich' and 'bilayer' assemblies	192–195, 202, 203
$[Ru^{II}(vipy)(bipy)_2L]^+$ $L = Cl^-$	Pt, Au, GC, SnO_2, TiO_2	Includes bilayers and copolymers with di-, tri-vinyl species.	186, 188

L = NO$_2^-$	Pt, Au, GC, SnO$_2$, TiO$_2$		188
L = N$_3^-$	Pt	Also formed by drop evaporation of pre-formed polymer (see Table 1).	16
[Ru(phen)$_2$(viphen)]$^{2+}$ [M(p-cinn)$_2$(bipy)$_2$]$^{2+}$ ^{a}	Pt, TiO$_2$		197
M = Ru	Pt	'Sandwiches' and 'bilayers'.	190
M = Os	Pt		196
M = Os, Ru	Pt	Os site dilution in homogeneous copolymers.	204, 205
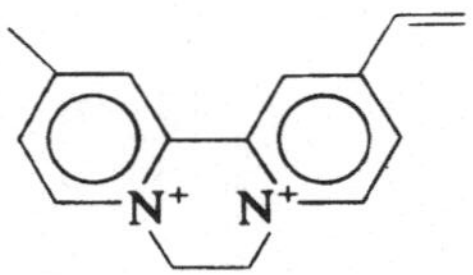	Pt, Au, SnO$_2$	Initiation by reduction to '0' state. Includes 'bilayers'.	5, 161, 190, 194
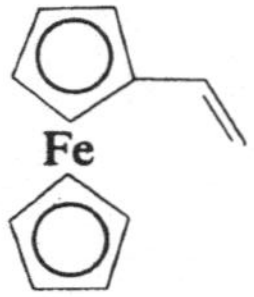	Graphite		206

^{a} p-cinn represents

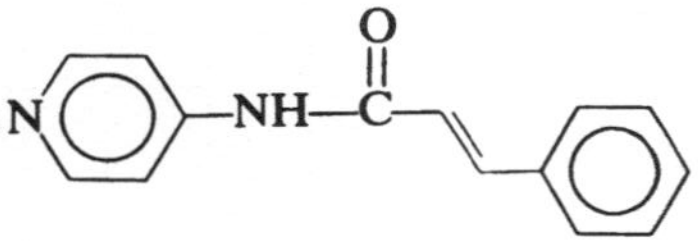

(continued)

TABLE 6—contd.

Monomer	Substrate	Comments	References
R = H (N-vinylcarbazole)	ITO, Pt		207–210, 329
R = Br	Pt		209
[M(protoporphyrin)] M = Ni^{II}, Zn^{II}, Co^{II}, Fe^{III}(Cl)	Pt, GC, SnO$_2$	Oxidative polymerisation of ring (metal oxidation insufficient).	211, 212
[M(protoporphyrin)]$_2$O, M = Fe^{III}, Cr^{III}	Pt, GC, SnO$_2$		211

Structure	Substituents	Electrode	Notes	Ref.
R'−⟨C=C⟩−C₆H₄−R (styrene derivative)	R' = H, R = CH=CH₂	Pt, Au	Oxidative polymerisation.	213
	R = N(p-C₆H₄—Br)₂	GC, Pt	Oxidative polymerisation.	214
	R' = CH₃, R = OCH₃	Pt	Oxidative polymerisation.	215
Vinylpyridine (N–R)	R = nothing	Pt	Anodic polymerisation.	213
	R = CH₃	Pt	Cathodic polymerisation.	213
	R = CH₂=CH—CO₂—CH₂OH	Fe	Probably indirect initiation by $S_2O_8^{2-}$.	216
	R = CH₂=CH—CN	Ni		217
C≡CH−C₆H₅ (phenylacetylene)		Cu	Simultaneous Cu corrosion.	218

(*continued*)

TABLE 6—contd.

Monomer	Substrate	Comments	References
OH–C₆H₅	Pt, Au		213, 219, 220
OH–C₆H₄–o,m,p-CHO	Fe, Cu, Pt, Ni		221
OH–C₆H₄–R:			
R = o,m,p-COCH₃	Fe, Cu, Pt, Ni		221
R = o,p-COC₂H₅	Fe, Cu, Pt, Ni		221
R = o-COC₆H₅	Fe, Cu, Pt		221
R = p-COC₆H₅	Fe, Cu, Ni		221
R = o,p-CO₂CH₃, o-CO₂C₆H₅, o-CONH₂, o-CONH·C₆H₅, p-CO₂C₃H₇	Steel		222
R = o,m,p-CH₂OH	Pt		223, 224
R = o-CH₂CO₂H	Pt		224

$R = p\text{-}CH_2CH_2COCH_3$	Cu, Fe, Ni, Pt	221
$R = p\text{-}C_2H_4\text{—}NH_2$	Pt, Fe, Cu	225, 226
$R = o\text{-}CH_2\text{—}NH\text{—}C_6H_5$	Pt, Fe, Cu	225
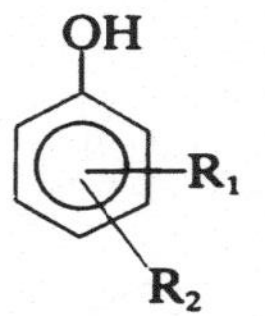		
$R_1 = 2\text{-}CH_3$, $R_2 = 3,5,6\text{-}CH_3$	Fe, Pt	226–228
$R_1 = 3\text{-}CH_3$, $R_2 = 4,5\text{-}CH_3$	Fe, Pt	224, 227
$R_1 = 2\text{-}CH_3$, $R_2 = 6\text{-}Cl$	Fe	227
$R_1 = 2\text{-}Cl$, $R_2 = 6\text{-}Cl$	Fe	227
$R_2 = 2\text{-}OCH_3$, $R_2 = 6\text{-}OCH_3$	Fe	227
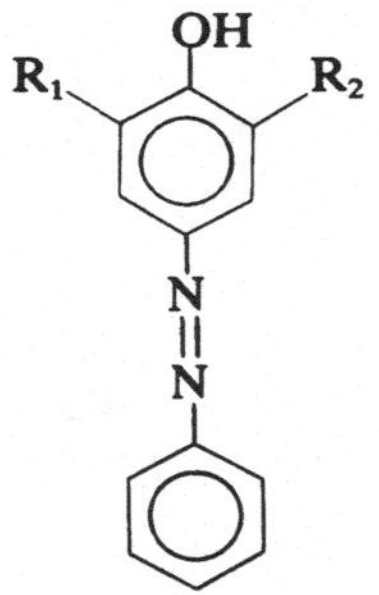	Fe, Cu	229
$R_1 = R_2 = H, CH_3$		
$R_1 = H$, $R_2 = CH\text{—}CH = CH_2$		

(continued)

TABLE 6—contd.

Monomer	Substrate	Comments	References
(2,4-dihydroxyphenylazo)pyridine structure	Pt		230
5-hydroxy-1,4-naphthoquinone structure	Graphite		231
1-hydroxyphenazine structure	Pt, Au SnO$_2$, GC SnO$_2$		232 233 234

(structure: tetraiodofluorescein/rose bengal-like with CO₂H)	n-Si, $\langle 11\bar{2}0\rangle n$-CdSe, $\langle 111\rangle$, $\langle 110\rangle$ p-GaAs, $\langle 110\rangle p$-GaP, p-Si		237
aniline	Pt, C		219, 235–242
1,2-phenylenediamine	Pt		219, 220, 243
1,4-phenylenediamine	n-WSe₂, n-MoSe₂	Single crystal, surface ⊥ to c-axis exposed.	244
substituted aniline	Pt		245
R = o,m-CF₃, p-CH₃, OCH₃, CO₂H	Pt		245

(*continued*)

TABLE 6—contd.

Monomer	Substrate	Comments	References
NH—SO$_2$R (phenyl) R = CH$_3$, C$_6$H$_5$	Cu		246
K$^+$ N$^-$—NO$_2$ (phenyl)	Cu	Polymerised in presence of allylamine.	247
(H$_2$N)—(phenyl)—O	Pt		219
N(CH$_3$)$_2$ (phenyl)	Pt	Fe(CN)$_6^{3-}$ ion-exchangeable into polymer.	248
[Ru(bipy)$_2$(ampyr)$_2$]$^{2+}$ [a] [Ru(bipy)$_n$(amphen)$_{3-n}$]$^{2+}$ [b] $0 \leq n \leq 2$	Pt Pt	Copolymerisation of $n = 2$ monomer with $n = 0$ monomer gives better films.	249 249

Pt, Au, GC, SnO$_2$	Oxidative polymerisation.	199, 232, 250–255
Pt, Au, GC, SnO$_2$	Oxidative polymerisation.	232
Pt, Au, GC, SnO$_2$	Oxidative polymerisation.	232
Pt, Au, GC, SnO$_2$	Oxidative polymerisation.	232

(continued)

^{a} ampyr =

^{b} amphen =

TABLE 6—contd.

Monomer	Substrate	Comments	References
(CH₃)₂N—[dimethylphenazine structure with CH₃, CH₃]	Pt, Au, GC, SnO₂	Oxidative polymerisation	232
[phenothiazine structure with N-H, S]	Au, Pt		256
[2,2'-bipyrazine structure] (bpz)	GC		257
Ru(bpz)$_3^{2+}$	GC		257
[pyrrole structure with N-H]	Pt	Range of 'dopants' includes ClO_4^-, BF_4^-, Cl^-, PF_6^-, AsF_6^-, $CF_3SO_3^-$, $CH_3-\phi-SO_3^-$, naphthalene—$(SO_3^-)_2$, SO_4^{2-}, $C_2O_4^{2-}$, $Fe(CN)_6^{3-}$ either directly during deposition or by subsequent ion exchange. Growth law studied.	85, 87, 258–277

Au	Dopants include ClO_4^-, BF_4^-, Cl^-, $Co^{II}(CH_3CO_2^-)_n$ ($n=2$ or 3), $CoTPP(SO_3^-)_4$.	274, 278–284
C	BF_4^-, ClO_4^-, $CF_3CO_2^-$, $Co(CH_3CO_2^-)_n$ ($n=2$ or 3), $Fe(CN)_6^{4-}$, glutamate, $Fe[Pc(SO_3^-)_4]$ dopants and with poly(L-valine) overlayer.	85, 274, 285, 286, 301
Ni	ClO_4^-, $Co(CH_3CO_2^-)_n$ ($n=2$ or 3) dopants.	274
Pd	BF_4^- dopant.	269
SnO_2	BF_4^-, ClO_4^- dopants.	275, 277, 287
In_2O_3	ClO_4^- dopant.	271, 288
Ta	BF_4^-, Cl^- dopants.	268, 289
PbO_2	BF_4^- dopant.	86
n-Si	BF_4^-, SO_4^{2-} dopant. (Also with prior Au/Pt island deposition.)	266, 290–294
n-CdX ($X=S$, Se, Te)	BF_4^- dopant; also with RuO_2 overlayer.	266, 295
n-GaAs	BF_4^- dopant.	266, 296
Pt/poly $[Ru(vibipy)_3^{2+}]$	ClO_4^- dopant; bilayer.	297
M/PVC (M = Pt), 'Nesa' glass	Composite films; ClO_4^-, BF_4^- dopants.	298, 299
n-Si/Cr-ferrocene	Bilayer, BF_4^- dopant.	292
$(CH)_x$	Morphology dependent on initial $(CH)_x$ doping state.	300

(continued)

TABLE 6—contd.

Monomer	Substrate	Comments	References
[pyrrole-N-R structure] R = CH$_3$, C$_6$H$_5$, C$_2$H$_5$, n-C$_3$H$_7$, n-C$_4$H$_9$, i-C$_4$H$_9$, —(CH$_2$)$_3$—N$^+$[pyridinium]—CH$_3$	Pt	BF$_4^-$ dopant. Growth law studied.	259, 264, 302–304
[pyrrole-N-phenyl-R structure] R = H, o-OCH$_3$, o-Cl, o-Br, o-F, p-F, p-Cl, p-Br, p-I, p-CH$_3$, p-OCH$_3$, p-CO$_2$H, p-CF$_3$, p-NO$_2$	Pt	BF$_4^-$ dopant; monomer oxidation potentials correlated with Hammett σ^+ constants.	305–307

Monomer	Electrode	Comments	Refs.
pyrrole / N-methylpyrrole mixtures	Pt	BF_4^- dopant	259
N-H pyrrole with trimethoxysilylpropyl group (N-M, M = Pt, n-Si, PbO_2, graphite)	Pt	BF_4^- dopant. Poly(L-valine), poly(L-leucine), poly(γ-benzyl-L-glutamate) dip coated on top.	85–87, 293
thiophene	Pt	BF_4^-, ClO_4^-, PF_6^-, $CF_3SO_3^-$ dopants.	263, 270, 272, 308–312
	In_2O_3/SnO_2	ClO_4^-, BF_4^-, AsF_6^-, PF_6^- dopants.	312–317
3-R-thiophene, R = CH_3	Pt	BF_4^-, ClO_4^-, PF_6^-, $CF_3SO_3^-$, picrate dopants.	272, 308–311, 318, 319
	n-Ge/thin Au overlayer	BF_4^-, ClO_4^- dopants.	320
R = Br	Pt	BF_4^-, ClO_4^- dopants.	309–311
R = CH_2—CN	Pt	BF_4^- dopant.	309, 311
R = S—CH_3	Pt	BF_4^- dopant.	310

(continued)

TABLE 6—*contd.*

Monomer	Substrate	Comments	References
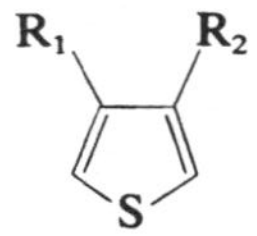			
$R_1 = R_2 = CH_3$	Pt	BF_4^-, ClO_4^-, $CF_3SO_3^-$ dopants.	272, 308, 310, 318
$R_1 = CH_3$, $R_2 = C_2H_5$	Pt	BF_4^-, ClO_4^- dopants.	310
$R_1 = R_2 = C_2H_5$	Pt	BF_4^-, ClO_4^- dopants.	310
$R_1 = R_2 = Br$	Pt	BF_4^-, ClO_4^- dopants.	309, 310
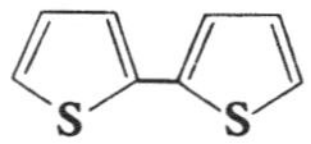	Pt	BF_4^-, ClO_4^- dopants.	308–310, 312
	In_2O_3/SnO_2	ClO_4^- dopant.	312
	GC	Cathodic polymerisation	321
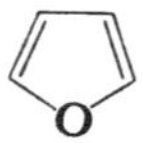	Pt	BF_4^- dopant.	270
	In_2O_3/SnO_2	ClO_4^- dopant.	322
	Pt, GC		323–326

Structure	Electrode	Notes	Ref.
biphenyl	Pt		323
azulene (R = H, CH₃)	Pt	BF_4^-, ClO_4^- dopants.	270, 327
indole	Pt n-MoSe₂	BF_4^-, ClO_4^- dopants. BF_4^- dopant.	270, 328 328
R–N⁺=⟨⟩=⟨⟩=N⁺–R, R = CH₃	Au	Ill-defined reductive polymerisation.	330–335
R = CH₂CH₂NH₂	Au	Ill-defined reductive polymerisation.	335
Br–⟨⟩–Ni(Pϕ_3)₂Br	Pt, GC	Reduction to Ni⁰ species to initiate polymerisation.	336

(continued)

TABLE 6—contd.

Monomer	Substrate	Comments	References
Phenazinium, R = CH₃, C₂H₅	Graphite	Anodic polymerisation, undefined polymer.	337
Crown ether with two catechol units, n = n' = 1, 2	Pt	Oxidation: trimeric fusion of aromatic rings.	338

electrolyte is of more than passing interest, for concomitant chain oxidation—the polymer is easier to oxidise than the monomer—requires the ingress of counter-ions ('dopants' in conducting polymer terminology) to maintain overall charge neutrality. On the one hand this may introduce further complexity or unexplained phenomena: Waltman[309] did not reproduce the data of Tourillon[270] for thiophene polymerisation in ClO_4^- media. On the other hand, this may be viewed as an additional degree of freedom, allowing one to incorporate additional redox groups, such as iron phthalocyaninesulphonate[285,286] or cobalt acetate[274] in polypyrrole, analogous to the ion-exchange systems. (More elaborate examples may be found in Table 7.)

As a general observation, the choice of substrate appears virtually unlimited, provided it is stable and not passivated in the potential range required for initiation. This is because all that is required is a source/sink of electrons and specific substrate/polymer, monomer interactions form no part of the strategy. An excellent illustration of this is the polymerisation of pyrrole on bulk noble metals (Au, Pt), metals in inert membranes (Pt/PVC), carbons (GC, graphite), semiconductors (n-CdS), polymer modified semi-conductors (n-Si/Cr–FeCp$_2$ complex), and silylated surfaces (I). Whilst control of initiation

$$M\text{—}O\text{—}Si\sim\sim N\diagdown\square, M = Pt, n\text{-Si}, PbO_2$$

I

via potential has received considerable attention—essentially all the entries in Table 6 quote optimum potential programs/values—control of propagation via mass transport of monomer to the surface has rarely been exploited;[250] perhaps this may come with time.

Table 7 contains examples of films produced by the elaboration of electrochemically produced polymers. In many instances workers have simply taken their chosen system (from Table 6) a stage further, using the established coordination, covalent attachment, ion-exchange or crosslinking techniques; again, the polypyrrole system illustrates the range of procedures well. It is particularly interesting to contrast two approaches to particulate incorporation: firstly, ion exchange and reduction of 'dopant' anion (RuO_4^{2-}) to produce RuO_2[344] in polypyrrole and, secondly, mediated reduction of cationic solution

TABLE 7
Electrochemically Polymerised Systems Subjected to Further Modification

Monomer	Substrate	Further treatment	References
[Ru(bipy)$_2$(vipy)$_2$]$^{2+}$	Pt	Ru^I or Ru^0 mediated reduction of metal ions to produce Pt/polymer/M sandwich; M = Cu, Ag, Co, Ni.	198
[Os(bipy)$_2$(vipy)$_2$]$^{2+}$	Pt	(a) Deposition of Au to form Pt/polymer/Au sandwich.	202
		(b) Further deposition of PVF-co-siloxane copolymer to produce 'sandwich–electrode diode'.	202
4-vinylpyridine (N-allyl)	GC, Pt	N-Alkylation (—CH$_3$, —CH=CH$_2$), extent unquantified; crosslinking suspected but unproven for alkyl derivative.	339
divinylbenzene, styrene	Graphite, Pt	(Indirect initiation via ϕ_2CO$^-$.) Sulphonation via ClSO$_3$H (also leads to some crosslinking via ~ϕ—S(=O)(=O)—ϕ~ formation), hydrolysis, and ion exchange with Ru(bipy)$_3^{2+}$, Ru(NH$_3$)$_6^{2+}$.	206
N-methyl-4-vinylpyridinium	Pt, Au	Ion exchange of Fe(CN)$_6^{4-}$.	213

Structure	Metal	Description	Ref.
OH-C6H4-R; R = o,m,p-CHO, o,m,p-COCH$_3$, o,p-COC$_2$H$_5$, o,p-COC$_6$H$_5$, p-C$_2$H$_4$COCH$_3$	Fe, Pt, Cu, Ni	2,4-Dinitrophenylhydrazine derivatives of polymer carbonyl function prepared, and observed using IR and XPS.	221
OH-C6H4-R; R = o,m,p-CH$_2$OH	Fe, Cu, Ni, Pt	Reaction of polymers with CH$_3$COCl to give esters; observed using IR, XPS.	223
2-allyl phenol (OH—C$_6$H$_4$—CH—CH=CH$_2$)	Fe	Thermally induced crosslinking via allyl functionality.	340
8-hydroxyquinoline derivative; R = H, CH$_3$	Pt	Coordination of Cu^{2+}.	341, 342

(*continued*)

TABLE 7—contd.

Monomer	Substrate	Further treatment	References
HO—C₆H₄—CH₂CH₂NH₂	Pt, Fe, Cu	(a) Condensation of ferrocene carboxaldehyde on to pendant amine groups.	225
	Pt, Fe	(b) Coordination of Cu^{II}, Co^{II}, Fe^{II}, Mn^{II}, Zn^{II} ions via O and/or N donor atoms.	226
NH₂—C₆H₄—R, R = H, NH₂	Pt	Application of Nafion/Ru(bipy)$_3^{2+}$ film subsequently on separate part of electrode to give reference/probe pair of redox couples.	343
pyrrole (N–H)	Pt	Nitration, electrochemical reduction of nitro group and condensation with TTF-CO₂H, to give [pyrrole-CH₂-NO·CO-TTF]$_n$ (BF$_4^-$ dopant).	261

174

![pyrrole structure]	Pt	(a) BF_4^- anion ion exchanged with $Fe(CN)_6^{3-}$.	266
	Pt, ⟨001⟩n-GaP	(b) SO_4^{2-} anion exchanged with RuO_4^{2-}, which is then reduced to particulate RuO_2. Oxidation of PP in ClO_4^- then gives M/PP^+, ClO_4^-·RuO_2.	344
![N-vinylimidazole structure]	PG	Coordination of $[Ru^{II}(bipy)_2X]$, $X = 2Cl^-$, CO_3^{2-}.	345
bpz	GC	Incorporation of $Ru(bipy)_3^{2+}$ by immersion.	257

TABLE 8
Miscellaneous Polymerisation/Coating Procedures

Polymer	Substrate	Deposition procedures	References
(ferrocene-containing vinyl polymer)	Pt	Plasma polymerisation.	6, 7, 346–351
	Graphite	Plasma polymerisation.	352, 353
	GC	Plasma polymerisation.	6, 349, 350
	TiO_2 (single crystal and thermally oxidised Ti)	Plasma polymerisation.	354
	RuO_2	Plasma polymerisation.	6
(poly(vinylpyridine)–$IrCl_5^{2-}$)	GC	Plasma polymerisation of vinylpyridine, then ligand displacement from $IrCl_5(acetone)^{2-}$.	355
(poly(vinylpyridine)–'Fe')	Pt, GC	Glow discharge polymerisation; in presence of $Fe(CO)_5$ vapour, Fe redox centres incorporated.	356, 357

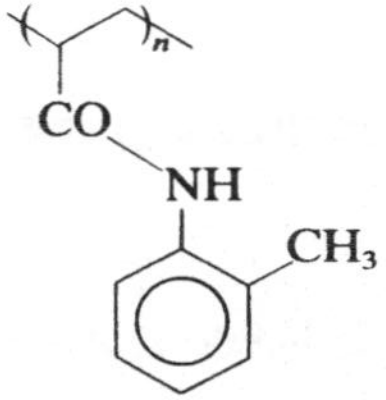

C (as CPE)	Plasma polymerisation of acrylic acid, conversion to acid chloride (SOCl$_2$), amide formation with o-toluidine.	358	
Pt, GC, steel	Glow discharge polymerisation. Also high-temperature copolymerisation with Co(acac)$_2$ incorporates metal redox sites.	359, 360	
Fe, Cu, Ni	Gas-phase polymerisation on metal surface, resulting in some metal species incorporation.	361	
Cu, Ni, Co, Fe, Ti	Thermal polymerisation of gas-phase 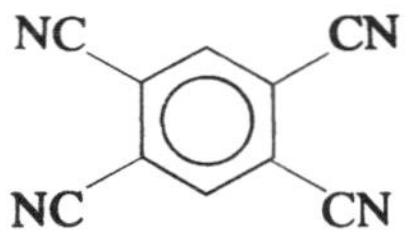 [tetracyanobenzene structure] on appropriate metal.	362	

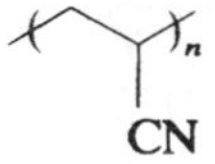

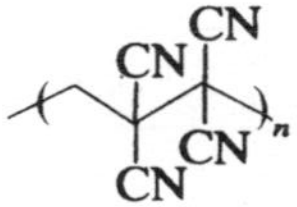

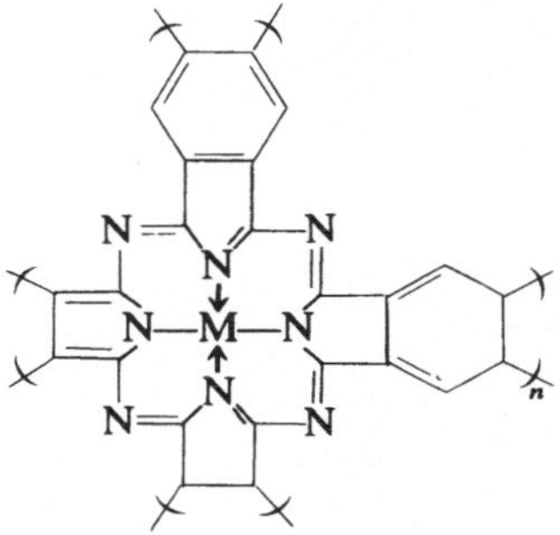

M = 2H, Co, Ni, Fe, Ti

(*continued*)

TABLE 8—contd.

Polymer	Substrate	Deposition procedures	References
⎡–CH₂–CH(CH₂–CO₂H)–⎤ₙ	GC	Solution-phase thermal polymerisation, followed by reduction of appropriate metal ions to give particulate Pt^0, Pd^0, Ag^0, Ni^0, Cd^0 and Pd^0–Pt^0 (bimetallic).	363, 364
Poly(TPPFe^{III}(Cl))	GC	Thermal polymerisation— probably involves phenyl group loss and some porphyrin ring degradation.	365
[Fe^{III}(TAPP)]$_2$O/SOCl$_2$ condensate	GC	Sulphinic diamide formation on electrode surfaces by amine/SOCl$_2$ condensation.	366

Structure	Substrate	Description	Ref.
![structure: C(=O)-NH-CH(CH2-NH-C(=O)-...)-(H3C)2CH-CH-NH]n amide polymer on surface	Graphite	Successive amide linkages of carboxylic surface functions with diaminopropane, modified surface with L-valine-N-carboxy-anhydride, etc.	51
'Poly[ferrocene–C₂H₄R]' R = NH₂, —N=CH₂, —N=CHφ	PG	Multilayer coverages (structure unspecified) result from reflux in monomer solution, but only with edge, not basal, plane graphite.	367

metal species by ruthenium centres in poly[Ru(bipy)$_2$(vipy)$_2$] to produce Cu, Ag, Co and Ni.[198] In the former instance, prior dispersion of the dopant throughout the film leads to an essentially homogeneous distribution of RuO_2 through the film. In the latter case, the relative rates of metal ion diffusion and reduction/nucleation govern the M^0 distribution: rapid reduction at the outer (film/solution) interface leads to a metal/polymer/electrodeposited metal sandwich. Further deposition of a vinylferrocene-co-siloxane polymer on the outside of the analogous osmium sandwich[202] was used to produce a device mimicking a diode in its electrical behaviour. The options seem to be limited only by the ingenuity of the operator.

Table 8 contains an assortment of synthetic procedures which do not fall into the previously discussed categories. A fairly common technique is plasma or glow discharge polymerisation, notably for vinylferrocene in the first case, the relative merits of which have been detailed by Murray[350] and Doblhofer,[360] respectively. From the available evidence (principally for PVF, where alternative syntheses have been well explored), it would appear that films produced by the plasma deposition procedure are broadly similar to those obtained by other means (see for example Table 1). On the credit side, coatings do tend to be uniform, but, on the debit side, plasma reactor conditions may cause redox centre damage[350,352] or unexpected substrate effects, such as the generation of surface states on TiO_2.[354] Gas and solution phase thermal polymerisations and polycondensations are also illustrated, although their use is not widespread and the products, particularly for the gas phase reactions, are not well characterised to date. Of the solution phase polymerisation systems, the poly(vinyl acetate)/M^0 films of Kuwana[363,364] are interesting: reduction of $PtCl_6^{2-}$, presumably at the glassy carbon/polymer interface since the polymer contains no obvious redox mediator sites, results in Pt^0 particle dispersion throughout the film, as evidenced by some very nice SEM photographs.

3. ELECTROCHEMICAL CHARACTERISATION

3.1. Objectives and Limitations

To obtain an understanding of the behaviour of an electrode coated with polymer towards solution species (covered in Chapter 6), and improvements in design and performance the resulting interface must be characterised as fully as possible. Prominent amongst the pieces of information desired are the following.

(i) polymer structure: the nature of polymer/electrode bonds (if present), interchain interactions and, particularly for *in situ* polymerised films, the nature of the monomer–monomer linkage;
(ii) film thickness and the degree of uniformity thereof;
(iii) the extent to which monomer properties (E', electron transfer rate constants, optical behaviour, etc.) are carried over to the polymer;
(iv) the local environment, notably extent of solvation and solution anion/cation incorporation;
(v) the redox composition/potential relation;
(vi) the presence (or otherwise) of electrostatic or other interactions between redox centres and the remainder of the polymer and possible resultant structural inhomogeneity;
(vii) the rates of electron and ion motion through the film, i.e. the conductivity, and the mechanism of charge transport.

It is clear that certain of these questions are amenable to electrochemical study whilst others, notably those concerned with structure, require spectroscopic or other physical probes. Ideally, these non-electrochemical techniques should be used *in situ*, since many properties may be intrinsic to the electrochemical environment (those where solvent, ionic strength, potential and concentration gradients are important). In practice, the various experimental conditions may be mutually exclusive—one cannot do XPS in the presence of bulk solution, for example—so that care must be exercised in the interpretation of *ex situ* data and the assessment of their relevance to *in situ* behaviour of polymer modified electrodes. This point, and the importance and correlation with electrochemical measurements of spectroscopic data, will be addressed in Section 4: here we shall concentrate on purely electrochemical measurements, focusing first on thermodynamic aspects and then moving on to kinetics.

3.2. Simple Models of Polymer Films on Electrodes and General Approach

The simplest model of a polymer modified electrode envisages a three-dimensional tangled network of polymer chains extending a uniform distance from the electrode, functionalised with homogeneously distributed non-interacting redox centres. This predicts a carry-over of solution monomer thermodynamic properties such as E', Nernstian redox composition/potential relation, pK_a and complexation constants, where appropriate. Whilst close approximations to these predictions have frequently been found for monolayer derivatised surfaces,[2,3] adherence to this model is far less common or close for polymeric systems. The principal reason is to be found in a comparison of the natures of solutions and immobilised polymer films. In the former case one usually deals with species at $\leq 10^{-3}$ mol dm^{-3} levels in excess background electrolyte, where activity effects can be dealt with reasonably comfortably, but in the latter instance the redox centres are frequently within ~1 nm of each other so that the effective concentration is in excess of 1 mol dm^{-3} and interaction/activity effects are liable to be more difficult to deal with. In addition to these redox centre–redox centre interactions, polymer–redox centre interactions may also preferentially stabilise one half of a redox couple, for example in ion-exchange systems where, since the two halves of the redox couple necessarily have different charges, their electrostatic interactions with the oppositely charged polymer will be different. If other (hydrophobic) interactions dominate, then the more highly charged partner will not be so favoured, so E' measurements immediately give a clue to binding forces. The simple model also does not account for the effects of coordinative changes or multiple binding on immobilisation, which may result in different or multiple E' values. Finally, a cautionary note: equality of solution and immobilised redox centre E' values does not signify identity of species, simply equal (not necessarily zero) changes in free energy of both partners.

3.3. Thermodynamic Parameters

The principal technique used to determine E'_{surf} values is cyclic voltammetry. More will be said about the details specific to modified electrodes in the kinetic discussion in Section 3.4, but for the moment it is sufficient to note that the value of E'_{surf} may be approximated by the average of the anodic and cathodic peak potentials $\frac{1}{2}(E_{p,a} + E_{p,c})$; at sufficiently low scan rates ($v \leq 10$ mV s^{-1}), the difference in peak potentials ΔE_p will be small so little error is incurred even if the peaks are not shifted symmetrically from E'_{surf}. Experimental simplicity and speed coupled with ease of (at least qualitative) data extraction have made the application of cyclic voltammetry essentially universal in this field. Exhaustive cataloguing is not warranted here: instead an attempt is made to present some of the more thoroughly studied phenomena, illustrating firstly the types of effects encountered, secondly the principles of data interpretation, and thirdly the information obtainable. Some emphasis will be placed on departures from solution species behaviour.

The ideal model has been considered by Laviron[370,371] and Aoki,[368] and the essential features of the cyclic voltammogram are summarised in Table 9, with the equivalent information for freely diffusing solution species exhibiting rapid kinetics. First we note two important features: zero peak separation and linearity of peak current with sweep rate. These provide important diagnostics for surface attachment, and even systems exhibiting non-ideality retain the latter characteristic, so it is a test almost universally applied. The reason for the linearity is simple: unlike the solution case, one oxidises and reduces the entire reservoir of redox centres, i.e. the charge to be passed is fixed, so decreases in the time taken (increases in scan rate) must be accompanied by proportionate increases in current for the integral to remain constant. This condition is violated if the kinetics cannot maintain equilibrium.

A frequent observation is that $|E'_{surf} - E'_{soln}| < 0.1$ V. Dominey[160] and Bruce[154] have shown equality of E'_{surf} and E'_{soln} for $Fe(CN)_6^{3-/4-}$,

TABLE 9
Cyclic Voltammetric Behaviour for an Ideal Redox Polymer Film on an Electrode and for a Freely Diffusing Solution Species, Both with Rapid Kinetics

E'	Polymer modified electrode[368] $E'_{surf} = E'_{soln}$	Solution redox couple[369] E'_{soln}
$i_{p,a}/i_{p,c}{}^a$	1	1
$\dfrac{\partial(\log i_p)}{\partial(\log v)}$	1	$\tfrac{1}{2}$
ΔE_p(mV)	0	$59/n^b$
$\Delta E_{p/2}{}^c$(mV)	$90.6/n$	$203/n$
Peak shape	Symmetric about $E = E'_{surf}$ and $i = 0$	Peak asymmetric: diffusional 'tailing'

^{a} Assuming equal diffusion coefficients.
^{b} n = number of electrons.
^{c} Half-peak width.

$Ru(CN)_6^{3-/4-}$, $Mo(CN)_8^{3-/4-}$ and $IrCl_6^{2-/3-}$ electrostatically bound to silylbenzyl- and silylpropyl-viologen polymers and Schneider[138] has found similar behaviour for $Cr(bipy)_3^{3+/2+}$ in an organosilane/styrene sulphonate copolymer, for example. However, non-ideality appears in other observations even with these systems. In the first case, Bookbinder[151] finds that E'_{surf} for the viologen in the propylsilane 'spine' is rather more negative than E'_{soln}, and suggested that aggregation might be responsible: this is a consequence of the different concentrations (~ 1 and 10^{-3} mol dm^{-3}) of viologen in the two cases. In the second case,[138] E'_{surf} did not equal E'_{soln} for $Ru(NH_3)_6^{3+/2+}$ in the film. A similar duplicity of behaviour was found by Oyama[110] for $Co(phen)_3^{3+/2+}$ ($E'_{surf} \simeq E'_{soln}$) and $Ru(NH_3)_6^{3+/2+}$ ($E'_{surf} \neq E'_{soln}$) in poly(styrene sulphonate) and poly(vinyl sulphonate). These examples provide evidence for dominance of electrostatic effects for small (high charge density) ions and of other interactions for larger (more hydrophobic) ions.

Two quantitative studies deserve mention here. Braun,[111] in agreement with Oyama's previous observation,[110] found that E'_{surf} for $Fe(CN)_6^{3-/4-}$ in quaternised PVP was shifted negatively, in contrast to the expected relative stabilisation of $Fe(CN)_6^{4-}$ on purely electrostatic grounds. He analysed the problem in terms of the work done (w^T) in transferring ions from bulk solution into the film, taking account of ion pairing, and found $w_{ox}^T \simeq -11$ kJ mol^{-1} and $w_{red}^T \simeq +13$ kJ mol^{-1} for PVP$^+$-C$_{12}$H$_{25}$, indicating accumulation of $Fe(CN)_6^{3-}$ and a deficit of $Fe(CN)_6^{4-}$. This effect is the opposite to the expected one, illustrated by the positive shift of E'_{surf} for $Fe(CN)_6^{3-/4-}$ in protonated poly('en'silane) by Kuo.[178] Tsou[125] studied cation incorporation in Nafion and a carboxylate derivative and found differences in E'_{surf} and E'_{soln} which varied with ionic strength: for example $E'_{surf} < E'_{soln}$ at low ionic strength but $E'_{surf} > E'_{soln}$ at high ionic strength for $Os(bipy)_3^{3+/2+}$. Here, both electrostatic and hydrophobic interactions are important, but their relative dominance is determined by ionic strength. Temperature dependences of E'_{surf} were used to obtain half-reaction entropies (and thus enthalpies).[125] Whilst the latter were usually unfavourable, large increases in entropy were able to provide the driving force for incorporation: this implies a major role for the solvent, in this case water. A more extreme manifestation of the way in which ionic strength can reduce electrostatic interactions was observed by Schneider[138] for the previously mentioned $Ru(NH_3)_6^{3+/2+}$

couple, in which an increase in electrolyte concentration caused the less tightly bound Ru^{II} species to be lost to the solution.

The dependence of E'_{surf} on pH has been studied for polymeric thionine films,[253] for a related polymer-bound thiazine species and some monoquaternised bipyridines.[47] For thionine,[253] a pK_a similar to the monomer in solution was found, but this was not found to be so for a methacrylate bound o-quinone,[54] or a polyethyleneimine-bound anthraquinone.[61] Indirect observation of pK_a for PVP,[99] by following the quantity of $Fe(CN)_6^{3-}$ bound to $PVP \cdot H^+$, showed an increase in basicity of a pyridine site in the presence of a $Fe(CN)_6^{3-}$ bound to a neighbouring protonated pyridine group. This is in contrast to the behaviour of PVP in solution in the absence of $Fe(CN)_6^{3-}$: apparently the multiply charged anions overcome repulsive forces which would otherwise predominate. The dependence of E'_{surf} on pH has also been used as a means of measuring the latter:[122] poly(o-phenylenediamine) and polyaniline films gave slopes of 58 and 115 mV/pH unit (corresponding to $1e^-,1H^+$ and $1e^-,2H^+$ processes) in the ranges pH 2–10 and pH 3–8, respectively. The importance of the underlying substrate on the observed E/pH relation has been discussed for a TCNQ polymer on Pt.[78]

Changes in E'_{surf} have also been used to detect chemical reactions following redox centre incorporation, such as ligand (L) exchange in PVP–Ru(bipy)$_2$(L),[16,21] and the presence of —Fe(CN)$_5^{2-/3-}$ redox centres in PVP in different environments[92] on partial protonation. The appearance of multiple peaks representing interconvertible sites in PVF film voltammetry has been discussed at length by Peerce and modelled numerically.[13] The 'scheme-of-squares' approach applied here could presumably be used to parameterise the PVP–Ru(EDTA) system, where it was found that, under equilibrium conditions, the Ru^{II} species was coordinated to two pyridine groups, but the Ru^{III} species to only one.[91] (Kinetic elements are, of course, also included here.)

The question of waveshape is an important and difficult one. The importance arises because the waveshape reflects the potential/composition relation, which governs the extent of 'titration' of mediator centres with potential in mediated charge-transfer-based applications or of chromophore in display applications, for example. The difficulty arises because the same effect may be the result of a variety of causes: wave broadening in cyclic voltammetry can be caused by repulsive site–site interactions (a thermodynamic effect, discussed here), slow charge transfer/propagation (kinetic effects, discussed later) or large uncompensated electrolyte resistance effects (an artefact). Here we restrict the discussion to the first case, where the kinetics are rapid and equilibrium prevails. Quantitatively, this means that the rates of charge transfer at the interface and of charge propagation through the film must be large. In the latter case, this corresponds to large values of $(D/L^2)/(vF/RT)$, where the quantity (D/L^2) represents an effective relaxation rate, and the means by which it may be measured are considered later. Values of this ratio greater than 3 result, for an 'ideal' film, with no interactive effects in (i) symmetrical waves about $E = E'_{surf}$ and $i = 0$ (i.e. $E_{p,a} = E_{p,c}$ and $i_{p,a} = i_{p,c}$), (ii) a linear i_p versus v relation and (iii) a peak half-width of $90.6 \, mV$.[368] The second result, which arises because the charge passed (q) is constant, provides, firstly, an important diagnostic test for surface attachment and, secondly, a means of determining the total number of electroactive centres $(\Gamma/mol \, cm^{-2})$:

$$\Gamma = q/nFA$$

The presence of interactive effects alters peak half-width and equality of peak currents (although they may still be individually proportional to scan rate if rapid kinetics prevail): repulsive (attractive) interactions result in peak broadening (narrowing). The effect was considered first by Brown[372] and later by Laviron.[373] Various models—usually evolved in response to the requirements of individual systems, but still generally applicable—have been developed to parameterise these effects, from one extreme of a small set of discrete E'_{surf} values[13,374] to a (continuous) Gaussian distribution,[253] whose width is related to the apparent Nernst slope in a composition/potential plot.[5,191,202,203,253] The range of systems for which these activity effects have been found to be important—from transition metal/pyridine systems[191,202,203] to aromatic redox centres[253] to silane-bound porphyrins[180] to ion exchange systems[374] to ferrocene[13]—clearly indicates an important common feature. The answer is very simple: within these films, the redox centres are present at very high concentrations ($\geq 1 \, mol \, dm^{-3}$) and dilute solution assumptions will not suffice. Indeed, apparently ideal behaviour is more likely to be the result of identical, rather than zero, interactions between all species.[13]

More complex manifestations include the shifting of bilayer trapping peaks[5] and the formation of new phases.[131,348] This latter effect is particularly interesting. In the case of Nafion/TTF$^+$, the presence of certain anions (e.g. Br$^-$) allows formation of a one-dimensional conducting (TTF)Br$_x$ phase ($x \simeq 0.7$), for the reduction/oxidation of which both chemical and electrochemical steps are required. The combination of strong attractive interactions and an 'EC' system (an electrochemical reaction step followed by a chemical reaction step) results in sharp but separated peaks. For PVF,[346,348] and in aminophenylferrocene silane,[180] striking changes with solvent revealed the important point that redox equilibrium does not necessarily imply solvation equilibrium. For the PVF system in water, the waveshape was consistent with constant ferrocene, and increasing ferricenium activity during oxidation. (The reduction proceeds normally with both activities varying.) The deduction here was that, in the former case, site segregation produces islands which grow or shrink, as necessary, and which, through slow establishment of equilibrium, differ in their solvent content. Segregation effects have also been found to be important for Fe(EDTA)$^{-/2-}$ and Co(C$_2$O$_4$)$_3^{2-}$ in poly(L-lysine),[116] where 'Donnan domains', in which ionic concentration and mobility are very different from those in the solution 'pools', are formed. Ionic clusters are also present in Nafion:[374] in a recent model, spheres of ~4 nm diameter containing solvent and attached ions are connected by narrow channels in a bulk (essentially fluorocarbon) phase. Oyama[133] has shown that whilst MV^{2+} is distributed randomly in Nafion, if the viologen centres are attached to a polystyrene spine this is no longer the case.

An important parameter is the partition coefficient of a solution species into a polymer film. The values in Table 10 are principally for ion exchange systems evaluated from the charge under cyclic voltammetric peaks, although data for a coordinative example and from other experiments are included for completeness. With the exception of the value for Fe(EDTA)/PLL·H$^+$ (where it is a solution 'pool', rather than 'Donnan domain', value) the values are large, often in excess of 10^2. Absolute comparisons are not necessarily helpful, firstly because background electrolyte ions compete (0·1 mol dm^{-3} SO$_4^{2-}$ would swamp 10^{-6} mol dm^{-3} IrCl$_6^{2-/3-}$ in silylviologen films[154]) and, secondly, because the values are concentration-dependent[121,133,138] (presumably a saturation effect related to the low absolute capacity of the films). 'Saturation' corresponding to 20–40% charge compensation is found for the poly(styrene sulphonate) system,[138] probably because of

TABLE 10
Partition Coefficients of Electroactive Species into Polymer Films

Polymer	Redox couple	Medium	κ	Comments	Ref.
PLL·H$^+$	[Fe^{III}(EDTA)]$^-$ (0·5–7·5 mM)	H$_2$O (pH 5·5); 0·2M-CF$_3$CO$_2^-$	1·0 ± 0·1	[Zn^{II}(EDTA)]$^{2-}$ at same concentration added to mimic Fe^{II} species without electroactivity. Spectroscopic determination.	114
PLL·H$^+$	Co(C$_2$O$_4$)$_3^{3-}$ (0·5–1 mM)	H$_2$O(pH 1·5); 0·2M-CF$_3$SO$_2$	17 ± 3	Value for 'bulk' solution in polymer (not Donnan domains) Chronocoulometry/RDE determination.	116
Benzylsilylviologen^{2+}	Various (as right) (50 μM)	H$_2$O; 0·1M-KCl	Mo(CN)$_8^{3-/4-}$ > Ru(Cn)$_6^{3-/4-}$ > Co(CN)$_6^{3-/4-}$ > Fe(CN)$_6^{3-/4-}$ > InCl$_6^{2-/3-}$ >> Cl$^-$	Relative ordering by competitive (pairwise) binding in CV determination.	160
p(Propylsilylviologen^{2+})	Various (as right)	H$_2$O; 0·1M-KCl	Mo(CN)$_8^{3-/4-}$ > Ru(CN)$_6^{3-/4-}$ > Co(CN)$_6^{3-/4-}$ > Fe(CN)$_6^{3-/4-}$ > IrCl$_6^{2-/3-}$ >> Cl$^-$ I$^-$ > SCN$^-$ ~ ClO$_4^-$ ~ SO$_4^{2-}$ > Br$^-$ > Cl$^-$ ~ p-CH$_3$—ϕ—SO$_3^-$ ($\kappa_{I^-}/\kappa_{Cl^-}$ ≃ 10; $\kappa_{IrCl_6^{2-}}/\kappa_{SO_4^{2-}}$ ≃ 10)	Relative ordering by competitive (pairwise) binding in CV determination. (N.B. Co-species 'silent', so 'absence' of response observed.) AES analysis for non-electroactive anions.	154
Nafion (from membrane 125)	34% viologen-functionalised polystyrene (~0·1–3 mM)	DMSO; 0·2M-NaClO$_4$	~10 to ~3, decreasing with c^{soln} (saturation?)	κ values not explicitly given; plot of (c_{poly}^{poly}) vs (c_{poly}^{soln}).	133

Polymer	Species	Electrolyte	K	Comments	Ref
Nafion (970 EW)	Cp_2FeTMA^+ (10^{-5}–10^{-3}M)	H_2O; 0.2M-Na_2SO_4	1500	At higher concentrations get saturation corresponding to > 100% charge compensation (not solely electrostatic binding).	121
Nafion (970 EW)	$Os(bipy)_3^{2+}$ (~10^{-3}M)	H_2O; 0.2M-Na_2SO_4	~500–750	$K_{Na^+}^{Mn^+} = 1.5 \times 10^4$	132
Nafion (970 EW)	$Ru(bipy)_3^{2+}$ (~10^{-3}M)	H_2O; 0.2M-Na_2SO_4	~500–750	7.3×10^4	
Nafion (1100 EW)	MV^{2+}	H_2O; 0.1M-$NaClO_4$	7.9×10^5	3.7×10^4	
	Cp_2FeTMA^+		1.1×10^6	5.7×10^6	
	$Ru(NH_3)_6^{3+}$		2.5×10^6	740	
	$Ru(bipy)_3^{2+}$		2.1×10^7		
	$Ru(NH_3)_6^{2+}$		12.6×10^4		
Poly(styrene sulphonate)-co-(methyl methacrylate)-silane	$Ru(NH_3)_6^{3+}$ 10^{-5}M	H_2O; 0.01M-KCl	2400	Electrostatic and hydrophobic interactions may occur. CV determination, using film flotation density.	138
	2.5×10^{-5}M		1800		
	5×10^{-5}M		1540		
	7.5×10^{-5}M		1240		
	10^{-4}M		1000^a		
	1.25×10^{-4}M		800^a		
Poly(styrene sulphonate)-co-(methyl methacrylate)-silane	MV^{2+}	H_2O; 0.01M-KCl			138
	4×10^{-5}M		100		
	8×10^{-5}M		81		
	10^{-4}M		87		
	2×10^{-4}M		76		
	4×10^{-4}M		55		
PVP	$[Ru^{II}(EDTA)]^{2-}$ (5×10^{-8}M)	H_2O (pH 3.4); 0.2M-$CF_3CO_2^-$	6000	Saturation effects. Coordinative attachment. Association constant approximately as for Ru/pyridine in solution.	91

^{a} Saturation effects at ~20–40% charge compensation.

inaccessible sulphonate groups. Conversely, 170% charge 'compensation' was found for Nafion:[121] this is indicative of additional binding mode(s). A good discussion of the treatment of data and an interesting account of structural effects, including differences between 970 and 1100 EW Nafion, have been given by Szentirmay.[132] Partitioning into poly(phenylene oxide) films was studied by Ohnuki:[219] the failure of the product of partition coefficient (κ) and film diffusion coefficient (D_{film}) to correlate with Stokes radius (as a measure of solution diffusion coefficient, D_{soln}) was taken as an indication of significant differences in κ values for $Fe(CN)_6^{3-}$ and Fe^{2+}. Whilst these values of κ provide useful indications of film concentration, most determinations are 'average' values for an assumed homogeneous film. Local (microscopic) concentration values in inhomogeneous films may quite clearly be somewhat larger.

3.4. Kinetic Parameters

3.4.1. Cyclic voltammetry

Cyclic voltammetry has been used in two principal ways to determine kinetic data. In the first, the scan time is sufficiently long for the entire film to have the opportunity to react and one may follow the course of comparatively slow film changes. In the second, one assumes that the film is not evolving in any way, and uses scan rate as the time domain variable to probe the rate(s) of process(es) associated with film oxidation state changes.

An obvious application of the first type of experiment is as a monitor of electroactive redox centre population during film production/elaboration by redox couple incorporation,[257] ion exchange,[135] silylation[182] or electropolymerisation.[151,211,249] In the latter case, 'observation' and polymerisation are obviously intimately linked, but the act of cycling the potential has also been shown to be important in the ion exchange of $Ru(bipy)_3^{2+}$ into poly(styrene sulphonate) films[135] (simple immersion was a far less effective incorporation procedure). Similarly, the loss of electroactive species has been followed by cyclic voltammetry[128,151,182,257] as well as being induced by it, in the case of H^+ reduction within a protonated pyridine iodide system.[185] In most cases the distinction between redox centre loss and isolation, and polymer loss, required more direct (e.g. spectroscopic) data. Shigehara's study of the time evolution of viologen centre populations (during incorporation and loss) for monomeric and polymeric species in Nafion is exemplary here.[128] Reaction (rather than loss) of redox centres may be followed by the appearance (and integration of) peaks at different E'_{surf}, or by the loss of reactant peak if the product is electroinactive.[212] Such processes observed include additional ligand coordination to a porphyrin,[212] and ligand exchange in Ru/bipy systems by thermal[16] or photochemical[21,345] processes. In one case,[16] it is interesting to note that fulfilment of the experimental requirement that the process be much slower than the scan time rested on the important result that environmental/structural consequences of film immobilisation slowed the exchange rate by a factor of approximately 5000.

In addition to the redox centre content of the film, the extent of solvent and background electrolyte ion incorporation may have important consequences. The importance of the solvent to the voltammogram ultimately obtained has been demonstrated for PVF[346] (in CH_3CN, C_2H_5OH/H_2O and H_2O), a silane-bound ferrocene[180] (in CH_3CN and H_2O) and PVP–Ru(EDTA)[375] [in CH_3CN, C_2H_5OH, CH_3OH, H_2O and $(CH_3)_2SO$]. Here, however, we are more interested in transient effects, which are likely to occur in 'medium transfer' experiments or on initial immersion following deposition from another solvent (e.g. by droplet evaporation). This phenomenon was first observed by Schroeder,[8,36] who saw the gradual appearance and then improvement in reversibility of peaks for a series of functionalised polystyrenes, and christened it 'break-in'. The effect has not been quantified, but clearly may involve ohmic potential drops in many cases, and frequent passing reference is made to it, even in room temperature molten salt media.[199] Electrodes are therefore often cycled until reproducible behaviour is observed. In the study of polyethyleneimine-bound anthraquinone and fluorenone systems[44] it has been claimed that the film structure was continuously undergoing potential cycling-induced changes, and was in fact kinetically, rather than thermodynamically, determined in the experiments reported.

Different equilibrium solvent/ion contents may also be associated with different oxidation states, resulting in more complex kinetic behaviour. An excellent example of this is provided by poly(styrene sulphonate)/Ru(bipy)$_3^{2+/3+}$, where the polymer ultimately configures itself to have two (three) sulphonate groups around the Ru centre in the 2+ (3+) oxidation states, but finds it kinetically more facile to move background electrolyte ions,[135] so that equilibrium ion populations are only found under certain circumstances. Ferrocene in

Nafion[127] exhibits similar behaviour, and both systems were discussed in 'scheme-of-squares' terms.

For the second type of experiment, several mathematical treatments have appeared, assuming that the potential drop occurs principally at the electrode/polymer interface. These models envisage potential-driven charge transfer at the interface, coupled with sequential self-exchange or physical movement of redox centres (along with the necessary polymer spine/solvent/ion motions in either case) to oxidise/reduce more distant redox centres. The simplest model envisages equilibrium[376] (Nernstian or quasi-Nernstian; see Section 3.3) at the electrode/film interface: this is frequently a good approximation, for the system is externally potential-driven here, but only concentration-gradient-driven in the bulk of the film. Then, whichever the rate-limiting step in the film, the process may be treated as a diffusive one,[376] even when there are very few layers present,[370] i.e. when the number of discrete electron 'hops' is small.

More elaborate treatments by Aoki,[377] Andrieux[376] and Laviron[371] include interfacial charge transfer kinetics and the dependence of apparent values on pH,[378] and the effect of interactions[379–381] (see Section 3.3) on waveshape. The numerical treatment of Aoki,[377] for which simple approximate expressions for i_p, E_p, $E_{p/2}$ and $E_{p/4}$ are given, is helpful: the analysis is in terms of two parameters comparing the interfacial kinetics ($\Lambda = k'_0 \, (RT/nFDv)^{1/2}$) and film charge propagation rates [$w = (nFv/RT) \cdot (L^2/D)$] with scan rate. Each of the two processes may be rapid, similar, or slow compared with scan rate, allowing construction of a kinetic zone diagram for each region for which simple diagnostic criteria were presented. For example, rapid interfacial kinetics and slow diffusional charge transport (large Λ and w—zone A1 in Aoki's case nomenclature[368,377]) result in the classical diffusional relationship[369,382]

$$i_p = 2 \cdot 69 \times 10^5 n^{3/2} (D^{1/2} c) v^{1/2}$$

which has been used by O'Connell[47] to evaluate $(D^{1/2}c)$, where c is concentration of redox centres in the film, for a monoquaternised bipyridine polymer and by Ghosh[257] to evaluate D for Ru(bipy)$_3^{2+}$ in polybipyrazine. Andrieux[376] used the value of v at which this square-root law first applies (cf. linear i_p, v dependence of Section 3.3) to estimate D for poly(p-nitrostyrene). The transition from equilibrium ($i_p \propto v$) to diffusional ($i_p \propto v^{1/2}$) behaviour has also been effected by variation of temperature,[347] redox centre spacing,[13] film thickness[350] and background electrolyte[13] for PVF, by solvent, electrolyte, temperature, film thickness and redox centre loading for PVP–Ru(EDTA)[375] and by film thickness for Nafion/Ru(bipy)$_3^{2+}$.[374] Digital simulations of voltammetric waves have been used to deduce values of D and the standard heterogeneous rate constant (k_0') for PVF[13] and Nafion/Ru(bipy)$_3^{2+}$.[119,374]

Cyclic voltammetry has also proved very useful for studying the bilayer films developed by Murray's group. Here, as illustrated in Fig. 2, outer film oxidation/reduction is mediated by the inner film, so that

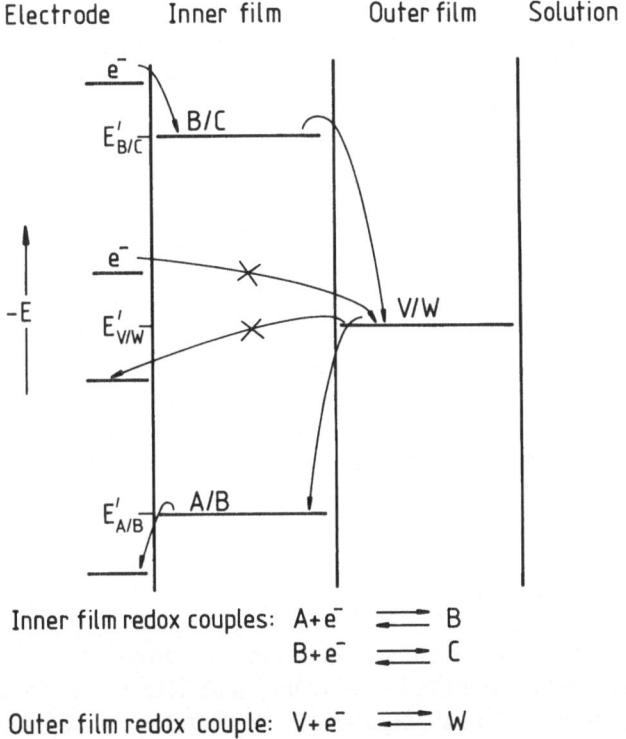

Inner film redox couples: $A + e^- \rightleftharpoons B$
$B + e^- \rightleftharpoons C$

Outer film redox couple: $V + e^- \rightleftharpoons W$

FIG. 2. Schematic representation of a bilayer modified electrode. Direct reduction (oxidation) of V(W) in the outer layer is prevented by the presence of the inner film. Reduction of V is mediated by the reduced partner C of the B/C inner film redox couple (when $E < E'_{B/C}$) and oxidation of W is mediated by the oxidised partner A of the A/B inner film redox couple (when $E > E'_{A/B}$).

outer film electrochemistry occurs at potentials chacteristic of the inner film ($E'_{inner,a}$ for oxidation, $E'_{inner,b}$ for reduction here) and a charge 'trapping'/'untrapping' device is created (see the next chapter for applications). Following early theoretical descriptions,[5,186] equations for rate-limiting inner film charge propagation and interfacial (polymer/polymer) kinetics have been developed.[193] In the companion paper[194] it was demonstrated that, except where the polymer/polymer interfacial reaction is thermodynamically 'uphill', inner film charge diffusion is most likely to be the rate-limiting step.

Finally we note some interesting variations on cyclic voltammetry. Degrand[60] has taken voltammograms of an azobenzene polymer on a hanging mercury drop before and after drop expansion and demonstrated that only at low azobenzene/polyethyleneimine functionalisation is the film flexible enough to bring more redox sites into contact with the increased mercury surface area upon expansion. Pickup[192,196,202] has studied charge transport rates in a series of polymeric Fe–, Ru– and Os–pyridine/bipyridine complexes by deposition of an outer metal layer. In this metal/polymer/metal sandwich, the response is now a steady state one, giving D/L via the concentration gradient. Values obtained are considered in the next section to facilitate comparison with those obtained from potential step experiments. Photogeneration of charge carriers in the underlying (p-Si) substrate has been demonstrated in the voltammetric behaviour of silylviologen films.[155,156] Semiconductor effects in the Ti/PVF system, where an intervening layer of TiO_2 was present, have been studied by Rolison[354] and the sensitivity of the results to trace contaminants from the plasma reactor (used for vinylferrocene polymerisation) noted.

3.4.2. Potential step methods

Continuous variation of the boundary condition (potential) and difficulties associated with uncompensated 'iR' drops and background currents can make quantitative modelling and data extraction in cyclic voltammetry difficult or ambiguous. It has therefore been found more useful to apply the simpler boundary condition of a potential step and monitor the resultant current or charge passed as a function of time as the film redox centres are oxidised/reduced by the 'electron bucket brigade'.[188] Current measurement (chronoamperometry) gives the rate of reaction and charge monitoring (chronocoulometry) the extent of reaction: the two are integrally related so the same information is present. Most workers apply large steps (across the voltammetric wave), in which case 'iR' losses may be prevented by addition of an appropriate voltage to ensure that the reactant at the metal/film interface is consumed rapidly. Under these conditions film charge propagation (rather than interfacial kinetics) is rate-limiting and a diffusionally limited response is observed (see Section 3.4.1). At sufficiently short times, when the depletion of reactant centres has not reached the outer interface, the Cottrell equation[383] is obeyed:

$$i = \frac{nFAD^{1/2}c}{\pi^{1/2}t^{1/2}} = \left(\frac{D}{L^2\pi t}\right)^{1/2} q \qquad (1)$$

where c and q represent the concentration change and total charge passed, respectively. (At longer times ($t \geqslant L^2/D$) the finite nature of the reactant reservoir results in the lower currents.[253,348] Thus a plot of i versus $t^{-1/2}$ gives a straight line, the slope of which is related to D, extrapolating to the origin. Both forms are given above, as some workers choose to use the $D^{1/2}c$ combination, others D/L^2; perhaps the latter is easier to visualise as it represents a relaxation rate in s^{-1}. It has been assumed here that the film is uniform in thickness. The response of an uneven film has been considered by Aoki,[384] who showed that the effect of a variation in L is a less abrupt fall-off from the linear Cottrell plot.

One now confronts two problems: firstly, extraction of D from the chosen combination and, secondly, its interpretation in terms of physical processes. In the first instance the difficulty revolves around *in situ* film thickness estimation, whichever combination is used ($c = \Gamma/L$, and Γ is known accurately): statements of film thickness should be inspected for *ex situ* (dry) determination, solvent swelling estimates or the use of circular arguments. Furthermore, solvent swelling (and thus L) may be oxidation-state-dependent.[36,348] As a result, caution should be exercised in using absolute values of D, although trends (associated with, for example, temperature or counter-ion changes) and orders of magnitude are unlikely to be seriously in error. A collection of values of 'D' obtained this way is given in Table 11.

An obvious point about the values in Table 11 is that, even allowing for uncertainty in the values of L used, they are orders of magnitude below those typical of species diffusing freely in solution ($10^{-5}\,cm^2\,s^{-1}$). This poses the second problem, then: what does 'D' represent? The possibilities include the intrinsic electron exchange

TABLE 11
Polymer Film Charge Transport Rates[a] Determined by Potential Step Transients Unless Otherwise Specified

System	Medium	$D\,(cm^2\,s^{-1})$	Species/process responsible	Comments	Refs
PVF (plasma polymerised)	CH_3CN, 0.1M-TEAT	9.8×10^{-12}		Anodic step.	350
	H_2O, 1M-LiClO$_4$	7×10^{-11}		Cathodic step.	348
		3.5×10^{-11}			
	C_3H_7CN, 0.1M-TBAP	10^{-13}–10^{-12}	Cooperative chain motion.	$-50 \leq T \leq -85°C$ $D_0 \sim (0.2$–$1.0) \times 10^{-8}\,cm^2\,s^{-1}$ $E_{act} \sim 15.5\,kJ\,mol^{-1}$ $\Delta S^\ddagger \sim -130$ to $-150\,J\,K^{-1}\,mol^{-1}$	347
PVF	CH_3CN, 0.1M-TBAP	$(2$–$3) \times 10^{-10}$		Values for three types of interconverting sites.[b]	13
PVF(silyl methacrylate copolymer)	CH_3CN, 0.1M-LiClO$_4$			$T = -36°C$ $35 \geq E_{act} \geq 17\,kJ\,mol^{-1}$, descending series.	146
$x_{Ferr} = 0.88$		3.4×10^{-10}			
$x_{Ferr} = 0.72$		6.2×10^{-10}			
$x_{Ferr} = 0.59$		10×10^{-10}		Apparent dispersion removed by correction for partial electroactivity.	
$x_{Ferr} = 0.38$		6.2×10^{-10}			
Polystyrene functionalised with TTF carboxylate	CH_3CN, 0.1M-TEA$^+$X$^-$				34
$x_{TTF} = 0.15$	X$^-$ = PF$_6^-$	10×10^{-10}	PF$_6^-$		
	X$^-$ = ClO$_4^-$	2.3×10^{-10}	ClO$_4^-$		
$x_{TTF} = 0.30$	X$^-$ = PF$_6^-$	6×10^{-10}	PF$_6^-$	[d] Ion size governs PTS$^-$ values, ion association with TTF$^+$ relative ClO$_4^-$, PF$_6^-$ values.	
	X$^-$ = CH_3–ϕ–SO_3^-	1.3×10^{-10}	CH_3–ϕ–SO_3^-		

Polystyrene functionalised with methylviologen				
$x_{viol} = 1.0$	H_2O, 1M-$LiClO_4$	3×10^{-12}	Extended area effects studied.	46
$x_{viol} = 0.5$	H_2O(pH 3·1), 0·2M-KCl		Multiply charged ions crosslink film and lower spinal flexibility (data are for viologen/ electrostatically bound ion charge transport processes as in column 1).	141
'V^{2+}' species		3.3×10^{-11}		
'V^{+}' species		1.3×10^{-11}		
'V^{2+}' species	($Fe(CN)_6^{4-}$ counter-ion)	1.1×10^{-11}		
'V^{+}' species	($Fe(CN)_6^{4-}$ counter-ion)	0.37×10^{-11}		
'$Fe(CN)_6^{3-}$' species		2.0×10^{-11}		
'$Fe(CN)_6^{4-}$' species		4.6×10^{-11}		
'$IrCl_6^{2-}$' species		1.4×10^{-11}		
'$IrCl_6^{3-}$' species		4.2×10^{-11}		
'$Mo(CN)_8^{3-}$' species		2.2×10^{-11}		
'$Mo(CN)_8^{4-}$' species		6.9×10^{-11}		
Poly(styrenesulphonate)/$Ru(bipy)_3^{2+}$	CH_3CN, 0·01M-TBAT	2×10^{-9}	'Dynamics of the motion of counter-ions'.	134 [d]
	CH_3CN, 0·01M-TBA^+X^-		'Uptake of anions'.	135 [d]
	$X^- = BF_4^-$	1.3×10^{-9}		
	$X^- = ClO_4^-$	1.3×10^{-9}		
	$X^- = PF_6^-$	0.78×10^{-9}		
	$X^- = CF_3SO_3^-$	1.3×10^{-9}		
	$X^- = CH_3{-}\phi{-}SO_3^-$	0.53×10^{-9}		
	CH_3CN, 0·01M-TEAT	1.2×10^{-9}		

[a] At room temperature, unless otherwise specified.
[b] Cyclic voltammetric determination.
[c] Chronopotentiometric determination.
[d] Chronocoulometric determination.

(continued)

TABLE 11—contd.

System	Medium	$D(cm^2 s^{-1})$	Species/ process responsible	Comments	Refs
Poly(styrene sulphonate)/ poly(p-xylylviologen)	H_2O(pH 7·2), 0·15M-phosphate buffer	$D/L^2 \simeq 1\,s^{-1}$ $(D \sim 10^{-10})^e$			70
Poly(p-xylylviologen) Nafion/M	H_2O, 1M-NaClO$_4$	$(1-2) \times 10^{-8}$			72
M = Ru(bipy)$_3^{2+}$	H_2O, 0·2M-Na$_2$SO$_4$	$2·7 \times 10^{-9}$	Not counter-ion motion.		119
	H_2O, 0·2M-Na$_2$SO$_4$	4×10^{-10}	Not physical motion.	For physical motion	121
	H_2O, 0·1M-K$_2$SO$_4$/H$_2$SO$_4$/ CH$_3\phi$SO$_3$H	$5·8 \times 10^{-10}$	Not counter-ion motion.	$D \simeq 0·2 \times 10^{-10}\,cm^2\,s^{-1}$. No concentration dependence.	374
	H_2O(pH 3·0), 0·2M-CF$_3$CO$_2$Na	$1·7 \times 10^{-10}$		'Electron hopping diffuser'.	120
M = Ru(bipy)$_3^{3+}$	H_2O, 0·2M-Na$_2$SO$_4$	$1·2 \times 10^{-9}$	Not counter-ion motion.		119
	H_2O, 0·1M-H$_2$SO$_4$	$6·8 \times 10^{-10}$	Not counter-ion motion.	No concentration dependence.	374
	H_2O, 0·1M-K$_2$SO$_4$/ CH$_3\phi$SO$_3$H	$5·8 \times 10^{-10}$			

M = Fe(bipy)$_3^{2+}$	H$_2$O(pH 3·0), 0·2M-CF$_3$CO$_2$Na	$1·8 \times 10^{-10}$		'Electron hopping diffuser'.	120
M = Fe(bipy)$_3^{3+}$	H$_2$O(pH 3·0), 0·2M-CF$_3$CO$_2$Na	$1·0 \times 10^{-10}$			120
M = Os(bipy)$_3^{2+}$	H$_2$O(pH 3·0), 0·2M-CF$_3$CO$_2$Na	$1·0 \times 10^{-10}$			120
	H$_2$O, 0·2M-Na$_2$SO$_4$	$0·7 \times 10^{-10}$		For physical motion $D \simeq 0·3 \times 10^{-10}$ cm^2 s^{-1}.	121
M = Fe(o-phen)$_3^{2+}$	H$_2$O(pH 3·0), 0·2M-CF$_3$CO$_2$Na	$0·4 \times 10^{-10}$			120
M = Fe(o-phen)$_3^{3+}$	H$_2$O(pH 3·0), 0·2M-CF$_3$CO$_2$Na	$0·4 \times 10^{-10}$			120
M = CpFeCp'—N(CH$_3$)$_3^+$	H$_2$O, 0·2M-Na$_2$SO$_4$	$1·7 \times 10^{-10}$		For physical motion $D \simeq (1·6–1·8) \times 10^{-10}$ cm^2 s^{-1}.	121
	H$_2$O(pH 3·0), 0·2M-CF$_3$CO$_2$Na	$1·3 \times 10^{-10}$	'True ionic diffuser'.		120
M = CpFeCp'N(CH$_3$)$_3^{3+}$	H$_2$O(pH 3·0), 0·2M-CF$_3$CO$_2$Na	$3·8 \times 10^{-10}$	'True ionic diffuser'.		120
M = Co(bipy)$_3^{2+}$	H$_2$O, 0·5M-NaSO$_4$	$(0·3–3·4) \times 10^{-11}$		[a] (For reduction to Co^{I}). Increasing with concentration.	123
	H$_2$O, 0·5M-NaSO$_4$	$(1·7–4·0) \times 10^{-12}$		[a] (For oxidation to Co^{III}). Decreasing with concentration.	123

[c] Using quoted value of L or c.

(*continued*)

TABLE 11—cond.

System	Medium	$D(\text{cm}^2\,\text{s}^{-1})$	Species/process responsible	Comments	Refs
M = Co(bipy)$_3^{3+}$	H$_2$O(pH 3·0), 0·2M-CF$_3$CO$_2$Na	2×10^{-12}		[a] (For oxidation to Co^{III}.) Decreasing with concentration.	120
M = MV$^{\cdot+}$	H$_2$O(pH 3·0), 0·2M-CF$_3$CO$_2$Na	2×10^{-11}			120
M = MV^{2+}	H$_2$O(pH 3·0), 0·2M-CF$_3$CO$_2$Na	4×10^{-10}	'True ionic diffusers'		120
	H$_2$O(pH 3·0), 0·2M-CF$_3$CO$_2$Na	18×10^{-10}			120
M = Ru(NH$_3$)$_6^{2+}$	H$_2$O(pH 3·0), 0·2M-CF$_3$CO$_2$Na	$3\cdot 4 \times 10^{-9}$			120
M = Ru(NH$_3$)$_6^{3+}$	H$_2$O(pH 3·0), 0·2M-CF$_3$CO$_2$Na	$2\cdot 3 \times 10^{-9}$			120
M = TTF^0	H$_2$O, 1M-KBr	8×10^{-7}		Reduction much faster.	385
PVP—M M = Ru(EDTA)	H$_2$O(pH 3·4), 0·2M-CF$_3$CO$_2$Na	—		From chronocoulometric slopes, $E_{act} \sim 19\cdot 2$ kJ mol^{-1}.	375
M = Fe(CN)$_5^{3-}$	H$_2$O(pH 3), 0·2M-CF$_3$CO$_2$Na	$(0\cdot 63 – 1\cdot 1) \times 10^{-7}$		Values increase to a plateau at high loading.	92
M = 'Fe$^{II/III}$'	H$_2$O, 5M-HCl	$(0\cdot 5 – 7\cdot 0) \times 10^{-11}$		Redox centres introduced into plasma as Fe(CO$_5$).	357

System	Conditions	D	Comments	Ref.
PVP·H$^+$, M M = Fe(CN)$_6^{3-}$ M = Fe(CN)$_6^{4-}$ M = W(CN)$_8^{4-}$	(H$_2$O, 0.2M-CF$_3$CO$_2$Na, 0.01M-CF$_3$CO$_2$H)	3.0×10^{-9} 3.9×10^{-9} 11.5×10^{-9}	^{cd} Chrono-amperometric, -coulometric, and -potentiometric data in good agreement. Decrease in D as Γ_M/Γ_{PVP} increases from 0·03 to 0·15, thereafter constant. Crosslinking or coulombic repulsion lowers D at higher loadings (≥ 0.3M). Rates enhanced by electron transfer in IrCl$_6^{3-}$/Fe(CN)$_6^{3-}$ mixtures.	97
M = Mo(CN)$_8^{4-}$		$(2.5–85) \times 10^{-9}$	Probably physical motion at low loadings.	
PVP·H$^+$ (copolymer with 6% crosslinking silane function), M M = $\frac{1}{3}$IrCl$_6^{3-}$ M = $\frac{1}{3}$Fe(CN)$_6^{3-}$	H$_2$O(pH 2·8), 2M-LiCl	$(2.8–0.05) \times 10^{-8}$ $(3.0–0.3) \times 10^{-8}$		108, 109
![structure: -(CH$_2$-CH(NH-R))$_n$-] R = —CO—Cp'CoCp (50% functionalised)	H$_2$O(pH 3·2), 0·2M-LiClO$_4$	$(0.4–1.6) \times 10^{-9}$		65
R = —CO—φ—CH$_2$—[N-pyridinium-φ-N] (50% functionalised)	H$_2$O(pH 1–6), 0·1M-LiClO$_4$	$(1.3–3.2) \times 10^{-12}$	Counter-ion or polymer motion. ^{be}	47

TABLE 11—contd.

System	Medium	$D(cm^2\ s^{-1})$	Species/process responsible	Comments	Refs
⟨N⁺(CH₃)⟩ₙ $\frac{1}{6}Fe^{II}[phen(-\phi SO_3^-)_2]_3$ [Ru^{II}(vibipy)₃]ₙ	H₂O, 1M-KCl/ NaCl/LiCl	$D/L^2 \simeq 10\ s^{-1}$ ($D \sim 3 \times 10^{-7}$)^e		Unexplained 'knee' on i-t transient.	95
	CH₃CN, saturated LiClO₄	$2 \cdot 2 \times 10^{-10}$	'Polymer lattice mobility'.		188
[M(bipy)₂(vipy)₂]ₙ M = Os	CH₃CN, 0·1M-TEAP	$(1-4) \times 10^{-9}$		^{f}	192
	CH₃CN, 0·1M-TEAP	$8(\pm 3) \times 10^{-9}$		Os$^{III/II}$ couple^f $E_{act} = 24\ kJ\ mol^{-1}$ $D_0 = 2 \cdot 7 \times 10^{-4}\ cm^2\ s^{-1}\ ^f$ Os$^{II/I}$ couple $E_{act} = 19\ kJ\ mol^{-1}$ $D_0 = 5 \cdot 4 \times 10^{-5}\ cm^2\ s^{-1}$ Os$^{I/0}$ couple^f $E_{act} = 14\ kJ\ mol^{-1}$ $D_0 = 4 \cdot 5 \times 10^{-5}\ cm^2\ s^{-1}$	202
		$24(\pm 9) \times 10^{-9}$			
		$200(\pm 70) \times 10^{-9}$			
M = Ru^{II}	CH₃CN, 0·1M-TEAP	8×10^{-10} $4 \cdot 6 \times 10^{-10}$		^{f}	202
[M(bipy)₂(cinn)₂]ₙ M = Os^{II}	CH₃CN, 0·1M-TEAP	$1 \cdot 5 \times 10^{-9}$		^{f}	196
M = Os/Ru mixtures	CH₃CN, 0·1M-TEAP			Ru is structurally similar 'diluent'.	204

System	Electrolyte	D	Comments	Ref.
$x_{os} = 0.074$		$\sim 0.3 \times 10^{-10}$		
$x_{os} = 0.45$		$\sim 1 \times 10^{-10}$		
$x_{os} = 0.74$		$\sim 6 \times 10^{-10}$		
Poly(ensilane·H$^+$, $\frac{1}{4}$Fe(CN)$_6^{4-}$)	H$_2$O(pH 3.2), 0.1M-KCl, 0.5M-glycine	3.4×10^{-9}	Probably electron hopping. D increasing with c, $E_{act} =$ 46.8 kJ mol^{-1}.	178
Poly(ethylpyridine silane·H$^+$, $\frac{1}{4}$Fe(CN)$_6^{4-}$)	H$_2$O(pH 2.5)	9×10^{-9}	D independent of c, $E_{act} =$ 26.8 kJ mol^{-1}. RDE experiment gives $D = 8 \times 10^{-9}$.	185
	0.4M-ClCH$_2$CO$_2$Na, 0.5M-KCl		D increasing with c, $E_{act} =$ 15.0 kJ mol^{-1}. D decreases at high loadings.	
Poly(propylviologensilane)	H$_2$O, xM-LiCl		Self-exchange, i.e. electron hopping, suggested. Little variation with electrolyte at fixed ionic strength.	151
	$x = 0.1$	0.37×10^{-10}		
	$x = 0.5$	0.97×10^{-10}		
	$x = 1.0$	1.40×10^{-10}		
	$x = 4.0$	2.10×10^{-10}		
Poly[L-lysine·H$^+$, $\frac{1}{3}$Co(C$_2$O$_4$)$_3^{3-}$]	H$_2$O(pH 1.5), 0.2M-CF$_3$CO$_2$Na/CF$_3$CO$_2$H	2×10^{-7}	Co-species inside^d Donnan domains.	116
		2×10^{-6}	Co-species outside^d Donnan domains. (RDE determination of κ.)	

^{f} 'Sandwich' device measurement.

(*continued*)

TABLE 11—contd.

System	Medium	$D (cm^2 s^{-1})$	Species/process responsible	Comments	Refs
Polyaniline	H_2O, 1M-HCl	$L^2/D \simeq 15$ ms ($D \simeq 2 \times 10^{-9}$)^e			239
Polythionine	H_2O(pH 1·3), 0·05M-HX		Counter-ion (X^-) motion.	$E_{act} = 35$ kJ mol^{-1}	253
	$X = HSO_4^-$	$9 \cdot 1 \times 10^{-13}$			
	$X = Cl^-$	$6 \cdot 5 \times 10^{-13}$			
	$X = CF_3CO_2^-$	$1 \cdot 7 \times 10^{-13}$			
	$X = CH_3\phi SO_3^-$	$0 \cdot 9 \times 10^{-13}$			
	H_2O(pH 5), 0·05M-KCl	$\sim 4 \times 10^{-12}$	Electron hopping.		
Azure A^g	H_2O(pH 1–7), 0·1M-KCl/ HCl/H_2SO_4/HClO$_4$	$1 \cdot 7 \times 10^{-9}$	Counter-ion or polymer motion.		47

^{g} N,N-Dimethylthionine.
TEAT = tetraethylammonium tetrafluoroborate.
TBAP = tetrabutylammonium perchlorate.
TEA$^+$ = tetraethylammonium cation.
TBAT = tetrabutylammonium tetrafluoroborate.
TBA$^+$ = tetrabutylammonium cation.

rate, diffusional encounter of redox centres (limited by local or large-scale polymer segmental motion), physical motion of redox or other species (solvent, counter-ions) necessary to maintain (solvation, electrostatic) equilibrium. Distinguishing between these requires analysis of the functional dependence of D on variables such as site–site separation, counter-ion size/charge, temperature, solvent and ionic strength, as illustrated in Table 11 and discussed below. Establishing a link between D and a physical process is important not only fundamentally but practically, for successful exploitation (see next chapter) of polymer modified electrodes relies on ensuring rapid charge transport.

Variation of D with redox site separation is interesting because a range of effects have been observed. Simple 'dilution' of sites (e.g. of $Os^{II/III}$ centres with structurally and electrostatically similar, but inactive, $Ru^{II/III}$ centres) results in a fall in D.[204] Increases in redox centre concentration, notably in electrostatically bound systems,[108,178] may result in crosslinking and thus restriction of polymer motion and a decrease in D. In other cases, for example $Ru(bipy)_3^{3+/2+}$ and $(Cp)Fe(Cp'\text{—}NMe_3)^{2+/+}$ in Nafion[121,374] and a ferrocene/silane copolymer,[146] D is relatively insensitive to c, although incomplete activity (if not accounted for) may give a misleading impression as the apparent concentration is varied.[140]

An interesting observation has been made by Mortimer[141] for a viologen polymer charge compensated with inactive (Cl^-) or electroactive [$Fe(CN)_6^{3-/4-}$, $IrCl_6^{3-/4-}$] anions: similar values of D are obtained for spinal (viologen) and electrostatically bound redox centre charge transport. The interpretation was a common rate-limiting step, probably polymer segmental motion. The exception was faster charge propagation when chloride was the counter-ion, probably because all the other (multiply charged) anions induced crosslinking.[141] Segmental motion has also been proposed as a rate-limiting step for suitable juxtapositioning of 'dilute' $Os^{III/II}$ sites in the Ru/Os copolymer of the preceding paragraph.[204]

Table 11 contains several clear examples of rate-limiting counter-ion diffusion. In all cases the restricted environment of the polymer film has a profound effect, lowering D by 4–8 orders of magnitude from typical free solution values. For a given polymer, counter-ion diffusion rates may be correlated with ionic size[34,135,253] or strength of interaction with charged polymer sites.[34]

In principle, parallel routes for transport exist and it is sometimes possible, for a given polymer medium, to alter the conditions in such a way as to change from one mechanism to another. In the ion exchange systems one may have parallel physical diffusion and self-exchange pathways: in free solution even rapid self-exchange does not allow this pathway to predominate,[121] but when physical diffusion rates (see preceding paragraph) are greatly reduced, it may do so. This has been illustrated by Co(bipy)$_3^{3+/2+}$ [123] and (Cp)FeCp'—N$^+$Me$_3$[121] (physical motion predominates), Os(bipy)$_3^{2+}$ [121] (both processes contribute) and Ru(bipy)$_3^{2+}$ (self-exchange predominates) charge transport rates in Nafion. The situation is complicated by the existence of parallel (electron- or place-exchange coupled in certain cases) pathways in hydrophylic and hydrophobic polymer phases,[123] and by competition for sites at high loadings when physical movement is required.[123] Solution-value based expectations of the relative rates of diffusion of two halves of a redox couple may not be realised if hydrophobic ion/polymer interactions are important, e.g. Ru(bipy)$_3^{2+/3+}$ in Nafion.[374] When self-exchange is rate-limiting, one might expect some dependence on ionic strength. This has indeed been found for a silane-bound viologen,[151] where an increase by a factor of ~8 in rate was found on raising the LiBr concentration from 0·1 to 4·0 mol dm^{-3}.

Comparatively few activation energies have been determined,[204,253,347,375] surprisingly perhaps, for significant activation energy differences in parallel pathways may lead to a switch from one to the other with temperature.[204] In the most detailed of these,[347] a very low pre-exponential factor, $D_0 \sim 10^{-8}$ cm^2 s^{-1}, was found for PVF, and attributed to a large negative entropy of activation ($\Delta S^{\ddagger} \sim -140$ J K^{-1} mol^{-1}). Using simple collision theory, a diffusion-controlled (probably via lattice mobility in this instance) rate constant of 10^2 mol^{-1} dm^3 s^{-1} was estimated—far below free-solution values. A similar 'levelling' effect (by polymer motion) of rapid electron exchange rates to a common (low) value has been noted for Nafion.[123]

In 'sandwich' systems, where a steady state current can flow, the transient (Cottrell) behaviour is superimposed on this background, as illustrated for Pt/poly(Os(bipy)$_2$(vipy)$_2$)$^{2+}$/Au assemblies.[202] Incorporation of particulates, such as carbon, in the film raises the apparent diffusion coefficient: for a thick viologen polymer, this was accounted for by inserting a 'roughness factor' into the Cottrell equation[46] to represent the increased area of electrode/polymer interface or decreased distance over which charge was required to diffuse.

Chronocoulometric measurements of D are included in Table 11, as the experiment/analysis is essentially the same: one plots charge versus $t^{1/2}$ and uses data at reasonably short $(t < L^2/D)$ times, for the same reason as in chronoamperometry. At very short times, uncompensated 'iR' losses result in negative intercepts on the charge axis and curvature, so it is usual practice[97,120] to take the slope of the straight line portion once this effect has decayed away, rather than force the data through the origin. A noteworthy example is the combination of chronocoulometric and rotating disc data for protonated poly(L-lysine)/Co(C$_2$O$_4$)$_3^{3-}$ [116] to deduce a partition coefficient as well as an effective charge transport diffusion coefficient (outside the Donnan domain).

3.4.3. Chronopotentiometry

This is the complementary technique to chronoamperometry/chronocoulometry: one measures the potential response to an applied step in current from zero to some chosen value. Application to polymer modified electrodes is not widespread, but an interesting comparison of all three techniques in this context has been given from a theoretical viewpoint and illustrated with consistent data from PVP·H$^+$/Mo(CN)$_8^{4-}$, W(CN)$_8^{3-/4-}$ and Fe(CN)$_6^{3-/4-}$ systems.[97] Non-uniformity of film thickness led to deviation from the Cottrell equation being 'smeared' in the time domain, but it is interesting to see that departure from the simple equation (1) occurs at different times for the three techniques, due to the way each probes the film redox state. Consistency with chronoamperometrically determined charge transport rates was found for Ru(bipy)$_3^{2+}$, Os(bipy)$_3^{2+}$ and (Cp)FeCp'—$\overset{+}{\text{N}}$Me$_3$ in Nafion.[121] Measurements of transition times (linearly related to charge injected/removed) have been used to determine the redox composition of Ru(NH$_3$)$_6^{2+/3+}$ and Os(bipy)$_3^{2+/3+}$ mixtures bound in carboxylate films[125] and to identify ruthenium centres kinetically isolated after an initial oxidative cycle in poly(styrene sulphonate)/Ru(bipy)$_3^{2+}$ films.[135] A rather earlier example of the technique was given by Schroeder,[8] who showed that TTF-functionalised polystyrene films underwent large resistive changes on oxidation/reduction. Constant current oxidation of the neutral species resulted in a large initial potential excursion before a return to values consistent with low film resistance. Prior partial oxidation (see reference to 'break-in' phenomena in Section 3.4.1), to introduce a few charged species, removed this uncompensated resistance effect completely.

3.4.4. Alternating current and pulse techniques

These more sophisticated techniques have been used by only a few groups, although the quality of data obtained imply that more widespread use may be expected in future. The impedance of PVP indicates a Warburg component in the absence of solution redox species: this has been attributed to quinone-like functionalities introduced by the plasma deposition conditions.[356] Polypyrrole impedance measurements were indicative of a porous film,[268] and the authors were able to fit their data to a simple equivalent circuit of resistive and capacitive elements, whose frequency dependences were explainable. Reversible and irreversible changes to insulating behaviour were found at -0.4 and 0.8 V, respectively. The reversible change was also studied by Burgmayer,[278,279] who determined both a.c. impedance and d.c. resistance as a function of oxidation state and used the in-phase impedance component as a measure of response rate for 'ion gate' devices.

Normal pulse voltammetry (NPV) has been used to study charge transfer processes in PVP–Fe(CN)$_5^{2-/3-}$,[93] PVP·H$^+$ Fe(CN)$_6^{3-/4-}$,[93] PVP·H$^+$ Mo(CN)$_8^{3-/4-}$ [100] and a series of viologen systems[48] (MV^{2+} in Nafion, a viologen functionalised polystyrene and a xylylviologen polymer). Analogously to d.c. polarography for solution species, one obtains kinetic information (k_0' and the transfer coefficient, α) for the metal/polymer interfacial charge transfer process and diffusional information (D) from the rising and plateau portions of the voltammogram, respectively. (The differences are that in NPV the plateau current depends on the sampling time and that in the polymer modified electrode context D represents the charge transport rate within the film.) In the PVP systems, k_0' and D showed similar dependences on coverage, whilst in the viologen systems only the Nafion/MV^{2+} system showed a dependence of k_0' and D on Γ. The reason is simple: in the styrene and xylene viologens, the redox centre is an integral part of the polymer and an increase in Γ is simply an increase in L, whilst in all the other systems the composition ($\Gamma_{\text{redox}}/\Gamma_{\text{polymer}}$) is varying. The facility to measure k_0' and D in one experiment and the apparent precision of the data indicate that this technique should be applied more widely.

Differential pulse voltammetry has been used to determine the amount of Cu^{2+} incorporated in poly(styrene sulphonate) modified carbon paste electrodes.[137] In this analytically oriented study the technique was chosen for its greater sensitivity and no mechanistic/

kinetic data were obtained, but peak current was found to be linear with Cu^{2+} concentration in the bathing solution and accumulation time under the conditions employed.

3.4.5. Rotating ring disc electrode (RRDE)

The RRDE and its simpler counterpart the rotating disc electrode (RDE) have been used extensively for mediated charge transfer studies, to be described in the next chapter, as a result of their steady state mass transport characteristics. The finite nature of polymer films means that only transient data can be obtained (without a bulk solution species present) so this advantage is not applicable here. However, two transient RRDE studies have been reported in which species ejected from a polymer film on the disc have been detected on the downstream ring electrode. Braun[113] found that on reduction of $PVP \cdot H^+/Fe(CN)_6^{3-}$ over 90% of the resulting $Fe(CN)_6^{4-}$ left the film, to be replaced by background electrolyte anions. In a similar experiment, Kobayashi[240] detected Br^- counter-ions ejected from polyaniline on reduction of the latter.

4. SPECTROSCOPIC AND OTHER NON-ELECTROCHEMICAL CHARACTERISATION TECHNIQUES

Advances in spectroscopic techniques have led to their increasing use in electrochemistry. This is well illustrated by the surface science techniques, whose high sensitivity is admirably suited to surface (e.g. polymer) modification studies. The approaches are complementary: electrochemistry is precise (the Faraday constant is accurately known) but structural information is indirect, whilst spectroscopy is not always precise (extinction coefficients, cross-sections, absolute instrumental sensitivities, etc., are not known with the precision of the Faraday) but, through choice of excitation species and energy, provides a wide range of direct structural information. Furthermore, current represents a rate of reaction of electroactive species, whilst spectroscopic techniques, representing concentration, allow one to observe also non-electroactive (or electroactive but kinetically isolated) species.

The techniques will be classified according to the entity to which they give access, starting with large collections of molecules and progressively increasing the (conceptual) magnification (not to be confused with physical resolution) through to individual atoms. Frequent attention will be drawn to comparison of *in situ* and *ex situ* data: the absence of field and solvent, and possible interactions during transfer with ambient species (O_2 and CO_2, for example) in the latter case mean that information is not always directly applicable. *Ex situ* conditions are sometimes a requirement of the technique—notable here are the UHV requirements of XPS (X-ray photoelectron-spectroscopy) and AES (Auger electron spectroscopy)—but, wherever possible, *in situ* measurement is to be recommended. A variety of non-spectroscopic techniques are also included in this section, as the information they impart is most suitably considered here.

4.1. Macroscopic Observations and Properties

The most popular technique for morphological observations is scanning electron microscopy (SEM), as evidenced by Table 12. The principal objectives are the ascertainment of film integrity (lack of pinholes) and homogeneity (absence of large-scale 'clustering'). In noting the comments in Table 12, it should be borne in mind that most observations were made at resolutions of the order of $0.1\,\mu$m, so contradictory evidence might be found from, for example, XPS or electrochemical experiments, where the electrons or ions need only much smaller holes to traverse the film unhindered. Loss of solvent on transfer to the vacuum chamber may also be important in SEM film thickness estimation.[160,350] Morphological changes induced by reduction/oxidation have been observed by SEM.[36] Very nice pictures of $(TTF)Br_{0.7}$ needles in Nafion, TTF^+ [131] and Pt^0 in poly(vinylacetic acid)[363] have been obtained. Many electron microscopes have EDAX attachments and this elaboration of the technique (strictly an 'atomic' method, but included here for completeness) has been elegantly used to demonstrate anion inclusion[160] and metal deposition.[14,131]

Optical microscopy has been coupled with X-ray diffraction to demonstrate the lack of crystallinity in electrochemically produced poly(p-phenylene) films on Pt.[325] This contrasts with the (at least partially) crystalline nature of chemically prepared material and was held to be the cause of observed differences in physical properties.

Friction coefficient measurement has been used to follow film formation accompanying substituted phenol and aniline oxidations,[221,223,227] as evidenced by 'slip-stick' behaviour. Unfortunately, the information is not very specific and the technique has not been widely used.

Film formation has also been studied by observing interference fringes during establishment of a diffusion layer during carbazole polymerisation.[325]

Contact angle changes for a droplet of aqueous solution on (chemically) oxidised and reduced plasma polymerised PVF (PPVF) films on Pt have been measured;[35] unfortunately, hysteresis effects prevented dynamic measurements under electrochemical redox state control.

4.2. 'Monomer'/Redox Centre Observation

Study of the redox group as an entity is of obvious interest, since this is the species deliberately introduced on to the surface. The principal effect monitored is electronic transition(s) within a suitable chromophore, which may be a pendant group (e.g. a metal/pyridine complex in a PVP film) or an integral part of the polymer spine (e.g. in polypyrrole). The most widely used technique is UV/VIS absorbance, firstly because of the extensive literature for monomeric analogues in solution, secondly because instrumentation is widely available, and thirdly because commercially available transparent conducting electrodes (e.g. doped SnO_2 on glass) allow transmission experiments with simple cells and optics. A simple Beer–Lambert law calculation shows the experimental requirement: for a typical detection limit of 0·01 absorbance units and a surface coverage (Γ) of 1 nmol cm^{-2}, we need $\varepsilon_{max} \geq 10^7$ mol^{-1} cm^2 (10^4 mol^{-1} dm^3 cm^{-1}). Whilst Γ is frequently 10 nmol cm^{-2} or more, the requirement of a good chromophore is clear.

Whilst a few measurements have been made in reflectance mode (for polyaniline,[239] a polysilylviologen,[151] a porphyrin polymer[180] and polypyrrole[267]), the majority of data have been acquired by transmission. In the simplest experiment, the absorption spectrum can be used to demonstrate successful immobilisation of a species of at least qualitatively similar nature to a known analogue.[38,106,107,145,147,160,181,224,314,327] Knowledge or assumption of Γ allows determination of extinction coefficient[151,253] or vice versa.[117] Electrochromic efficiency determination[70,95] and demonstration of charge trapping in bilayer films[5] have been accomplished. The absorbance may be used as a redox state monitor in steady

TABLE 12
SEM Examination of Polymer Films on Electrodes

System	Observation(s)	Refs
GC/(plasma)–PVF	No gross cracks or pinholes. Film thickness estimated by cutting film.	350
Pt/PVF	Small pinholes detected, although continued exposure to electron beam was observed to cause damage.	12
GC/PVF	Cu^0 deposition on underlying GC is even (not in islands), characteristic of membrane (not pinhole) diffusive behaviour. EDAX data support this view.	14
GC/PVP	Film uneven—more so when Ru(EDTA) coordinated.	375
Pt/polystyrene (functionalised with TTF, pyrazoline)	Morphological changes accompanying electrochemical cycling observed. Polymer swelling correlated with electroactivity.	36
SnO_2/poly(benzylviologensilane)	Film thickness estimation. Detection of counter-ions by EDAX.	160
SnO_2/poly(ethylpyridinesilane·H^+, $\frac{1}{4}Fe(CN)_6^{4-}$)	Film thickness estimation.	185
SnO_2/Nafion, TTF	Observation of $TTF^+Br_{0.7}^-$ crystals. EDAX used to show selective Cu^0 deposition on these conducting 'needles'.	131
Pt/poly[Ru(bipy)$_2$(vipy)$_2^{2+}$]	Study of Cu^0, Ni^0, Co^0 deposition: ~0·1–2 μm spheres observed, dependent on conditions used.	198
Pt/poly(N-R-pyrrole)$_n$, R = H, CH$_3$	Surface compact, not fibrous. Smoother films grown from CH_3CN solution containing 1% H_2O. No difference in morphology (at 0·1 μm scale) between BF_4^- doped and undoped films.	261, 264, 265
$(CH)_x$/poly(pyrrole)$_n$	Distinct 'layered' structure found only when polyacetylene substrate is undoped.	300

Substrate / Polymer	Description	Ref.
SnO₂ / —[thiophene-S⁺ ClO₄⁻]ₙ—	~1 μm clusters.	313
Pt / —[R-thiophene-S⁺ BF₄⁻]ₙ—, R = H, Br, —CH₃	~1–5 μm clusters; no evidence of crystallinity.	309
Pt / —[thiophene-S⁺ CF₃SO₃⁻]ₙ—	Fibrous structure, diameter 150 nm, increasing on deposition of Cu⁰. More dilute plating solutions give particulate Cu⁰.	319
Pt / —(C₆H₄)ₙ— 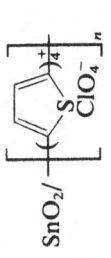	Amorphous structure (unlike chemically prepared material).	325
GC / —[Ni(C₆H₄)ₓ]ₙ—, x = 6–7	More compact films formed at less cathodic polymerisation potentials.	336
Pt / —(C₆H₄=NH)ₙ— 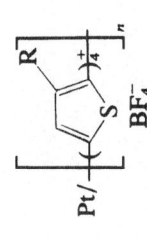	Continuous, fairly smooth film.	236

(continued)

TABLE 12—contd.

System	Observation(s)	Refs
Pt/ azulenium structure with R groups; R = CH$_3$, X$^-$ = ClO$_4^-$; R = H, X$^-$ = BF$_4^-$	~2 μm clusters.	327
Pt/ carbazole polymer	Film slightly 'rippled'.	329
Pt/polyacrylonitrile (by GDP) (by electrochemical coating)	No pores or cracks detected. Non-uniform coverage, with preferential deposition on substrate defects, such as polishing scratches.	359 386
GC/ –(CH(CH$_2$CO$_2$H))$_n$– · Pt^0	Observation of particulate Pt^0 deposition by electroreduction of PtCl$_6^{2-}$.	363

state[117,239–241,253,277] or transient[76,89,161,207,253,271,272] modes, as a measure of redox centre incorporation/loss/reaction by chemical, electrochemical or photochemical means,[5,16,89,207,212,275,385,387] as a measure of fractional electroactivity,[37,89] as a monitor of film growth,[271] for observation of transient species,[34] and as an environmental probe. The latter is interesting because the high concentration conditions prevailing in the film would be expected to promote aggregation: whilst this has been observed for variously immobilised TTF-based species,[34,40,385] TCNQ-based species[75,76] and a pyrazoline,[39] the contrary has been found for one viologen polymer[160] because polymer rigidity prevents the otherwise favoured dimerisation. Band broadening is sometimes seen and this may be correlated with broadening of cyclic voltammetric waves as a result of interactions between redox centres; derivative absorbance/potential plots for a TCNQ polymer demonstrate the point.[74,76] In these latter studies,[75,77] absorbance data were correlated with the concentrations of radical chromophores measured using electron spin resonance (ESR). Visible absorption data also provide the key to successful Raman spectroscopy when resonance enhancement is relied upon.[211,255] In mediated charge transfer based applications (see next chapter), the extent to which one 'titrates' in the mediator with potential, i.e. the redox state/potential relation, is crucial. Since it may well not be Nernstian (see Section 3.3) an independent measure of this relation via absorbance[117,253] is extremely useful. Whilst structural information is not detailed, some interesting observations have been made for polyaromatics, such as polyphenylene[336] and substituted polythiophenes,[272,310] where electron delocalisation is intimately related to the goal of high conductivity.

As well as absorption, one can also obtain information from emission of visible radiation. Fluorescence from Croconate Violet in $PVP \cdot H^+$,[107] and thionine[255] has been observed. Given the possibilities of double electron transfer with the electrode and of self-quenching it is noteworthy that the effect is observed at all: presumably, polymer-imposed constraints on energy/charge transfer to neighbouring redox centres are the key, although no definite evidence exists. Luminescence from $Ru(bipy)_3^{2+*}$ centres produced either by electrochemically generated Ru^I/Ru^{III} conproportionation[117,134,189,388] or reaction with oxalate[118,119] has been observed in several films and Stern–Volmer plots have been used to extract kinetic parameters. Data on charge propagation rates[134,388] and slow structural changes[117] have been obtained this way. The stimulus/probe relationship of electrochemistry and spectroscopy is to be noted here.

Another established technique applied to polymer modified electrodes is ESR, using special cell arrangements so that the electrode can be placed in the cavity. By definition, the technique is specific for radical species. In contrast with the complex spectra of radicals in solution, surface immobilised species give single-line spectra as a result of rapid exchange: this difference is both an advantage (as a surface versus solution diagnostic) and a disadvantage (as much information is lost). Dimerisation in TCNQ[74,75,77] and TTF[35] polymers has been studied: unequivocal evidence for $TCNQ_2^-$ and TTF_2^{+} was obtained, correlation with UV/VIS absorbance was used to construct distribution diagrams for the various species present as a function of potential, and, in addition to the thermodynamic (E', K) data, slow dimerisation kinetics[75] and trapping of radical sites[77] were revealed for the TCNQ polymer. Polynitrostyrene and polyvinylanthraquinone radicals have been observed,[30] and the ESR signal has been used as a concentration monitor in potential step experiments. Whilst the Cottrell equation (in integrated form, as for absorbance or chronocoulometric data) was found to be a good approximation to the time dependence of the signal, deviations due to lineshape variation with concentration were observed. Polythiophene films have been studied[318] and their behaviour described in metallic terms with a 'skin effect'; depending on the skin depth, different lineshapes, correlatable with observed conductivity, were seen. From the g-value, it was deduced that the sulphur p-orbitals contribute little to the π ground state.

A technique well established in thin film analysis, but hitherto not widely applied to polymer modified electrodes, is ellipsometry. Whilst the data pertaining to refractive index do not lead to detailed structural information, the method does allow *in situ* determination of film thickness without the need for assumptions based on estimated swelling factors, i.e. L is determined directly, not as a combination with D or c as it is in chronoamperometry/chronocoulometry or cyclic voltammetry. Electropolymerisation of acrylonitrile on Ni[217] and thionine on Pt[389] have been followed this way: reduction/formation of surface oxides were found to be necessary in these cathodically/anodically initiated processes, respectively.

4.3. Sub-molecular Units
The objective here is to determine the bonding in the polymer, the monomer–monomer linkage and the mode of attachment of subsequently introduced entities being of primary interest. Molecular

fragments may be studied directly by SIMS. This technique has been used for polyacrylonitrile[390] and poly(THF) films,[391] but suffers from being destructive and *ex situ*. Far more useful is to look at the 'fragments' without removing them, by vibrational (infrared or Raman) spectroscopy.

Solvent and/or electrode absorbance problems in the infrared region have been overcome by using an *ex situ* (multiple reflectance where necessary) method for polymeric phenols,[221–223,225–227,231] polymeric anilines[245] and polypyrrole[267] or, more crudely, by removing the polymer from the electrode and preparing a pellet for poly(3-methylthiophene)[272] and a Ru/aminophenanthroline polymer.[249] Fourier transform methodology has been used in transmission[56,159,160] (on Si substrates) and reflectance[219,320] (on Au and Ge/Au substrates) to permit the more desirable *in situ* measurements. Dominey showed ion exchange of electro-active and -inactive cyanide complexes[160] and adsorption of CO on to Pt^0 particles (including a ^{13}C isotopic shift)[159] in two polymeric viologen silane films. Dubois[56] showed coordination of N_2 to a polymeric Mo^0/phosphine complex and its loss on molybdenum oxidation. These elegant pieces of work indicate that the technique will find wide application in this area.

Raman spectroscopy has the advantage that one can select the exciting wavelength to avoid solvent absorbance problems, but its lower sensitivity places reliance on resonance and/or surface enhancement. Resonance Raman has been used to study electropolymerised protoporphyrin films on Pt:[211] approximately half of the vinyl groups become saturated on polymerisation, clearly identifying the linkage mode. A combination of resonance and surface enhancement, via choice of oxidation state and substrate, respectively, was used to determine the structure of polymeric thionine films on Au^{255} with the aid of selective deuteration. PVP-bound ruthenium bipyridyl complexes on Ag have been studied with the aid of the large surface enhancement factor associated with Ag;[24,25] it is interesting to note that the nitrogen lone pair bonding envisaged for pyridine adsorption on Ag cannot be operative here as the nitrogen is coordinated to the metal.

4.4. Atoms

The principal techniques here are XPS and AES; although EDAX strictly speaking belongs here, it was discussed in Section 4.1 as it is

TABLE 13
Polymer Modified Electrodes Studied by AES (Where Specified) and XPS (All Other Examples)

System	Observation(s)	Refs
(ferrocene polymer, variously deposited)	No pinholes.	12, 14, 146
	Fe oxidation state changes accompanying deposition.	5, 6, 9, 352
	Uneven thickness (by angular dependence of signal).	7, 9
	Bilayer 'charge trapping' demonstration.	5
	Substrate transfer in plasma reactor.	354
	Some segregation of silane copolymer: more silane at surface.	146
Poly[M(vibipy)$_3$]$^{2+}$		
M = Ru, Fe	Bilayer film structure demonstrated.	5, 186
M = Ru	Correlation of XPS and electrochemical film thickness measurements.	7
Poly[Ru(bipy)$_2$(vipy)$_2$]$^{2+}$	Observation of Cu^0, Ni^0 and Co^0 deposited on outer surface (some charging effects).	198
Poly[Ru(bpz)$_3$]$^{2+}$		257
Poly[Ru(amphen)$_3$]$^{2+}$	Counter-ion (ClO$_4^-$) observed.	249

Deactivation associated with Hg impurity (from SCE?). Presence and oxidation state of Pd firstly as ion-exchanged $PtCl_4^{2-}$, then as Pd^0 by reduction.	149
Auger depth profiles of C, N, O, Si and Br in 'as prepared' films and Pt, Pd, Cl in $PtCl_4^{2-}$ and $PdCl_4^{2-}$ ion-exchanged and Pt^0 and Pd^0 dispersed particulate derivatives.	149, 154, 156–158
Competitive binding of anions followed by Auger spectroscopy (and cyclic voltammetry where ions electroactive): kinetics and thermodynamics of exchange studied.	154
Absence of 'Cl' implies complete hydrolysis of $Fe(Cp_2)\!\!>\!\!SiCl_2$ monomer (also by Auger).	164, 171
Some Fe loss/Si surface accumulation.	164
Correlation with cyclic voltammetry implies complete electroactivity.	164

(*continued*)

TABLE 13—contd.

System	Observation(s)	Ref.
(—O—Si(—O—)$_n$)—C$_2$H$_4$—pyridinium—N$^+$—H X$^-$	Tertiary and quaternised N in neutral and protonated states.	185
X$^-$ = Cl$^-$, $\frac{1}{2}$IrCl$_6^{2-}$, $\frac{1}{4}$Fe(CN)$_6^{4-}$; (—CH—CH$_2$—)$_n$—C$_6$H$_4$—SO$_3^-$ As$^+\phi_4$	AES depth profiles imply homogeneity and Cl/N in accord with complete charge neutralisation.	185
(—C(CH$_3$)—CH$_2$—)$_m$—CO—O—C$_3$H$_6$—Si(OCH$_3$)$_3$	H$_2$O exposure raises surface Si population.	138

Structure	Observations	Refs
[pyrrole-N-H]$_n^+$·X$^-$, X$^-$ = BF$_4^-$, PF$_6^-$	Dopant observation.	264, 392
	Probable formation of carbonyl groups on exposure to oxygen.	264, 392
	Auger depth profiles for parent and silane surface bonded polymer.	293
[3-methylthiophene]$_n^+$·PF$_6^-$	Dopant observation.	272, 392
	C oxidation states.	221, 224
phenylene oxide with R substituents (various substituents)	Condensation of FeCp$_2$ moiety observed through Fe signal.	225
	Presence of corroding substrate in film.	222
	Metal complexation by amine function.	226
	Film growth law by intensities of film (C) and substrate (Pt) peaks as a function of polymerisation time.	228

(continued)

TABLE 13—contd.

System	Observation(s)	Ref.
poly(8-hydroxyquinoline) type, R = H, CH$_3$	Co^{II}, Cu^{II} coordination.	342
poly(naphthoquinone) type	Observation of $>$C=O, C—O—C and C=C functional groups.	231
poly(aniline) type (various substituents)	Electrolyte ions found in films produced by pulsed (but not constant or swept) potential polymerisation.	245
poly(thionine) type	Oxidation states of film and substrate species. Intensity/film thickness correlation.	254

$\left[\underset{Ni^{II}}{\bigcirc}\!\!-\!\!\bigcirc \right]_m \Big]_n$ Nickel present as Ni^{II}. No P or Br from precursors.   323

$m \simeq 6\text{–}7$

$\left[-NH-C_3H_6-N^+\bigcirc\!\!-\!\!\bigcirc N^+-C_3H_6-NH-\underset{N}{\bigcirc}\!\!-Cl \right]_n$ Loss of electroactivity does not result from polymer desorption.   73 usually coupled with electron microscopy. Whilst all three techniques involve interaction with atomic entities, their spatial resolutions are very different: less than a micron for EDAX but of the order of a millimetre for XPS and AES. The latter two are also surface-sensitive, whilst EDAX is not. In all three cases, of necessity, measurements are obtained *ex situ*.

X-ray photon spectroscopy has, as a result of its surface sensitivity, been used for characterising modified electrodes since their monolayer beginnings.[1-3] In all cases (see Table 13) the technique has been used at least qualitatively to establish successful surface immobilisation, usually via suitable 'tag' elements (e.g. Ru, Fe, S, N) or the disappearance of selected elements as a consequence of the synthetic procedure (e.g. hydrolytic Cl loss in silane polymerisation[164]). (Care must be taken with C or O species as they may, at least in part, arise from surface contamination.) More specifically, the technique has been used to establish approximate oxidation states,[5,6,9,149,185,198,221,254,323,342,352] to check film integrity via absence of pinholes,[12,14,254] to study film growth rates[228] and to distinguish between electroinactivity and polymer desorption.[73] Quantitative measurements have established elemental ratios following redox centre incorporation[225] and demonstrated surface segregation effects.[138,146,164] Correlation with electrochemical coverage data has been used to estimate (dry) film thickness[7,164,254] and establish complete electroactivity.[7,164,254] Variation of photoelectron pathlength via measurement at different angles was used to demonstrate the segregated nature of bilayer films[186] and, through deviation of estimated film thickness from electrochemical estimates, the presence of non-uniformity.[7,9]

Auger electron spectroscopy with depth profiling has been used by Wrighton's group to study the distribution of immobilised Pt^0 [156] and Pd^0 [149] particles (from reduction of ion-exchanged $PtCl_6^{2-}$ and $PdCl_4^{2-}$, respectively) in polysilylviologen films and to probe variously structured interfaces. Relative populations of transition metal complex and simple anions in these films were established in competitive binding experiments:[154] where species were electroactive, agreement with cyclic voltammetric estimation was good. Elemental ratios were used to demonstrate charge neutralisation of a protonated silylpyridine polymer.[185] Polypyrrole[293] and polysilylferrocene[171] films on Si have also been studied.

4.5. Other Techniques

Several spectroscopic techniques have been used to characterise surface modifying polymers, but not on the electrode or in the electrochemical environment. These include electron energy loss spectroscopy (EELS) of BF_4^--doped polypyrrole films stripped from SnO_2,[287] ^{13}C solid state magic angle spinning NMR of polyazulene,[327] and ^{31}P NMR of a molybdenum/phosphine polymer.[88]

5. CONCLUSIONS

A range of procedures has been developed for the attachment of controlled amounts of polymer-bound redox centres to electrode surfaces. Preformed or *in situ* generated polymers may be used and the resulting surface further modified by thermally, photochemically or electrochemically induced coordination, covalent bond formation, ion exchange or oxidation/reduction. Combinations of these procedures allow the construction of more complex arrangements, such as bilayer, sandwich and immobilised particulate systems. The range of polymers and procedures available allows the surface attachment of a given redox couple (ferrocenes and viologens being the best examples) in a variety of ways (spinal, variously pendant or ion-exchanged), allowing one to vary systematically the site–site separation (i.e. concentration or 'loading') and the environmental/solvation characteristics experienced by the redox centres. The latter have important consequences for the rates of charge and mass transport through the film, both of which are of crucial importance with respect to virtually all applications.

Characterisation techniques (principally electrochemical) are well established for the determination of thermodynamic and charge injection/transport kinetic parameters. The former (notably E' and the redox composition/potential relation) seem to be largely transferred from monomeric solution analogues to the surface-bound species, with important qualifications. These centre principally around the difference in concentration (typically in excess of 1 mol dm^{-3} in the polymer film, compared with $10^{-3} \text{ mol dm}^{-3}$ for solution studies), resulting in significant activity effects. Kinetic parameters are affected more substantially: encounter rates of both covalently and electrostatically bound redox centres are much lower than for solution species, because of a combination of slow spinal motion, inter-/intra-chain interactions and the spatially restricted environment. For systems where the redox species are able to move through the film (ion-exchange films), the high redox centre concentration may mean that the electron-exchange contribution to charge propagation is more important than in solution. Additionally, changes in ion and solvent populations may be so slow as to limit the observed behaviour.

Structural information of a more direct nature has largely been obtained by the application of conventional spectroscopic techniques, notably visible absorbance, IR and photoelectron spectroscopies. It is unfortunate that the number of systems where reasonably complete characterisation has been carried out are small: on occasions undue faith seems to have been placed in the ideality of synthetic strategies. Whilst many of the techniques are not routinely available to electrochemists, the fruits of collaborative studies are well worthwhile and have frequently provided explanations to unexpected behaviour in electrochemical experiments. It is clear, however, that procedures do exist for the synthesis and detailed characterisation of polymer films on electrodes, opening the way to their use for mediated charge transfer to solution species and their exploitation in a range of applications discussed in the following chapter.

REFERENCES

1. R. W. Murray, *Acc. Chem. Res.*, 1980, **13**, 135.
2. W. J. Albery and A. R. Hillman, *Roy. Soc. Chem. Ann. Rep. C*, 1981, **78**, 377.
3. R. W. Murray, in *Electroanalytical Chemistry*, Vol. 13, ed. A. J. Bard, Marcel Dekker, New York, 1984, p. 191.
4. C. P. Andrieux and J. M. Saveant, *J. Electroanal. Chem.*, 1978, **93**, 163.
5. P. Denisevich, K. W. Willman and R. W. Murray, *J. Amer. Chem. Soc.*, 1981, **103**, 4727.
6. M. Umana, D. R. Rolison, R. Nowak, P. Daum and R. W. Murray, *Surf. Sci.*, 1980, **101**, 295.
7. M. Umana, P. Denisevich, D. R. Rolison, S. Nakahama and R. W. Murray, *Anal. Chem.*, 1981, **53**, 1170.
8. A. H. Schroeder, F. B. Kaufman, V. Patel and E. M. Engler, *J. Electroanal. Chem.*, 1980, **113**, 193.
9. D. R. Rolison, M. Umana, P. Burgmayer and R. W. Murray, *Inorg. Chem.*, 1981, **20**, 2996.
10. A. Merz and A. J. Bard, *J. Amer. Chem. Soc.*, 1978, **100**, 3222.
11. P. J. Peerce and A. J. Bard, *J. Electroanal. Chem.*, 1980, **108**, 121.
12. P. J. Peerce and A. J. Bard, *J. Electroanal. Chem.*, 1980, **112**, 97.

13. P. J. Peerce and A. J. Bard, *J. Electroanal. Chem.*, 1980, **114**, 89.
14. J. Leddy and A. J. Bard, *J. Electroanal. Chem.*, 1983, **153**, 223.
15. J. M. Calvert and T. J. Meyer, *Inorg. Chem.*, 1981, **20**, 27.
16. J. M. Calvert and T. J. Meyer, *Inorg. Chem.*, 1982, **21**, 3978.
17. G. J. Samuels and T. J. Meyer, *J. Amer. Chem. Soc.*, 1981, **103**, 307.
18. J. M. Calvert, J. V. Caspar, R. A. Binstead, T. D. Westmoreland and T. J. Meyer, *J. Amer. Chem. Soc.*, 1982, **104**, 6620.
19. T. D. Westmoreland, J. M. Calvert, R. W. Murray and T. J. Meyer, *J. Chem. Soc. Chem. Comm.*, 1983, 66.
20. O. Haas and J. G. Vos, *J. Electroanal. Chem.*, 1980, **113**, 139.
21. O. Haas, M. Kriens and J. G. Vos, *J. Amer. Chem. Soc.*, 1981, **103**, 1318.
22. O. Haas, N. Müller and H. Gerischer, *Electrochim. Acta*, 1982, **27**, 991.
23. H. Kido and C. H. Langford, *J. Chem. Soc. Chem. Comm.*, 1983, 350.
24. J. A. Chambers and R. P. Buck, *J. Electroanal. Chem.*, 1982, **140**, 173.
25. J. A. Chambers and R. P. Buck, *J. Electroanal. Chem.*, 1984, **163**, 297.
26. J. A. Cox and P. J. Kulesza, *J. Electroanal. Chem.*, 1984, **175**, 105.
27. T. Kikuchi, H. Sasaki and S. Toshima, *Chem. Lett.*, 1980, 5.
28. J. B. Kerr and L. L. Miller, *J. Electroanal. Chem.*, 1979, **101**, 263.
29. J. M. Kerr and L. L. Miller, *J. Amer. Chem. Soc.*, 1980, **102**, 3383.
30. W. J. Albery, R. G. Compton and C. C. Jones, *J. Amer. Chem. Soc.*, 1984, **106**, 469.
31. M. R. Van de Mark and L. L. Miller, *J. Amer. Chem. Soc.*, 1978, **100**, 3223.
32. N. Oyama, S. Yamaguchi, M. Kaneko and A. Yamada, *J. Electroanal. Chem.*, 1982, **139**, 215.
33. K. Rajeshwar, M. Kaneka and A. Yamada, *J. Electrochem. Soc.*, 1983, **130**, 38.
34. J. Q. Chambers, F. B. Kaufman and K. H. Nichols, *J. Electroanal. Chem.*, 1982, **142**, 277.
35. G. Inzelt, J. Q. Chambers and F. B. Kaufman, *J. Electroanal. Chem.*, 1983, **159**, 443.
36. A. H. Schroeder and F. B. Kaufman, *J. Electroanal. Chem.*, 1980, **113**, 209.
37. F. B. Kaufman and E. M. Engler, *J. Amer. Chem. Soc.*, 1979, **101**, 547.
38. F. B. Kaufman, A. H. Schroeder, E. M. Engler and V. V. Patel, *Appl. Phys. Lett.*, 1980, **36**, 422.
39. F. B. Kaufman, A. H. Schroeder, V. V. Patel and K. H. Nichols, *J. Electroanal. Chem.*, 1982, **132**, 151.
40. F. B. Kaufman, A. H. Schroeder, E. M. Engels, S. R. Kramer and J. Q. Chambers, *J. Amer. Chem. Soc.*, 1980, **102**, 483.
41. L. L. Miller, A. N. K. Lau and E. K. Miller, *J. Amer. Chem. Soc.*, 1982, **104**, 5242.
42. A. N. K. Lau and L. L. Miller, *J. Amer. Chem. Soc.*, 1983, **105**, 5271.
43. A. N. K. Lau, L. L. Miller and B. Zinger, *J. Amer. Chem. Soc.*, 1983, **105**, 5278.
44. L. L. Miller, B. Zinger and C. Degrand, *J. Electroanal. Chem.*, 1984, **178**, 87.

45. C. Degrand, L. Roullier, L. L. Miller and B. Zinger, *J. Electroanal. Chem.*, 1984, **178**, 101.
46. P. Burgmayer and R. W. Murray, *J. Electroanal. Chem.*, 1982, **135**, 335.
47. K. M. O'Connell, E. Waldner, L. Roullier and E. Laviron, *J. Electroanal. Chem.*, 1984, **162**, 77.
48. T. Ohsaka, H. Yamamoto, M. Kaneka, A. Yamada, M. Nakamura, S. Nakamura and N. Oyama, *Bull. Chem. Soc. Jap.*, 1984, **57**, 1844.
49. B. L. Funt and P. M. Hoang, *J. Electroanal. Chem.*, 1983, **154**, 229.
50. R. S. Lawton and A. M. Yacynych, *Anal. Chim. Acta*, 1984, **160**, 149.
51. S. Abe and T. Nonaka, *Chem. Lett.*, 1983, 1541.
52. M. Fukui, A. Kitani, C. Degrand and L. L. Miller, *J. Amer. Chem. Soc.*, 1982, **104**, 28.
53. R. Tamilarasan and P. Natajaran, *Nature (London)*, 1981, **292**, 224.
54. C. Degrand and L. L. Miller, *J. Amer. Chem. Soc.*, 1980, **102**, 5728.
55. H. Yoneyama, Y. Murao and H. Tamura, *J. Electroanal. Chem.*, 1980, **108**, 87.
56. D. L. Dubois and J. A. Turner, *J. Amer. Chem. Soc.*, 1982, **104**, 4989.
57. A. Bettelheim, R. J. H. Chan and T. Kuwana, *J. Electroanal. Chem.*, 1980, **110**, 93.
58. F. J. Davis, H. Block and R. G. Compton, *J. Chem. Soc. Chem. Comm.*, 1984, 890.
59. Y. Morishima, M. Isono, Y. Itoh and S.-I. Nozakura, *Chem. Lett.*, 1981, 1149.
60. C. Degrand and E. Laviron, *J. Electroanal. Chem.*, 1981, **117**, 283.
61. C. Degrand and L. L. Miller, *J. Electroanal. Chem.*, 1981, **117**, 267.
62. C. Degrand and L. L. Miller, *J. Electroanal. Chem.*, 1982, **132**, 163.
63. C. Degrand, *J. Electroanal. Chem.*, 1984, **169**, 259.
64. Y. Morishima, M. Isono and S.-I. Nozakura, *Chem. Lett.*, 1982, 1427.
65. L. Roullier, E. Waldner and E. Laviron, *J. Electroanal. Chem.*, 1982, **139**, 199.
66. V. J. Razumas, A. S. Samulius and J. J. Kulys, *J. Electroanal. Chem.*, 1984, **164**, 195.
67. L. L. Miller and M. R. Van de Mark, *J. Amer. Chem. Soc.*, 1978, **100**, 639.
68. L. L. Miller and M. R. Van de Mark, *J. Electroanal. Chem.*, 1978, **88**, 437.
69. N. Oyama, N. Oki, H. Ohno, Y. Ohnuki, H. Matsuda and E. T. Tsuchida, *J. Phys. Chem.*, 1983, **87**, 3642.
70. H. Akahoshi, S. Toshima and K. Itaya, *J. Phys. Chem.*, 1981, **85**, 818.
71. A. Factor and T. O. Rouse, *J. Electrochem. Soc.*, 1980, **127**, 1313.
72. P. Martigny and F. C. Anson, *J. Electroanal. Chem.*, 1982, **139**, 383.
73. P. Jander, J. Weber and L. Kavan, *J. Electroanal. Chem.*, 1984, **180**, 109.
74. R. W. Day, G. Inzelt, J. F. Kinstle and J. Q. Chambers, *J. Amer. Chem. Soc.*, 1982, **104**, 6804.
75. G. Inzelt, R. W. Day, J. F. Kinstle and J. Q. Chambers, *J. Phys. Chem.*, 1983, **87**, 4592.

76. G. Inzelt, R. W. Day, J. F. Kinstle and J. Q. Chambers, *J. Electroanal. Chem.*, 1984, **161,** 147.
77. G. Inzelt, J. Q. Chambers, J. F. Kinstle and R. W. Day, *J. Amer. Chem. Soc.*, 1984, **106,** 3396.
78. G. Inzelt, J. Q. Chambers, J. F. Kinstle, R. W. Day and M. A. Lange, *Anal. Chem.*, 1984, **56,** 301.
79. H. Behret, H. Binder, G. Sandstede and G. G. Scherer, *J. Electroanal. Chem.*, 1981, **117,** 29.
80. L. Kreja and R. Dabrowski, *J. Power Sources*, 1981, **6,** 35.
81. R. Larsson, L. Y. Johansson and L. Jönssen, *J. Appl. Electrochem.*, 1981, **11,** 489.
82. J. Blomquist, U. Helgeson, L. C. Moberg, L. Y. Johansson and R. Larsson, *Electrochim. Acta*, 1982, **27,** 1453.
83. S. Abe, T. Nonaka and T. Fuchigami, *J. Amer. Chem. Soc.*, 1983, **105,** 3630.
84. S. Abe, T. Fuchigami and T. Nonaka, *Chem. Lett.*, 1983, 1033.
85. T. Komori and T. Nonaka, *J. Amer. Chem. Soc.*, 1983, **105,** 5690.
86. T. Komori and T. Nonaka, *Chem. Lett.*, 1984, 509.
87. T. Komori and T. Nonaka, *J. Amer. Chem. Soc.*, 1984, **106,** 2656.
88. D. L. Dubois, *Inorg. Chem.*, 1984, **23,** 2047.
89. N. S. Scott, N. Oyama and F. C. Anson, *J. Electroanal. Chem.*, 1980, **110,** 303.
90. N. Oyama and F. C. Anson, *J. Amer. Chem. Soc.*, 1979, **101,** 739.
91. N. Oyama and F. C. Anson, *J. Amer. Chem. Soc.*, 1979, **101,** 3450.
92. K. Shigehara, N. Oyama and F. C. Anson, *J. Amer. Chem. Soc.*, 1981, **103,** 2552.
93. N. Oyama, T. Ohsaka, M. Kaneko, K. Sato and H. Matsuda, *J. Amer. Chem. Soc.*, 1983, **105,** 6003.
94. R. Jasinski, *J. Electrochem. Soc.*, 1983, **130,** 834.
95. K. Itaya, H. Akahoshi and S. Toshima, *J. Electrochem. Soc.*, 1982, **129,** 762.
96. N. Oyama and F. C. Anson, *J. Electrochem. Soc.*, 1980, **127,** 247.
97. N. Oyama, S. Yamaguchi, Y. Nishiki, K. Tokuda, H. Matsuda and F. C. Anson, *J. Electroanal. Chem.*, 1982, **139,** 371.
98. K. Shigehara, N. Oyama and F. C. Anson, *Inorg. Chem.*, 1981, **20,** 518.
99. H.-R. Zumbrunnen and F. C. Anson, *J. Electroanal. Chem.*, 1983, **152,** 111.
100. K. Sato, S. Yamaguchi, H. Matsuda, T. Ohsaka and N. Oyama, *Bull. Chem. Soc. Jap.*, 1983, **56,** 2004.
101. N. Oyama, K. Sato and H. Matsuda, *J. Electroanal. Chem.*, 1980, **115,** 149.
102. N. Oyama and F. C. Anson, *Anal. Chem.*, 1980, **52,** 1192.
103. F. C. Anson, J. M. Saveant and K. Shigehara, *J. Electroanal. Chem.*, 1983, **145,** 423.
104. J. A. Cox and P. J. Kulesza, *Anal. Chim. Acta*, 1983, **154,** 71.
105. J. A. Cox and P. J. Kulesza, *J. Electroanal. Chem.*, 1983, **159,** 337.
106. P. V. Kamat and M. A. Fox, *J. Electroanal. Chem.*, 1983, **159,** 49.

107. P. V. Kamat, M. A. Fox and A. J. Fatiadi, *J. Amer. Chem. Soc.*, 1984, **106,** 1191.
108. J. Facci and R. W. Murray, *J. Electroanal. Chem.*, 1981, **124,** 339.
109. J. Facci and R. W. Murray, *J. Phys. Chem.*, 1981, **85,** 2870.
110. N. Oyama, J. Shimomura, K. Shigehara and F. C. Anson, *J. Electroanal. Chem.*, 1980, **112,** 271.
111. H. Braun, W. Storck and K. Doblhofer, *J. Electrochem. Soc.*, 1983, **130,** 807.
112. G. J. Samuels and T. J. Meyer, *J. Amer. Chem. Soc.*, 1981, **103,** 307.
113. H. Braun, F. Decker, K. Doblhofer and H. Sotobayashi, *Ber. Bunsenges. Phys. Chem.*, 1984, **88,** 345.
114. F. C. Anson, J. M. Saveant and K. Shigehara, *J. Amer. Chem. Soc.*, 1983, **105,** 1096.
115. F. C. Anson, T. Ohsaka and J. M. Saveant, *J. Amer. Chem. Soc.*, 1983, **105,** 4883.
116. F. C. Anson, T. Ohsaka and J. M. Saveant, *J. Phys. Chem.*, 1983, **87,** 640.
117. D. A. Buttry and F. C. Anson, *J. Amer. Chem. Soc.*, 1982, **104,** 4824.
118. I. Rubinstein and A. J. Bard, *J. Amer. Chem. Soc.*, 1980, **102,** 6641.
119. I. Rubinstein and A. J. Bard, *J. Amer. Chem. Soc.*, 1981, **103,** 5007.
120. C. R. Martin and K. A. Dollard, *J. Electroanal. Chem.*, 1983, **159,** 127.
121. H. S. White, J. Leddy and A. J. Bard, *J. Amer. Chem. Soc.*, 1982, **104,** 4811.
122. I. Rubinstein, *Anal. Chem.*, 1984, **56,** 1135.
123. D. A. Buttry and F. C. Anson, *J. Amer. Chem. Soc.*, 1983, **105,** 685.
124. F. C. Anson, Y.-M. Tsou and J. M. Saveant, *J. Electroanal. Chem.*, 1984, **178,** 113.
125. Y.-M. Tsou and F. C. Anson, *J. Electrochem. Soc.*, 1984, **131,** 595.
126. D. A. Buttry, J. M. Saveant and F. C. Anson, *J. Phys. Chem.*, 1984, **88,** 3086.
127. I. Rubinstein, *J. Electroanal. Chem.*, 1984, **176,** 359.
128. K. Shigehara, E. Tsuchida and F. C. Anson, *J. Electroanal. Chem.*, 1984, **175,** 291.
129. D. A. Buttry and F. C. Anson, *J. Amer. Chem. Soc.*, 1984, **106,** 59.
130. T. P. Henning, H. S. White and A. J. Bard, *J. Amer. Chem. Soc.*, 1981, **103,** 3937.
131. T. P. Henning, H. S. White and A. J. Bard, *J. Amer. Chem. Soc.*, 1982, **104,** 5862.
132. M. N. Szentirmay and C. R. Martin, *Anal. Chem.*, 1984, **56,** 1898.
133. N. Oyama, T. Ohsaka, K. Sato and H. Yamamoto, *Anal. Chem.*, 1983, **55,** 1429.
134. M. Majda and L. R. Faulkner, *J. Electroanal. Chem.*, 1982, **137,** 149.
135. M. Majda and L. R. Faulkner, *J. Electroanal. Chem.*, 1984, **169,** 77.
136. C. D. Ellis, J. A. Gilbert, W. R. Murphy and T. J. Meyer, *J. Amer. Chem. Soc.*, 1983, **105,** 4842.
137. J. Wang, B. Greene and C. Morgan, *Anal. Chim. Acta*, 1984, **158,** 15.
138. J. R. Schneider and R. W. Murray, *Anal. Chem.*, 1982, **54,** 1508.
139. C. D. Ellis and T. J. Meyer, *Inorg. Chem.*, 1984, **23,** 1748.

140. N. Oyama and F. C. Anson, *J. Electrochem. Soc.*, 1980, **127**, 249.
141. R. J. Mortimer and F. C. Anson, *J. Electroanal. Chem.*, 1982, **138**, 325.
142. H. D. Abruna and A. J. Bard, *J. Amer. Chem. Soc.*, 1981, **103**, 6898.
143. P. Ghosh and A. J. Bard, *J. Amer. Chem. Soc.*, 1983, **105**, 5691.
144. A. Yamagishi and A. Aramata, *J. Chem. Soc. Chem. Comm.*, 1984, 452.
145. P. V. Kamat and M. A. Fox, *J. Electrochem. Soc.*, 1984, **131**, 1032.
146. S. Nakahama and R. W. Murray, *J. Electroanal. Chem.*, 1983, **158**, 303.
147. E. S. Decestro, D. A. Smith, J. E. Mark and W. R. Heineman, *J. Electroanal. Chem.*, 1982, **138**, 197.
148. K. Itaya and A. J. Bard, *Anal. Chem.*, 1978, **50**, 1487.
149. J. A. Bruce, T. Murahashi and M. S. Wrighton, *J. Phys. Chem.*, 1982, **86**, 1552.
150. C. J. Stalder, S. Chao and M. S. Wrighton, *J. Amer. Chem. Soc.*, 1984, **106**, 3673.
151. D. C. Bookbinder and M. S. Wrighton, *J. Electrochem. Soc.*, 1983, **130**, 1080.
152. N. S. Lewis and M. S. Wrighton, *Science*, 1981, **211**, 944.
153. N. S. Lewis and M. S. Wrighton, *J. Phys. Chem.*, 1984, **88**, 2009.
154. J. A. Bruce and M. S. Wrighton, *J. Amer. Chem. Soc.*, 1982, **104**, 74.
155. D. C. Bookbinder and M. S. Wrighton, *J. Amer. Chem. Soc.*, 1980, **102**, 5123.
156. D. C. Bookbinder, J. A. Bruce, R. N. Dominey, N. S. Lewis and M. S. Wrighton, *Proc. Nat. Acad. Sci. USA*, 1980, **77**, 6280.
157. R. N. Dominey, N. S. Lewis, J. A. Bruce, D. C. Bookbinder and M. S. Wrighton, *J. Amer. Chem. Soc.*, 1982, **104**, 467.
158. D. J. Harrison and M. S. Wrighton, *J. Phys. Chem.*, 1984, **88**, 3932.
159. R. N. Dominey, *J. Electrochem. Soc.*, 1982, **129**, 300C.
160. R. N. Dominey, T. J. Lewis and M. S. Wrighton, *J. Phys. Chem.*, 1983, **87**, 5345.
161. K. W. Willman and R. W. Murray, *J. Electroanal. Chem.*, 1982, **133**, 211.
162. P. Denisevich, K. W. Willman and R. W. Murray, *J. Amer. Chem. Soc.*, 1981, **103**, 4729.
163. M. S. Wrighton, M. C. Palazzotto, A. B. Bocarsly, J. M. Bolts, A. B. Fischer and L. Nadjo, *J. Amer. Chem. Soc.*, 1978, **100**, 7264.
164. A. B. Fischer, M. S. Wrighton, M. Umana and R. W. Murray, *J. Amer. Chem. Soc.*, 1979, **101**, 3442.
165. M. S. Wrighton, *Acc. Chem. Res.*, 1979, **12**, 303.
166. M. S. Wrighton, A. B. Bocarsly, J. M. Bolts, M. G. Bradley, A. B. Fischer, N. S. Lewis, M. C. Palazzotto and E. G. Walton, *ACS Adv. Chem. Ser. I*, 1980, **184**, 269.
167. J. M. Bolts, A. B. Bocarsly, M. C. Palazzotto, E. Walton, N. S. Lewis and M. S. Wrighton, *J. Amer. Chem. Soc.*, 1979, **101**, 1378.
168. N. S. Lewis, A. B. Bocarsly and M. S. Wrighton, *J. Phys. Chem.*, 1980, **84**, 2033.
169. A. B. Bocarsly, E. W. Walton and M. S. Wrighton, *J. Amer. Chem. Soc.*, 1980, **102**, 3390.
170. N. S. Lewis and M. S. Wrighton, *ACS Symp. Ser.*, 1981, **146**, 37.

171. J. A. Bruce and M. S. Wrighton, *J. Electroanal. Chem.*, 1981, **122**, 93.
172. J. M. Bolts and M. S. Wrighton, *J. Amer. Chem. Soc.*, 1978, **100**, 5257.
173. J. M. Bolts and M. S. Wrighton, *J. Amer. Chem. Soc.*, 1979, **101**, 6179.
174. M. S. Wrighton, R. G. Austin, A. B. Bocarsly, J. M. Bolts, O. Haas, K. D. Legg, L. Nadjo and M. C. Palazzotto, *J. Electroanal. Chem.*, 1978, **87**, 429.
175. S. Chao, J. L. Robbins and M. S. Wrighton, *J. Amer. Chem. Soc.*, 1983, **105**, 181.
176. A. B. Bocarsly, E. G. Walton, M. G. Bradley and M. S. Wrighton, *J. Electroanal. Chem.*, 1979, **100**, 283.
177. J. R. Lenhard and R. W. Murray, *J. Amer. Chem. Soc.*, 1978, **100**, 7870.
178. K.-N. Kuo and R. W. Murray, *J. Electroanal. Chem.*, 1982, **131**, 37.
179. G. S. Calabrese, R. M. Buchanan and M. S. Wrighton, *J. Amer. Chem. Soc.*, 1983, **105**, 5594.
180. K. W. Willman, R. D. Rocklin, R. Nowak, K.-N. Kuo, F. A. Schultz and R. W. Murray, *J. Amer. Chem. Soc.*, 1980, **102**, 7629.
181. P. K. Ghosh and T. G. Spiro, *J. Amer. Chem. Soc.*, 1980, **102**, 5543.
182. P. K. Ghosh and T. G. Spiro, *J. Electrochem. Soc.*, 1981, **128**, 1281.
183. A. B. Bocarsly, S. A. Galvin and S. Sinha, *J. Electrochem. Soc.*, 1983, **130**, 1319.
184. R. M. Buchanan, G. S. Calabrese, T. J. Sobieralski and M. S. Wrighton, *J. Electroanal. Chem.*, 1983, **153**, 129.
185. D. J. Harrison, K. A. Daube and M. S. Wrighton, *J. Electroanal. Chem.*, 1984, **163**, 93.
186. H. D. Abruna, P. Denisevich, M. Umana, T. J. Meyer and R. W. Murray, *J. Amer. Chem. Soc.*, 1981, **103**, 1.
187. T. Ikeda, C. R. Leidner and R. W. Murray, *J. Amer. Chem. Soc.*, 1981, **103**, 7422.
188. P. Denisevich, H. D. Abruna, C. R. Leidner, T. J. Meyer and R. W. Murray, *Inorg. Chem.*, 1982, **21**, 2153.
189. H. D. Abruna and A. J. Bard, *J. Amer. Chem. Soc.*, 1982, **104**, 2641.
190. T. Ikeda, R. Schmehl, P. Denisevich, K. Willman and R. W. Murray, *J. Amer. Chem. Soc.*, 1982, **104**, 2683.
191. T. Ikeda, C. R. Leidner and R. W. Murray, *J. Electroanal. Chem.*, 1982, **138**, 343.
192. P. G. Pickup, W. Kutner, C. R. Leidner and R. W. Murray, *J. Amer. Chem. Soc.*, 1984, **106**, 1991.
193. P. G. Pickup, C. R. Leidner, P. Denisevich and R. W. Murray, *J. Electroanal. Chem.*, 1984, **164**, 39.
194. C. R. Leidner, P. Denisevich, K. W. Willman and R. W. Murray, *J. Electroanal. Chem.*, 1984, **164**, 63.
195. A. G. Ewing, B. J. Feldman and R. W. Murray, *J. Electroanal. Chem.*, 1984, **172**, 145.
196. P. G. Pickup and R. W. Murray, *J. Amer. Chem. Soc.*, 1982, **105**, 4510.
197. P. K. Ghosh and T. G. Spiro, *J. Electrochem. Soc.*, 1981, **128**, 1281.
198. P. G. Pickup, K. N. Kuo and R. W. Murray, *J. Electrochem. Soc.*, 1983, **130**, 2205.

199. P. G. Pickup and R. A. Osteryoung, *J. Electrochem. Soc.*, 1983, **130**, 1965.
200. C. D. Ellis, W. R. Murphy and T. J. Meyer, *J. Amer. Chem. Soc.*, 1981, **103**, 7480.
201. L. D. Margerum, T. J. Meyer and R. W. Murray, *J. Electroanal. Chem.*, 1983, **149**, 279.
202. P. G. Pickup and R. W. Murray, *J. Electrochem. Soc.*, 1984, **131**, 833.
203. C. R. Leidner and R. W. Murray, *J. Amer. Chem. Soc.*, 1984, **106**, 1606.
204. J. S. Facci, R. H. Schmehl and R. W. Murray, *J. Amer. Chem. Soc.*, 1982, **104**, 4959.
205. R. H. Schmehl and R. W. Murray, *J. Electroanal. Chem.*, 1983, **152**, 97.
206. B. R. Shaw, G. P. Haight and L. R. Faulkner, *J. Electroanal. Chem.*, 1982, **140**, 147.
207. P. C. Lacaze, J. E. Dubois, A. Desbene-Monvernay, D. L. Desbene, J. J. Basselier and D. Richard, *J. Electroanal. Chem.*, 1983, **147**, 107.
208. A. Desbene-Monvernay, P. C. Lacaze, J. E. Dubois and P. L. Desbene, *J. Electroanal. Chem.*, 1983, **152**, 87.
209. H. Kanega, Y. Shirota and H. Mikawa, *J. Chem. Soc. Chem. Comm.*, 1984, 158.
210. Y. Shirota, N. Noma, H. Kanega and H. Mikawa, *J. Chem. Soc. Chem. Comm.*, 1984, 470.
211. K. A. Macor and T. G. Spiro, *J. Amer. Chem. Soc.*, 1983, **105**, 5601.
212. K. A. Macor and T. G. Spiro, *J. Electroanal. Chem.*, 1984, **163**, 223.
213. H. O. Finklea and R. S. Vithanage, *J. Electroanal. Chem.*, 1984, **161**, 283.
214. J.-C. Moutet, *J. Electroanal. Chem.*, 1984, **161**, 181.
215. P. Cerrai, G. Guerra, M. Tricoli and L. Nucci, *Eur. Polym. J.*, 1980, **16**, 867.
216. G. Mengoli, A. Martina, F. Furlanetto and M. M. Musiani, *Chim. Ind. (Milan)*, 1980, **62**, 16.
217. Y. Bouizem, F. Chao, M. Costa, A. Tadjeddine and G. Lecayon, *J. Electroanal. Chem.*, 1984, **172**, 101.
218. T. M. Abrantes, L. M. Castillo, M. Fleischmann, I. R. Hill, L. M. Peter, G. Mengoli and G. Zotti, *J. Electroanal. Chem.*, 1984, **177**, 129.
219. Y. Ohnuki, H. Matsuda, T. Ohsaka and N. Oyama, *J. Electroanal. Chem.*, 1983, **158**, 55.
220. G. Cheek, C. P. Wales and R. J. Nowak, *Anal. Chem.*, 1983, **55**, 380.
221. M. C. Pham, P.-C. Lacaze and J.-E. Dubois, *J. Electroanal. Chem.*, 1978, **86**, 147.
222. P. Mourcel, M. C. Pham, P.-C. Lacaze and J.-E. Dubois, *J. Electroanal. Chem.*, 1983, **145**, 467.
223. M.-C. Pham, J.-E. Dubois and P.-C. Lacaze, *J. Electroanal. Chem.*, 1979, **99**, 331.
224. J.-E. Dubois, G. Tourillon, M.-C. Pham and P.-C. Lacaze, *Thin Solid Films*, 1980, **69**, 141.
225. J.-E. Dubois, P.-C. Lacaze and M.-C. Pham, *J. Electroanal. Chem.*, 1981, **117**, 233.

226. M.-C. Pham, P.-C. Lacaze and J.-E. Dubois, *J. Electrochem. Soc.*, 1984, **131,** 777.
227. F. Bruno, M.-C. Pham and J.-E. Dubois, *Electrochim. Acta,* 1977, **22,** 451.
228. M. Delamar, M. Chehimi and J.-E. Dubois, *J. Electroanal. Chem.*, 1984, **169,** 145.
229. M. M. Musiani, G. Mengoli, B. Pelli and C. Folonari, *Makromol. Chem.*, 1982, **183,** 1869.
230. M. Lapkoswki, J. Zak and J. W. Strojek, *J. Electroanal. Chem.*, 1983, **145,** 173.
231. M.-C. Pham, A. Hachemi and J.-E. Dubois, *J. Electroanal. Chem.*, 1984, **161,** 199.
232. J. M. Bauldreay and M. D. Archer, *Electrochim. Acta,* 1983, **28,** 1515.
233. O. Haas and H.-R. Zumbrunnen, *Helv. Chim. Acta,* 1981, **64,** 854.
234. A. K. Mesmaeker and R. Dewitt, *Electrochim. Acta,* 1981, **26,** 297.
235. M. Breitenbach and K.-H. Heckner, *J. Electroanal. Chem.*, 1973, **343,** 267.
236. A. F. Diaz and J. A. Logan, *J. Electroanal. Chem.*, 1980, **111,** 111.
237. R. Noufi, A. J. Nozik, J. White and L. F. Warren, *J. Electrochem. Soc.*, 1982, **129,** 2261.
238. N. Oyama, Y. Ohnuki, K. Chiba and T. Ohsaka, *Chem. Lett.*, 1983, 1759.
239. T. Kobayashi, H. Yoneyama and H. Tamura, *J. Electroanal. Chem.*, 1984, **161,** 419.
240. T. Kobayashi, H. Yoneyama and H. Tamura, *J. Electroanal. Chem.*, 1984, **177,** 281.
241. T. Kobayashi, H. Yoneyama and H. Tamura, *J. Electroanal. Chem.*, 1984, **177,** 293.
242. A. Kitani, J. Izumi, Y. Hiromoto and K. Sasaki, *Bull. Chem. Soc. Jap.*, 1984, **57,** 2254.
243. W. R. Heineman, H. J. Wieck and A. M. Yacynych, *Anal. Chem.*, 1980, **52,** 345.
244. H. S. White, H. D. Abruna and A. J. Bard, *J. Electrochem. Soc.*, 1982, **129,** 265.
245. A. Volkov, G. Tourillon, P.-C. Lacaze and J.-E. Dubois, *J. Electroanal. Chem.*, 1980, **115,** 279.
246. G. Mengoli, M. M. Musiani, S. Daolio and C. Folonari, *Makromol. Chem.*, 1980, **181,** 1909.
247. G. Mengoli, M.-T. Munari and C. Folonari, *J. Electroanal. Chem.*, 1981, **124,** 237.
248. A. Kitani, J. Yano and K. Sasaki, *Chem. Lett.*, 1984, 1565.
249. C. D. Ellis, L. D. Margerum, R. W. Murray and T. J. Meyer, *Inorg. Chem.*, 1983, **22,** 1283.
250. W. J. Albery, W. R. Bowen, F. S. Fisher, A. W. Foulds, K. T. Hall, A. R. Hillman, R. G. Egdell and A. F. Orchard, *J. Electroanal. Chem.*, 1980, **111,** 295.
251. W. J. Albery, A. W. Foulds, K. J. Hall and A. R. Hillman, *J. Electrochem. Soc.*, 1980, **127,** 654.

252. W. J. Albery, A. W. Foulds, K. J. Hall, A. R. Hillman, R. G. Egdell and A. F. Orchard, *Nature (London)*, 1979, **282,** 793.
253. W. G. Albery, M. G. Boutelle, P. J. Colby and A. R. Hillman, *J. Electroanal. Chem.*, 1982, **133,** 135.
254. W. J. Albery, A. R. Hillman, A. R. Egdell and H. Nutton, *J. Chem. Soc., Faraday I*, 1984, **80,** 111.
255. W. J. Albery, A. R. Hillman, R. E. Hester and K. Hutchinson, *J. Chem. Soc., Faraday I*, 1984, **80,** 2053.
256. G. Mengoli, M. M. Musiani and G. Zotti, *J. Electroanal. Chem.*, 1984, **175,** 93.
257. P. K. Ghosh and A. J. Bard, *J. Electroanal. Chem.*, 1984, **169,** 113.
258. A. F. Diaz, K. K. Kanazawa and G. P. Gardini, *J. Chem. Soc., Chem. Comm.*, 1979, 635.
259. K. K. Kanazawa, A. F. Diaz, R. H. Geiss, W. D. Gill, J. F. Kwak, J. A. Logan, J. F. Rabolt and G. B. Street, *J. Chem. Soc. Chem. Comm.*, 1979, 854.
260. A. F. Diaz and J. I. Castillo, *J. Chem. Soc. Chem. Comm.*, 1980, 397.
261. A. F. Diaz, W.-Y. Lee, J. A. Logan and D. G. Green, *J. Electroanal. Chem.*, 1980, **108,** 377.
262. A. F. Diaz, J. M. V. Vallejo and A. M. Duran, *IBM J. Res. Dev.*, 1981, **25,** 42.
263. A. F. Diaz, J. Crowley, J. Bargon, G. P. Gardini and J. B. Torrance, *J. Electroanal. Chem.*, 1981, **121,** 355.
264. A. F. Diaz, J. I. Castillo, J. A. Logan and W.-Y. Lee, *J. Electroanal. Chem.*, 1981, **129,** 115.
265. A. F. Diaz, *Chem. Scripta*, 1981, **17,** 145.
266. R. Noufi, D. Tench and L. F. Warren, *J. Electrochem. Soc.*, 1981, **128,** 2596.
267. W. Watanabe, M. Tanaka and J. Tanaka, *Bull. Chem. Soc. Jap.*, 1981, **54,** 2278.
268. R. A. Bull, F. R. F. Fan and A. J. Bard, *J. Electrochem. Soc.*, 1982, **129,** 1009.
269. J. Prejza, I. Lundström and T. Skotheim, *J. Electrochem. Soc.*, 1982, **129,** 1685.
270. G. Tourillon and F. Garnier, *J. Electroanal. Chem.*, 1982, **135,** 173.
271. E. M. Genies, G. Bidan and A. F. Diaz, *J. Electroanal. Chem.*, 1983, **149,** 101.
272. G. Tourillon and F. Garnier, *J. Phys. Chem.*, 1983, **87,** 2289.
273. A. S. N. Murthy and K. S. Reddy, *Electrochim. Acta*, 1983, **28,** 473.
274. O. Ikeda, K. Okabayashi and H. Tamura, *Chem. Lett.*, 1983, 1821.
275. T. Inoue and T. Yamase, *Bull. Chem. Soc. Jap.*, 1983, **56,** 985.
276. S. Asavapiriyanont, G. K. Chandler, G. A. Gunawardena and D. Pletcher, *J. Electroanal. Chem.*, 1984, **177,** 229.
277. S. Kuwabata, H. Yoneyama and H. Tamura, *Bull. Chem. Soc. Jap.*, 1984, **57,** 2247.
278. P. Burgmayer and R. W. Murray, *J. Amer. Chem. Soc.*, 1982, **104,** 6139.
279. P. Burgmayer and R. W. Murray, *J. Electroanal. Chem.*, 1983, **147,** 339.

280. K. Okabayashi, O. Ikeda and H. Tamura, *J. Chem. Soc. Chem. Comm.*, 1983, 684.
281. P. Burgmayer and R. W. Murray, *J. Phys. Chem.*, 1984, **88**, 2515.
282. H. S. White, G. P. Kittlesen and M. S. Wrighton, *J. Amer. Chem. Soc.*, 1984, **106**, 5375.
283. K. Santhanam and R. N. O'Brien, *J. Electroanal. Chem.*, 1984, **160**, 377.
284. R. N. O'Brien and K. S. V. Santhanam, *J. Electrochem. Soc.*, 1983, **130**, 1114.
285. R. A. Bull, F.-R. Fan and A. J. Bard, *J. Electrochem. Soc.*, 1983, **130**, 1636.
286. R. A. Bull, F.-R. Fan and A. J. Bard, *J. Electrochem. Soc.*, 1984, **131**, 687.
287. J. J. Ritsko, J. Funk and G. Crecelius, *Sol. St. Comm.*, 1983, **46**, 477.
288. E. M. Genies and J. M. Pernant, *Syn. Metals*, 1984/5, **10**, 117.
289. G. Cooper, R. Noufi, A. J. Frank and A. J. Nozik, *Nature (London)*, 1982, **295**, 578.
290. F.-R. F. Fan, R. L. Wheeler, A. J. Bard and R. N. Noufi, *J. Electrochem. Soc.*, 1981, **128**, 2042.
291. R. Noufi, A. J. Frank and A. J. Nozik, *J. Amer. Chem. Soc.*, 1981, **103**, 1849.
292. R. E. Malpas and B. Rushby, *J. Electroanal. Chem.*, 1983, **157**, 387.
293. R. A. Simon, A. J. Ricco and M. S. Wrighton, *J. Amer. Chem. Soc.*, 1982, **104**, 2031.
294. T. Skotheim, I. Lundström and J. Prejza, *J. Electrochem. Soc.*, 1981, **128**, 1625.
295. A. J. Frank and K. Honda, *J. Phys. Chem.*, 1982, **86**, 1933.
296. R. Noufi, T. Tench and L. F. Warren, *J. Electrochem. Soc.*, 1980, **127**, 2310.
297. K. Murao and K. Suzuki, *J. Chem. Soc. Chem. Comm.*, 1984, 238.
298. O. Niwa and T. Tamamura, *J. Chem. Soc. Chem. Comm.*, 1984, 817.
299. M. De Paoli, R. T. Waltman, A. F. Diaz and J. Bargon, *J. Chem. Soc. Chem. Comm.*, 1984, 1015.
300. G. Ahlgren and B. Krische, *J. Chem. Soc. Chem. Comm.*, 1984, 946.
301. B. Zinger and L. L. Miller, *J. Amer. Chem. Soc.*, 1984, **106**, 6861.
302. A. F. Diaz, J. Castillo, K. K. Kanazana, J. Logan, M. Salmon and O. Fajardo, *J. Electroanal. Chem.*, 1982, **133**, 233.
303. S. Asavapiriyanont, G. K. Chandler, G. A. Gunawardena and D. Pletcher, *J. Electroanal. Chem.*, 1984, **177**, 245.
304. G. Bidan, A. Deronzier and J.-C. Moutet, *J. Chem. Soc. Chem. Comm.*, 1984, 1185.
305. M. Salmon, M. Aguilar and M. Saloma, *J. Chem. Soc. Chem. Comm.*, 1983, 570.
306. M. Salmon, M. E. Carbajal, M. Aguilar, M. Saloma and J. C. Juarez, *J. Chem. Soc. Chem. Comm.*, 1984, 1532.
307. M. Salmon, M. E. Carbajal, J. C. Juarez, A. F. Diaz and M. C. Rock, *J. Electrochem. Soc.*, 1984, **131**, 1802.
308. F. Garnier, G. Tourillon, M. Gazard and J. C. Dubois, *J. Electroanal. Chem.*, 1983, **148**, 299.

309. R. J. Waltman, J. Bargon and A. F. Diaz, *J. Phys. Chem.*, 1983, **87**, 1459.
310. G. Tourillon and F. Garnier, *J. Electroanal. Chem.*, 1984, **161**, 51.
311. R. J. Waltman, A. F. Diaz and J. Bargon, *J. Electrochem. Soc.*, 1984, **131**, 1452.
312. J. H. Kaufman, T.-C. Chung, A. J. Heeger and F. Wudl, *J. Electrochem. Soc.*, 1984, **131**, 2092.
313. K. Kaneto, K. Yoshino and Y. Inuishi, *Jap. J. Appl. Phys.*, 1982, **21**, L567.
314. K. Kaneto, K. Yoshino and Y. Inuishi, *Sol. St. Comm.*, 1983, **46**, 389.
315. K. Kaneto, Y. Kohno, K. Yoshino and Y. Inuishi, *J. Chem. Soc. Chem. Comm.*, 1983, 382.
316. K. Yoshino, K. Kaneto and Y. Inuishi, *Jap. J. Appl. Phys.*, 1983, **22**, L157.
317. K. Kaneto, K. Yoshino and Y. Inuishi, *Jap. J. Appl. Phys.*, 1983, **22**, L412.
318. G. Tourillon, D. Gourier, P. Garnier and D. Vivien, *J. Phys. Chem.*, 1984, **88**, 1049.
319. G. Tourillon, E. Dartyge, H. Dexpert, A. Fontaine, A. Jucha, P. Lagarde and D. E. Sayers, *J. Electroanal. Chem.*, 1984, **178**, 357.
320. H. Neugebauer, G. Nauer, A. Neckel, G. Tourillon, F. Garnier and P. Lang, *J. Phys. Chem.*, 1984, **88**, 652.
321. G. Zotti and G. Schiavon, *J. Electroanal. Chem.*, 1984, **163**, 385.
322. T. Ohsawa, K. Kaneto and K. Yoshino, *Jap. J. Appl. Phys.*, 1984, **23**, L663.
323. M. Delamar, P.-C. Lacaze, J.-Y. Dumonsseau and J.-E. Dubois, *Electrochim. Acta*, 1982, **27**, 61.
324. I. Rubinstein, *J. Electrochem. Soc.*, 1983, **130**, 1506.
325. I. Rubinstein, *J. Polym. Sci.*, 1983, **21**, 3035.
326. K. Kaeriyama, M. Sato, K. Someno and S. Tanaka, *J. Chem. Soc. Chem. Comm.*, 1984, 1199.
327. J. Bargon, S. Mohmand and R. J. Waltman, *Mol. Cryst. Liq. Cryst.*, 1983, **93**, 279.
328. L. Fornarini, F. Stirpe and B. Scrosati, *J. Electrochem. Soc.*, 1983, **130**, 2184.
329. R. N. O'Brien, N. S. Sundaresan and K. S. V. Santhanam, *J. Electrochem. Soc.*, 1984, **131**, 2028.
330. H. L. Landrum, R. T. Salmon and F. M. Hawkridge, *J. Amer. Chem. Soc.*, 1977, **99**, 3154.
331. J. F. Stargardt, F. M. Hawkridge and H. L. Landrum, *Anal. Chem.*, 1978, **50**, 930.
332. E. F. Bowden, F. M. Hawkridge and H. N. Blount, *Bioelectrochem. Bioenerg.*, 1980, **7**, 447.
333. J. F. Castner and F. M. Hawkridge, *J. Electroanal. Chem.*, 1983, **143**, 217.
334. C. D. Crawley and F. M. Hawkridge, *J. Electroanal. Chem.*, 1983, **159**, 313.

335. V. J. Razumas, A. V. Gudavicius and J. J. Kulys, *J. Electroanal. Chem.*, 1983, **151**, 311.
336. G. Schiavon, G. Zotti and G. Bontempelli, *J. Electroanal. Chem.*, 1984, **161**, 323.
337. A. Torstensson and L. Gorton, *J. Electroanal. Chem.*, 1981, **130**, 199.
338. V. LeBerre, R. Cartier, A. Tallec and J. Simonet, *J. Electroanal. Chem.*, 1983, **143**, 425.
339. C. M. Elliott and W. S. Martin, *J. Electroanal. Chem.*, 1982, **137**, 377.
340. G. Mengoli, P. Bianco, S. Daolio and M. T. Munari, *J. Electrochem. Soc.*, 1981, **128**, 2276.
341. M.-C. Pham, G. Tourillon, P.-C. Lacaze and J.-E. Dubois, *J. Electroanal. Chem.*, 1980, **111**, 385.
342. M.-C. Pham, J.-E. Dubois and P.-C. Lacaze, *J. Electrochem. Soc.*, 1983, **130**, 346.
343. I. Rubinstein, *Anal. Chem.*, 1984, **56**, 1135.
344. R. Noufi, *J. Electrochem. Soc.*, 1983, **130**, 2126.
345. S. Geraty and J. G. Vos, *J. Electroanal. Chem.*, 1984, **176**, 389.
346. P. Daum and R. W. Murray, *J. Electroanal. Chem.*, 1979, **103**, 289.
347. P. Daum, J. R. Lenhard, D. Rolison and R. W. Murray, *J. Amer. Chem. Soc.*, 1980, **102**, 4649.
348. P. Daum and R. W. Murray, *J. Phys. Chem.*, 1981, **85**, 389.
349. R. Nowak, F. A. Schultz, M. Umana, H. Abruna and R. W. Murray, *J. Electroanal. Chem.*, 1978, **94**, 219.
350. R. J. Nowak, F. A. Schultz, M. Umana, R. Lam and R. W. Murray, *Anal. Chem.*, 1980, **52**, 315.
351. K. W. Willman and R. W. Murray, *Anal. Chem.*, 1983, **55**, 1139.
352. M. F. Dautartas and J. F. Evans, *J. Electroanal. Chem.*, 1980, **109**, 301.
353. M. F. Dautartas, K. R. Mann and J. F. Evans, *J. Electroanal. Chem.*, 1980, **110**, 379.
354. D. R. Rolison and R. W. Murray, *J. Electrochem. Soc.*, 1984, **131**, 337.
355. J. Facci and R. W. Murray, *Anal. Chem.*, 1982, **54**, 772.
356. K. Doblhofer, *Electrochim. Acta*, 1980, **25**, 871.
357. K. Doblhofer, W. Dürr and M. Jauch, *Electrochim. Acta*, 1982, **27**, 677.
358. G. R. Heider, M. B. Gelbert and A. M. Yacynych, *Anal. Chem.*, 1982, **54**, 324.
359. K. Doblhofer, D. Nölte and J. Ulstrup, *Ber. Bunsenges. Phys. Chem.*, 1978, **82**, 403.
360. K. Doblhofer and W. Dürr, *J. Electrochem. Soc.*, 1980, **127**, 1041.
361. K. Hiratsuka, H. Sasaki and S. Toshima, *Chem. Lett.*, 1979, 751.
362. M. Wöhrle, R. Bannehr, B. Schumann and N. Jaeger, *Angew. Makromol. Chem.*, 1983, **117**, 103.
363. W.-H. Kao and T. Kuwana, *J. Amer. Chem. Soc.*, 1984, **106**, 473.
364. D. E. Weisshaar and T. Kuwana, *J. Electroanal. Chem.*, 1984, **163**, 395.
365. O. Ikeda, H. Fukuda and H. Tamura, *J. Chem. Soc. Chem. Comm.*, 1984, 567.
366. C. M. Elliott and C. A. Marrese, *J. Electroanal. Chem.*, 1981, **119**, 395.
367. N. Oyama, K. B. Yap and F. C. Anson, *J. Electroanal. Chem.*, 1979, **100**, 233.

368. K. Aoki, K. Tokuda and H. Matsuda, *J. Electroanal. Chem.*, 1983, **146**, 417.
369. R. S. Nicholson and I. Shain, *Anal. Chem.*, 1964, **36**, 706.
370. E. Laviron, *J. Electroanal. Chem.*, 1980, **112**, 1.
371. E. Laviron, L. Roullier and C. Degrand, *J. Electroanal. Chem.*, 1980, **112**, 11.
372. A. P. Brown and F. C. Anson, *Anal. Chem.*, 1977, **49**, 1589.
373. E. Laviron and L. Roullier, *J. Electroanal. Chem.*, 1980, **115**, 65.
374. C. R. Martin, I. Rubinstein and A. J. Bard, *J. Amer. Chem. Soc.*, 1982, **104**, 4817.
375. N. Oyama and F. C. Anson, *J. Electrochem. Soc.*, 1980, **127**, 640.
376. C. P. Andrieux and J.-M. Saveant, *J. Electroanal. Chem.*, 1980, **111**, 377.
377. K. Aoki, K. Tokuda and H. Matsuda, *J. Electroanal. Chem.*, 1984, **160**, 33.
378. E. Laviron, *J. Electroanal. Chem.*, 1980, **109**, 57.
379. E. Laviron, *J. Electroanal. Chem.*, 1979, **100**, 263.
380. E. Laviron and L. Roullier, *J. Electroanal. Chem.*, 1980, **115**, 65.
381. E. Laviron, *J. Electroanal. Chem.*, 1981, **122**, 37.
382. A. J. Bard and L. R. Faulkner, *Electrochemical Methods: Fundamentals and Applications*, Wiley, New York, 1980, p. 218.
383. A. J. Bard and L. R. Faulkner, *Electrochemical Methods: Fundamentals and Applications*, Wiley, New York, 1980, p. 143.
384. K. Aoki, K. Tokuda, H. Matsuda and N. Oyama, *J. Electroanal. Chem.*, 1984, **176**, 139.
385. T. P. Henning and A. J. Bard, *J. Electrochem. Soc.*, 1983, **130**, 613.
386. F. Bruno, M.-C. Pham and J.-E. Dubois, *J. Chim. Phys.*, 1975, **72**, 490.
387. N. Oyama, H. Yamamoto, T. Ohsaka and M. Kaneko, *Bull. Chem. Soc. Jap.*, 1984, **57**, 2942.
388. M. Majda and L. R. Faulkner, *J. Electroanal. Chem.*, 1984, **169**, 97.
389. A. Hamnett and A. R. Hillman, *J. Electroanal. Chem.*, 1985, **195**, 189.
390. P. C. Lacaze and G. Tourillon, *J. Chim. Phys. Phys.—Chim. Biol.*, 1979, **76**, 371.
391. G. Tourillon, J.-E. Dubois and P.-C. Lacaze, *J. Chim. Phys. Phys.—Chim. Biol.*, 1979, **76**, 369.
392. G. Tourillon and F. Garnier, *J. Electrochem. Soc.*, 1983, **130**, 2042.

CHAPTER 6

Reactions and Applications of Polymer Modified Electrodes

A. ROBERT HILLMAN

School of Chemistry, University of Bristol, UK

1. INTRODUCTION

This chapter describes how the chemical, optical and electronic properties of polymer modified electrodes may be harnessed in potentially useful ways. Mediated charge transfer to solution species forms the basis of many of these applications. A brief resumé of the theory of this phenomenon and some experimental tests of the predictions arising from it are therefore the starting points. From a fundamental viewpoint this theory allows the elucidation of mechanism and determination of kinetic parameters. From an operational viewpoint, it allows one to define the criteria for optimum behaviour and performance under these conditions. At the outset, two unfortunate consequences of theoretical treatments post-dating much of the experimental work are to be noted. Firstly, clear judgements are frequently impossible as a result of insufficient or inappropriate data being reported; this chapter will make clear what is desirable in this respect. Secondly, hindsight will reveal that many systems were always doomed to failure. Nevertheless, analysis of some of these cases is fruitful in that it highlights the requirements, leads to a qualitative understanding of some underlying physical processes and provides a source of quantitative kinetic/diffusion parameters for use in derivative systems. The second half of this chapter surveys some of the applications of polymer modified electrodes. The presentation follows the ultimate objective, rather than the underlying mechanism, to facilitate comparison of different approaches to the same goal.

2. THEORETICAL TREATMENTS OF MEDIATED CHARGE TRANSFER

2.1. Basic Model and Concepts

Reduction of a solution species Y by the reduced partner of the surface attached A/B couple was illustrated in Fig. 1 of Chapter 5. The processes involved are injection of charge and substrate at the electrode/polymer and polymer/solution interfaces, respectively, their diffusion within the film and reaction upon meeting. Y enters the film according to a partition coefficient κ. In all cases but one,[1] this process has been assumed to be at equilibrium. The diffusion coefficient of Y in the film, D_Y, is expected to be rather less than the value in solution, D_S, for all but small species in highly swollen polymers. Charge injection at the electrode/polymer interface will be to redox centres A in close proximity (≤ 1 nm) to the electrode according to the usual Butler–Volmer relation, characterised by k_0' and α.[2] At this interface A/B interconversion is potential-driven, whilst within the film charge propagation (electron 'diffusion', D_E) is concentration-gradient-driven. Assuming Marcusian A/B self-exchange behaviour, this leads to the important result that charge injection will be more rapid than charge propagation, a point noted early on by Anson.[3] This assumption is likely to be a good one at the high overpotentials for the A/B couple usually associated with limiting currents in RDE voltammetry, described in the following sections. Finally, Y and B in the same region of space react to produce A and the product Z:

$$B + Y \xrightarrow{k} A + Z \qquad (1)$$

Depending on the relative rates of the kinetic and transport parameters, reaction may occur throughout or only in a restricted region of the film, under kinetic (k) or electron/substrate (t_E/t_Y) diffusional control. Two further possibilities exist: reaction at the outer interface without the requirement for partition ('surface' cases) and direct (unmediated) reaction at the underlying electrode ('electrode' cases). Again each of these may be under kinetic [k'' and $k_E'(E)$, respectively] or diffusional (t_E and t_Y, respectively) control. A summary of treatments of differing complexity is given in Table 1. Those approaches limited to 'surface' or 'electrode' cases are rather less generally useful for, as pointed out in earlier work,[4] if modified

TABLE 1

Theoretical Treatments of Mediated Charge Transfer at Modified Electrodes

Case(s) considered	Rapid partition?	Irreversible mediation reaction?	Other features/limitations/comments	Refs
Surface	Not applicable	Yes	Kinetic zone diagram for transport/kinetic control. Calculation of i/E curves and $E_{1/2}$.	5
Electrode	Yes ($\kappa = 1$)	Not applicable	i, E curves and $E_{1/2}$ calculated.	6
Layer	Yes	Yes	k case only (D_E, D_Y large).	4
	Yes	Yes	Numerical solution for kinetic/transport effects as a function of film thickness.	7
	Yes	Yes	Marcusian ideas applied to charge transfer. k'_{ME} as a function of $(E'_{A/B} - E_{Y/Z})$. Transport and kinetic effects.	3
	Yes	Yes	Transport and kinetic flux limitations.	8
	—	—	'Typical' (PVF) parameters used for economic modelling.	9
Layer, surface	Yes	Yes	Diagnostics for transport/kinetic control. Kinetic zone diagram.	10
Layer, electrode	Yes	No ($K = 1$)	Charge- and place-exchange contributions to self-exchange.	11
	Yes	Yes	Solution concentration polarisation treated. Kinetic zone diagrams.	12–14
	Yes	Yes	'CE' mechanism. Kinetic zone diagrams.	15
	Yes	No	Kinetic zone diagram.	16
	No	Yes	Procedure for interfacial mass transport rate determination.	17
Layer, electrode, surface	Yes	Yes	Kinetic zone diagram and optimisation of L.	18

electrodes are to compete with homogeneous catalysts, monolayer derivatisation is unlikely to be adequate simply because there are not enough mediator sites present. By expanding from a two to a three-dimensional reaction zone with multilayers of mediator this deficiency may be overcome—hence the desire for polymer modification of electrode surfaces.

This raises an important concept to which frequent reference will be made, the reaction layer. Its importance lies in the facts that its thickness conveys what fraction of the film is usefully employed and that its location reflects the relative transport rates of charge and substrate. For Y entering a film of B, the reaction layer thickness at the polymer/solution interface ($x = L$, where x is the distance from the electrode/polymer interface) is given by

$$X_L = (D_Y/kb_0)^{1/2} \qquad (2)$$

(b_0, the concentration of B at $x = 0$, is used rather than b_L because under conditions where X_L enters expressions for k'_{ME}, B is not concentration polarised[10]). Analogously, for electrons injected at the electrode/film interface ($x = 0$) diffusing into a region containing Y,

$$X_0 = (D_E/k\kappa y_s)^{1/2} \qquad (3)$$

where inclusion of κ allows the known polymer/solution interfacial concentration of Y to be used and similar arguments for concentration polarisation apply.[10]

In the extreme case, when k is zero, Y has to diffuse across the film to react directly at the electrode. When the kinetics are slow, eqns (2) and (3) show that reaction layers will be thick and mediated reactions will take place throughout the film. However, when k is large, B and Y cannot co-exist in the same region of space and separate regions dominated by each will be found, the relative sizes of which will be governed by their diffusional fluxes ($D_E b_0/L$) and $\kappa D_Y y_s/L$). When k is sufficiently large (or κD_Y sufficiently small) reaction may be confined to molecular dimensions at the outer interface and Y will not have the opportunity to partition into the film. This very qualitative consideration of the interplay of the transport and kinetic processes is placed on a rather more quantitative footing in the next section.

2.2. Analysis of Mediated Charge Transfer

The object of this section is to arrive at expressions which will firstly allow mechanistic diagnosis, secondly facilitate determination of parameters, and thirdly point the way towards optimisation. The principal technique used for quantitative data acquisition (see Section 2.3) is the rotating disc electrode: this is simply because its controlled steady-state solution mass transport properties give straightforward access to the concentration of Y at the polymer/solution interface (y_s) from its bulk solution value (y_∞).[19] Consequently, most theoretical treatments are cast in this form, as is done here.

The analysis presented here is that developed separately by Albery and Saveant (see Table 1), which includes the possibility of reaction throughout at least some of the film ('layer' cases) because these are likely to be of greatest practical importance. (The solely 'electrode' and 'surface' analyses of Delamar[6] and Laviron[5] are sub-cases, of rather less general applicability.) As both have noted,[10,14] the approaches of Albery and Saveant are essentially equivalent. Whilst a slightly different choice of descriptive parameters is used (reaction layer/film thickness comparison in Albery's case and 'characteristic currents' describing maximum diffusional and kinetic fluxes in Saveant's case), the ultimate combination of parameters, the conceptual 'kinetics versus diffusion' approach and the results obtained are basically the same. Saveant's analysis produces a result in terms of a current, described by a notation system denoting the rate-limiting process. Albery's analysis produces an effective rate constant (k'_{ME}), for comparison with the 'bare electrode' value (k'_E), and describes the situation in a notation conveying the site of reaction and the rate-limiting process. This is summarised in Table 2. Now that Saveant's expressions[12] have been modified to include solution concentration polarisation,[14] the one remaining difference is the treatment of 'surface' reactions, to which we will return later.

We need to solve the steady-state diffusion equations for B and Y, subject to their mutual consumption with rate constant k, to give their concentration profiles within the film:

$$D_E \frac{d^2 b}{dx^2} - kby = 0 \qquad (4)$$

$$D_Y \frac{d^2 y}{dx^2} - kby = 0 \qquad (5)$$

The boundary condition for b at $x = 0$ is imposed by the potential (usually in a Nernstian-like fashion, see Section 3.3 of Chapter 5). The boundary condition for y at $x = L$ is imposed by the partition relation

$$y_L = \kappa y_s \tag{6}$$

which is assumed to be in equilibrium here. The observed current, i_F, is related to the sum of the fluxes of direct (j_Y) and mediated (j_B) charge transfer (i.e. charge injection into Y and A) and the concentration gradients at the electrode/polymer interface by

$$(i_F/nFA) = j_0 = j_B + j_Y = -D_E\left(\frac{db}{dx}\right)_0 + D_Y\left(\frac{dy}{dx}\right)_0 = k'_{ME}y_s \tag{7}$$

Here the effective heterogeneous rate constant for the modified electrode, k'_{ME}, has been introduced and one can see the simple relationship between the Albery and Saveant formulations in terms of rate constants and currents.

Whilst more general numerical[12] and analytical[18] solutions to these equations have been demonstrated, the presentation here will be restricted to a series of more specific cases resulting from various inequalities between the rates of the kinetic and transport processes. As pointed out,[12,18] this is far more informative and leads to straightforward treatment of experimental data, as will be demonstrated in the next section. Furthermore, the simple expressions which result contain the parameters in such combinations that their effects may be studied individually, opening the way to mechanistic diagnosis and overall optimisation.

For illustrative purposes, the kinetic zone diagram[14,15] for the various 'layer' cases is shown in Fig. 1. (Inclusion of 'surface' and 'electrode'[18] cases adds a third dimension and additional parameters, but the principle is the same.) The axes and zones have been labelled according to both the Albery and Saveant notations, and the 'signpost' shows the effect of varying individual parameters on the location of a given system on the diagram. The expressions for the rate constant[18]/current[12] in each of the zones are given in Table 2 and are in accord with each other. Near the borders, the fuller expressions must be used (Saveant broadens the borders into mixed-case zones), but it should be noted that there are no discontinuities: e.g. in the expression for k'_{ME} at the Lk/LSk border, L is replaced by X_L where the two are equal. It should be borne in mind that certain parameters (notably b_0, y_s, L) are readily variable within a system, but that others (D_E, D_Y, κ, k) require more drastic steps (or a different system altogether) to effect their variation.

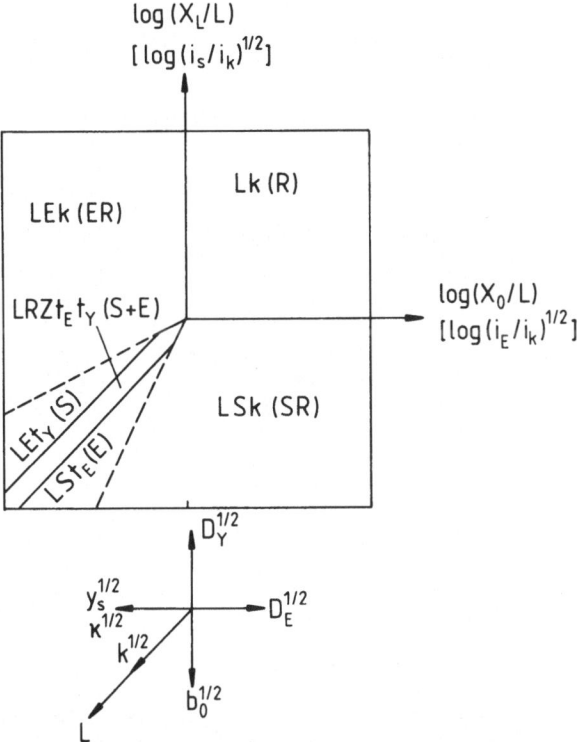

FIG. 1. Kinetic zone diagram for the 'layer' cases, showing the location and rate-limiting step in the mediated charge transfer process to a solution species as a function of the values of the normalised kinetic/diffusion parameters for the system. Axes and zones are labelled according to the Albery (Saveant) notations of ref. 18 (14), with the concentration of Y corrected for the concentration polarisation in the solution outside the film in both cases. The 'signpost' shows the effect on the position of a system within the diagram as the various parameters are increased. (Reproduced by courtesy of Elsevier Science Publishers.)

Inspection of Table 2 shows that establishment of the functional dependence of i_F or k'_{ME} on each of the members of the first set is, with the exception of the St_E/LSt_E and Et_Y/LEt_Y pairs, sufficient to define the mechanistic case. Evaluation of X_L or X_0, if necessary, resolves these two pairs.

There are four possibilities to consider when optimisation of film thickness is desired.[18] The most important pair of these is when both

TABLE 2
Notation and Expressions Describing Behaviour of Different Cases for Faradaic Reactions at Polymer Modified Electrodes

Case notation		Expression for charge transfer rate	
Ref. 10	Ref. 14	k'_{ME} [10]	i_F [14,c,d]
Sk''	—[a]	$k''b_0$	—
St_E	—[a]	$D_E b_0/Ly_s$	—
LSk	SR	$\kappa k b_0 X_L$	$(i_s i_k)^{1/2}$
LSt_E	E	$D_E b_0/Ly_s$	i_E
Lk	R	$\kappa k b_0 L$	i_k
$LRZt_E t_Y$	S+E	$D_E b_0/Ly_s + \kappa D_Y/L$	$i_E + i_s$
LEk	ER	$\kappa k b_0 X_0$	$(i_E i_k)^{1/2}$
LEt_Y	S	$\kappa D_Y/L$	i_s
Ek'_E	—[b]	$\kappa k'_E$	—
Et_Y	—[b]	$\kappa D_Y/L$	—

[a] 'Surface' cases treated as limiting cases of 'LS' situations.
[b] Expressions for total (mediated and direct) currents given.
[c] Characteristic currents corrected for solution concentration polarisation effects.[14]
[d] $i_F = nFAk'_{ME}y_s$.

charge and Y are able to pass through the film. Here, the value of k'_{ME} (or i_F) will initially increase with L (all sites mediate under kinetic control), then pass through a maximum, before decreasing with L (only some sites mediate under transport control). Depending on the ratio $D_E b_0/\kappa D_Y y_s$, this corresponds to the sequence of cases $Lk \rightarrow LEk \rightarrow LEt_Y$ or $Lk \rightarrow LSk \rightarrow LSt_E$. The optimum cases here, then, are LEk or LSk, with $L \simeq 3X_0$ or $3X_L$, respectively. In qualitative terms, this provides sufficient polymer film (i.e. mediator sites) for consumption of B (or Y), without constituting a barrier to substrate (or charge) diffusion. The remaining two eventualities are those where charge (substrate) is unable to penetrate the film, in which case the film should be thin, so that it presents no diffusional barrier. Here, then, the optimum cases are 'electrode' (or 'surface') reaction under kinetic control, Ek'_E (or Sk''), depending on which species is able to penetrate the film. The former case corresponds to no mediation, the latter to mediation without partition. At first sight, these latter two cases would appear uninteresting to the polymer chemist. This is not necessarily so in the Ek'_E case because the rate constant k'_E is enhanced by κ in k'_{ME}, and κ is likely to be dependent on the polymer/substrate combination. It is also not necessarily so in the Sk'' case, because the rate constant k'' may not be related to k'_E in simple outer-sphere Marcusian terms, i.e. B may be a specific catalyst for the reduction of Y.

Returning to the more likely LEk and LSk cases, the expressions for k'_{ME} can be used[18] to deduce target values of parameters for successful modification of the electrode surface. Typically, a useful current density would be $1\,A\,cm^{-2}$ or above, for a concentration, y_s, of the order of $1\,mol\,dm^{-3}$—this corresponds to $k'_{ME} \sim 10^{-2}\,cm\,s^{-1}$. Additionally, it is required that k'_{ME} be greater than k'_E for an advantage to accrue from modification. Illustrating the argument for the LSk case,[18]

$$k'_{ME} = \kappa(D_Y k b_0)^{1/2} \simeq \left(\frac{\kappa X_L}{l}\right) k'_E \qquad (8)$$

where l represents the distance over which direct electron transfer can occur ($\sim 1\,nm$). Clearly, even if κ is not large, k'_{ME}/k'_E can be made considerably greater than unity through manipulation of X_L. Taking a typical sluggish reaction with $k \sim 10^2\,mol^{-1}\,dm^3\,s^{-1}$, a film diffusion coefficient, D_Y, of $10^{-7}\,cm^2\,s^{-1}$ would lead to a reaction layer thickness, X_L, of $100\,nm$ and a catalytic advantage (for $\kappa = 1$) of 10^2. Slower rate constants require thicker films, and thus more rapid diffusion, but give the promise of greater catalytic advantage. The message is simple: transport rates must be maximised to gain the greatest advantage. This implies use of highly swollen films with rapid charge transport.

Saveant has extended this basic analysis to include additional features such as chemical reversibility of the mediation reaction,[16] self-exchange by charge- or place-exchange,[11] a finite interfacial mass transfer rate[17] and an additional chemical step.[15] This last possibility corresponds to the 'CE' scheme:

$$A + e^- \rightleftharpoons B$$

$$Y \underset{k_{-1}}{\overset{k_1}{\rightleftharpoons}} X$$

$$B + X \overset{k_2}{\to} Z$$

frequently found in organic and biological systems and thus likely to be of considerable importance from an applications viewpoint. The kinetics are now characterised by two parameters, describing the competition for X between the forward (k_2) and reverse (k_{-1}) processes (Saveant's choice[15] of parameters is $\kappa k_1 y_0 L$ and $k_2 b_0/k_{-1}$ for the generation and fate of X). The kinetic zone diagram for layer cases alone (analogous to Fig. 1) now has a third dimension provided by the competition parameter. Whilst inclusion of an extra dimension for the possibility of surface reactions would make graphical representation difficult, there may be an important point here, for large biological substrates likely to follow this 'CE' scheme may be obliged to undergo surface reaction as a result of an unfavourable partition coefficient.

2.3. Experimental Test of the Analysis

In this section, the behaviour of some 'model' redox couples at polymer modified electrodes is classified. The couples, usually single-electron outer-sphere systems, were chosen to probe the permeation and mediation characteristics of polymer films towards substrates of varying size, charge and reactivity. The principal technique used is rotating disc electrode voltammetry, where the steady-state hydrodynamic conditions result in a solution mass transport limited current given by the Levich equation:[20]

$$i_{Lev} = 1\cdot554 n F A D^{2/3} v^{-1/6} y_\infty W^{1/2} = B W^{1/2} \tag{9}$$

Inclusion of an additional possible rate-limiting step, characterised by the current i_F, in the presence of the polymer film results in an observed limiting current given by the Koutecky–Levich equation:[21]

$$\frac{1}{i_{Lim}} = \frac{1}{i_F} + \frac{1}{i_{Lev}} = \frac{1}{i_F} + \frac{1}{BW^{1/2}} \tag{10}$$

A plot of observed limiting current/rotation speed data according to eqn (10) (i_{Lim}^{-1} versus $W^{-1/2}$) allows extrapolation to infinite solution mass transport rates ($W^{-1/2} \to 0$) and thus extraction of i_F. It is the functional dependence of i_F or k'_{ME} on b_0, y_s and L which distinguishes the cases summarised in Table 2: this forms the basis of the classification procedure here.

Whilst some authors present clear evidence for the variation of k'_{ME} (or i_F) with b_0, y_s and L, they are by no means in the majority. Frequent cases exist of assumption of mechanism followed by (partial) 'confirmatory' evidence, as well as cases where the information is given but not exploited. The reviewer has attempted here to classify broadly these 'model' systems into unmediated (direct) charge transfer ('electrode'), mediated polymer/solution interfacial charge transfer ('surface') and partially or wholly film mediated charge transfer ('layer') cases.

Examples identified as 'electrode' cases are collected in Table 3. In the case of a reduction (oxidation), increasingly negative (positive) potentials results in a shift from kinetic to transport control. Table 3 contains limiting current data, corresponding to t_Y control. The initially extracted quantity is thus $\kappa D_Y/L$. Provided an estimate of L is available, one can then determine κD_Y, as in the table. In some cases,[26,27] 'dry' film thicknesses have been used so there may be slight errors in absolute values (though not trends). Since κ is not readily obtainable the process usually stops there, although the assumption that κ is unity[22] is sometimes made, albeit implicitly. Quoted values of D_Y should therefore be inspected for assumptions as the values of L and κ.

Direct reaction may result because the film contains no redox sites and merely acts as an inert membrane,[22,23] because the mediation process is sluggish (often as a consequence of uphill thermodynamics)[24-27] or because the potential is not at an appropriate value to generate the necessary film oxidation state, i.e. mediator sites are not 'switched on'.[28,31,32] In the latter instance certain circumstances, described by Saveant,[12] give rise to two waves in the current/voltage curve, and one sees direct and mediated charge transfer at potentials characteristic of the solution and mediator couples, respectively. Clearly both size and charge are important in permeation of Y. Neutral species diffuse through poly[Ru(vibipy)$_3^{2+}$] at rates inversely related to their size: quinone > FeCp$_2$ > Ru(bipy)$_2$Cl$_2$ > Fe(bipy)$_2$(CN)$_2$.[27] Furthermore, the wide divergence in permeation rates implies that these species must be diffusing through spaces with dimensions comparable to their own. More quantitatively, diffusion through PVF was shown[24] to fit a membrane, rather than a pinhole, model.

Charge effects may well be manifested through the κ component of the pair, for replacement of a Cl$^-$ by pyridine in Ru(bipy)$_2$Cl$_2$ to give a unipositive ion results in a decrease in κD_Y through poly[Ru(vibipy)$_3^{2+}$] by over two orders of magnitude.[27] Selectivity in polyphenol films has been shown[23] by correlating κD_Y values with Stokes' ionic radii to separate out the trend of the D_Y component.[23] A more complex

TABLE 3
Direct 'Electrode' Charge Transfer Cases

Polymer	Solution species	κD_Y $(cm^2\,s^{-1})$	Comments	Refs
‑(‑O‑)ₙ‑ with OCH₃	FeCp₂	$(2\text{–}5) \times 10^{-8}$	κ not explicitly referred to, but quoted 'D' contains it.	22
	HO‑C₆H₄‑OH	$1\cdot 3 \times 10^{-8}$		
‑(‑O‑)ₙ‑ with CH₃, CH₃	FeCp₂	$1\cdot 3 \times 10^{-7}$		22
H₂NC₂H₄‑C₆H₃‑(O)ₙ‑	FeCp₂	$1\cdot 0 \times 10^{-9}$		22
	HO‑C₆H₄‑OH	$1\cdot 8 \times 10^{-9}$		
‑(‑O‑)ₙ‑ with CN	HO‑C₆H₄‑OH	$\sim 3\cdot 7 \times 10^{-8}$		22

Polymer	Species	Value	Notes	Ref.
poly(phenoxide) -(-C6H4-O-)n-	H^+	4.1×10^{-7}	κD_Y for $H^+ > Br^- > Fe^{2+}$.	23
poly(anilide) -(-C6H4-NH-)n-	H^+	8.5×10^{-8}		23
poly(o-aminoaniline) (with NH2)	H^+	3.5×10^{-7}		23
poly(o-aminoanilide)	H^+	$>2 \times 10^{-7}$		23
poly(4-nitrophenyl-NH-)	MV^{2+}	2.1×10^{-7}	Using $\kappa = 6$ (chronoamperometry), $D_Y \sim 4 \times 10^{-8}$.	24
PVF	hydroquinone (HO–C6H4–OH)	7.2×10^{-7}	Using $\kappa = 4$ (chronoamperometry), $D_Y \sim 2 \times 10^{-7}$.	24

(continued)

TABLE 3—contd.

Polymer	Solution species	κD_Y $(cm^2 s^{-1})$	Comments	Refs
PVP·H$^+$, (cyclopentadienone derivative with NC, CN, O, O$^-$ groups)	CpFeCp'—$\overset{+}{\text{N}}$(CH$_3$)$_3$	3×10^{-10}	Using $\kappa D_Y/L$ and lower quoted estimate of L.	25
PVP	CpFeCp'—$\overset{+}{\text{N}}$(CH$_3$)$_3$	4×10^{-11}	Using $\kappa D_Y/L$ and lower quoted estimate of L.	25
Poly[Ru(bipy)$_2$(vipy)$_2^{2+}$]	FeCp$_2$	6.4×10^{-8}		26
Poly[Ru(bipy)$_2$(vipy)$_2^{2+}$]/Co^0	FeCp$_2$	'Apparent' reduction by ~2–3 times	Co^0 partially obscures film surface.	26
Poly[Ru(vibipy)$_3^{2+}$]	Br$^-$	$>4 \times 10^{-7}$		27, 28
	HO—⟨⟩—OH	5.8×10^{-8}		
	FeCp$_2$	1.3×10^{-8}		
	Ru(bipy)$_2$Cl$_2$	1.3×10^{-9}		
	Fe(bipy)$_2$(CN)$_2$	3.3×10^{-10}		
	Diquat^{2+}	2.0×10^{-9}		
	Ru(bipy)$_2$(py)(Cl)$^+$	$<7 + 10^{-12}$		

Polymer	Substrate	Value	Notes	Ref
Poly[Ru(bipy)$_2$ (cinn)$_2^{2+}$]	FeCp$_2$	9.0×10^{-8}		27
	Ru(bipy)$_2$(Cl)$_2$	1.0×10^{-8}		27
Poly(VDQ^{2+})	FeCp$_2$	1.0×10^{-9}		
	Ru(bipy)$_2$Cl$_2$	3.3×10^{-10}		
	Co(tpy)$_2^{2+}$	3×10^{-6}		
Polylysine·H$^+$, $\frac{1}{4}$Mo(CN)$_8^{4-}$			Some dependence on Mo(CN)$_8^{4-}$ loading via film thickness and diffusion coefficient effects.	29
Polylysine·H$^+$, $\frac{1}{3}$Co(C$_2$O$_4$)$_3^{3-}$	Co(C$_2$O$_4$)$_3^{3-}$	$\sim 4 \times 10^{-5}$ (outside Donnan domains) ($\kappa \sim 17$)	Two independent pathways for reduction of Co^{III} inside and outside Donnan domains (no in- or cross-phase electron exchange) ($D_s = 6.6 \times 10^{-6}$ cm^2 s^{-1}, $D_{\text{Donnan}} \sim 2 \times 10^{-7}$ cm^2 s^{-1}, $D_Y \sim 2 \times 10^{-6}$ cm^2 s^{-1}).	30

arrangement, in which (partly coalesced) particulate Co^0 was deposited on the outside of the polymer, was studied by Pickup.[26] In this case information on polymer diffusion (κD_Y) was obtained from the intercepts and on the particulate morphology from the slopes of Koutecky–Levich plots. A general observation is that, even allowing for values of κ rather less than unity, the implied values of D_Y are well below their solution analogues, D_s. (Exceptions to this are the values in polylysine films;[29,30] however, these are values for species in the large electrolyte 'pools' in films swollen by a factor of about 10^2.[30]) These low values of D_Y have been rationalised using Walden's rule and by describing the solvated polymer film as an extremely viscous medium[24,27,33] through which Y must move.

At the other extreme are the 'surface' cases, where mediated reaction is so rapid (compared with permeation) that it occurs at the polymer/solution interface without partition. This corresponds to the reaction layer thickness at the outer interface being reduced to submolecular dimensions.

$$X_L = (D_Y/kb_0)^{1/2} \leq 0.5 \text{ nm} \tag{11}$$

Depending on the magnitudes of the kinetic (k'') and diffusive (D_E) parameters, the reaction may be under transport (St_E) or kinetic (Sk'') control. The examples collected in Table 4 show that the clearly identified surface cases to date concern rapid (usually outer-sphere) electron transfer reactions[28,34–38,40] or very impermeable films.[39] Complete diagnosis is, unfortunately, rare, as the gaps in columns 3–5 of Table 4 show. Whilst the reviewer believes many of the conclusions reached by authors, the assumption rather than determination of reaction order is not condoned. Particularly dangerous is the transfer of parameters between even rather similar systems. This is well illustrated by the differences in polysilylpropyl-[37] and polysilylbenzyl-[38] viologens, where there are significant differences in electroactivity 'attributable to differences in the nature of the polysiloxanes that result'.[38] As illustrated for a series of transition metal complexes reacting at polypyridyl metal complex films[28,34,35] and for Fe^{3+} reduction at a polymeric thionine film[39] (see Fig. 2) clear reaction orders of unity for y_s and zero for L were established from Koutecky–Levich plots. According to Table 2, this establishes the case as Sk'' or LSk. Distinction between these rests on whether the order with respect to b_0 is unity or one-half, respectively. It was made by considering not the plateau, but the rising portion of the current/

voltage curves[28,35,39] (where B is 'titrated in' with potential as described in Section 3.3 of Chapter 5), the resulting value of unity implying a surface reaction.

A check on diagnosis is available by location of the system on the kinetic zone diagram,[39] or at least partially so by a calculation of X_L to see if inequality (11) is satisfied. Values quoted (or derived by the reviewer) and collected in Table 4 are reassuring. The value for the oxidation of $IrCl_6^{3-}$ [40] deserves further discussion. Original calculations, based on assumption of an Lk ('R') case, gave apparent values of k increasing with decreasing film thickness implying incomplete accessibility ($X_L < L$) until a coverage of $\sim 10^{-10}$ mol cm^{-2} was reached. This is in accord with monolayer accessibility, i.e. an Sk'' case, and the value of X_L given.

Correlation of the observed rate constants with those calculated using appropriate values in the expressions in Table 2 is satisfactory for both charge transport[28,36,37] and kinetically[35,39] limited cases. In the latter instance, adherence to Marcusian predictions for the variations of k with $(E'_{A/B} - E'_{Y/Z})$ was demonstrated,[34,35] suggesting, at least for the films studied, that the solvating energetics are the same as for the monomeric analogues in solution. Whether this is true throughout the film—this case only probes the outer layer—is not proven.[35]

Sufficient increase in the rate constant results in a shift to transport control, as exemplified by the change in Y from Fe^{3+} to $Fe(CN)_6^{3-}$ as the species to be reduced by a polymeric thionine film.[39] Under these circumstances $E_{1/2}$ is much closer to $E'_{A/B}$.[37] This is rather different from the Sk'' case: there, small quantities of B generated well positive of $E'_{A/B}$ (in the case of a reduction) are able to handle the incoming flux of Y, with the result that $E_{1/2}$ may be several hundred millivolts positive of $E'_{A/B}$.

There are other instances where the reaction is probably a 'surface' one but diagnosis is incomplete. Oxidation of $Fe(CN)_6^{4-}$ by poly(silylpyridinium)-[41] and crosslinked poly(pyridinium)-[42] bound $Fe(CN)_6^{3-}$ has been found to be transport-limited. In the first case, t_E control (with a value of D_E of 9×10^{-9} cm^2 s^{-1}) was diagnosed.[41] In the second,[42] the process was analysed as an Et_Y case (i.e. no mediation), although this seems less likely; furthermore the value of 'D_Y' derived, 10^{-9} cm^2 s^{-1}, is very similar (given the probable effects of crosslinking here) to that assigned to D_E in the previous example.[41] Given the likely rapid electron-exchange contribution alone to exchange here, these are probably St_E cases.

TABLE 4
'Surface' Cases of Mediated Charge Transfer at Polymer Modified Electrodes

Polymer	Solution reactant, Y	Order[a] with respect to:			X_L (nm)	Case	Refs
		y_s	L	b_0			
Poly[Ru(vibipy)$_3^{3+}$]	Ru(bipy)$_2$(L^1)(L^2)$^{2+}$, [L^1 = L^2 = 4,4'-bipy, CH$_3$CN, pyrazine; L^1 = pyr, L^2 = CH$_3$CN; L^{1}L^2 = (bipy)]	1	0	—	0·01 (L^1 = L^2 = bipy)	Sk''	28,34
	Ru(bipy)$_2$(Cl)$_2$ Fe(bipy)$_2$(CN)$_2$	0	−1	1	<0·3 (using $\kappa \sim 1$ and lowest estimate of k_2)	St_E (only measurable at low b_0)	28
Poly[Fe(vibipy)$_3^{2+}$]	FeL$_3^{2+}$ (L = bipy, vibipy, 4,4'-dimethylbipy)	1	0	—		Sk''	34
Poly[Os(bipy)$_2$ (vipy)$_2^{3+}$]	FeL$_3^{2+}$ [L = phen,4-Me(phen), 5-Me(phen), 5,6-(Me)$_2$(phen), 4,7-(Me)$_2$(phen), 3,4,7,8-(Me)$_4$(phen), 4,7-(ϕ)$_2$(phen), bipy, 4,4'-(Me)$_2$(bipy), 4,4'-(ϕ)$_2$(phen)]	1	0	1	<0·03	Sk''	35

Poly[M(bipy)$_2$ cinn)$_2^{3+}$] M = Ru/Os mixture[b]	[Ru(i-pr-bipy)$_2$Cl$_2$] i-pr-bipy = 4,4'-di-iso-propylcarboxy-2,2'-bipyridine (≥0.4 mM)	0	−1	—	0.02 (using $\kappa \sim 1$)	St_E	36
Poly[silylpropyl-viologen$^{·+}$] $\frac{1}{4}$ Fe(CN)$_6^{4-}$	Ru(NH$_3$)$_6^{3+}$ (>6 mM)	0	—	—	<0.1 using lower limit for k	St_E	37
Poly[silylbenzyl-viologen$^{·+}$]	Ru(NH$_3$)$_6^{3+}$ (≥30 mM)	0	−1	—	<0.1 (using lower limit for k)	St_E	38
Poly(thionine)	Fe^{3+}	1	0	1	0.06	Sk''	39
PVP·H$^+$,IrCl$_6^{3-}$	IrCl$_6^{2-}$	1	≠1	—	~0.1	Sk''	40
PVP·H$^+$,Fe(CN)$_6^{4-}$	Fe(CN)$_6^{3-}$	1	≠1	—	~0.1	Sk''	40

[a] For $k''(t_E)$ control these columns should read 1,0,1 (0,−1, 1).
[b] Mediation by Os^{III} only, here, as combined Ru^{III}, Os^{III} mediation is solution mass transport limited.

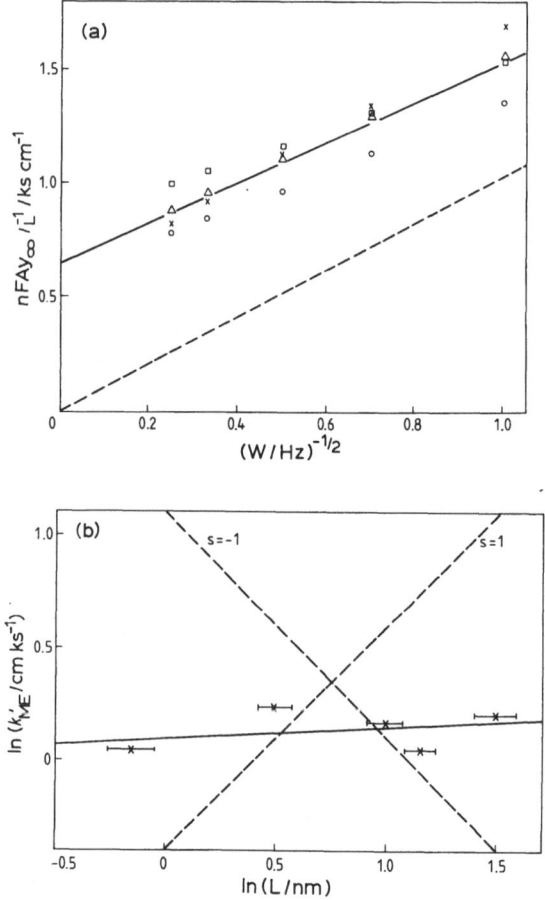

FIG. 2. (a) Koutecky–Levich plots for the reduction of Fe^{3+} at a polymeric thionine coated electrode. Normalisation of limiting currents with respect to bulk Fe^{3+} concentration [y_∞(mmol dm^{-3}) = 0·01 (×), 0·25 (○), 0·50 (△) and 1·00 (□)] on to a single line indicates first-order behaviour with respect to y_s. The dashed line represents the Levich limiting current at a bare electrode. (b) Plot of k'_{ME}, obtained from plots such as those in (a), for films of various thickness. Comparison with lines of ±unit slope clearly shows zero-order behaviour with respect to L. (Reproduced by courtesy of Elsevier Science Publishers.)

A rather more complex 'surface' reaction is the reduction of $Fe(CN)_6^{3-}$ at conducting islands of $(TTF)Br_{0.7}$ on the surface of Nafion, TTF.[43] Here information on the inter-site spacing and site size were obtained from Koutecky–Levich plots, analogous to that described above for Co^0 deposited on polymer films.[26]

The 'layer' cases (see Table 5) are arguably the most interesting, for multilayers of sites are active in the mediation reaction. Simplest to visualise is the Lk ('R') case, where rapid transport (large D_E/L, $\kappa D_Y/L$ or i_E, i_s) compared with cross-reaction kinetics (small k or i_k) results in reaction layers X_L and X_0 larger than the film thickness and consequent mediation throughout the film without significant reactant depletion. First-order behaviour in y_s, b_0 and L or X_L/L, X_0/L ratios above unity can be used to diagnose this case[31,32] (see Tables 2, 5). Frequently, results are discussed in terms of variation with coverage, Γ, rather than b_0 and L individually, and an important distinction should be made here. In cases where the composition is fixed, e.g. polyhydroxyphenazine,[45] it is necessarily L which is varied, but in certain cases, notably the coordinative[44] and ion-exchange[29,31,44] systems, the loading level (b_0) and thickness may be varied separately.

Sufficient increase in k, b_0 or L results in consumption of B or Y before traversing the layer and reaction is confined to X_L or X_0 (LS and LE cases, respectively). This is nicely demonstrated in Table 5, by the oxidation of Fe^{2+} by $PVP \cdot H^+$(lysine copolymer), $IrCl_6^{2-}$:[31] low loadings result in reaction throughout the layer, but high loadings reduce the value of $X_L/L((i_s/i_k)^{1/2})$ and cause a shift from the Lk ('R') to the LSk ('SR') zone (see Fig. 1). Use of the reaction order diagnosis on data from the earlier study[44] leads to the same result: increasing b_0 by a factor of 13 resulted in a decrease of the 'apparent' rate constant (assuming complete participation) by a factor of 3·3. The LSk case is half-order in b_0, predicting a change by a factor of 3·6, in good agreement.

The work on the oxidation of $Co(tpy)_2^{2+}$ by polymer-bound $Mo(CN)_8^{3-}$ and $W(CN)_8^{3-}$ [29] illustrates a general point concerning the likelihood of various cases. The authors compared i_E and i_s (equivalent to X_0^2 versus X_L^2 comparison), i.e. decided on which side of the 45° line $X_L = X_0$ (or $i_s = i_E$) in Fig. 1 their system lay. This corresponds to determining which interface (LS or LE) will be the site of reaction when it no longer occurs throughout the film. It is likely that

$$\frac{i_E}{i_s} = \frac{D_E b_0}{\kappa D_Y y_s} > 1 \tag{12}$$

TABLE 5
'Layer' Cases of Mediated Charge Transfer by Polymer Modified Electrodes[a]

Polymer	Solution species	Order with respect to:			X_0/L $=(i_E/i_k)^{1/2}$	X_L/L $=(i_s/i_k)^{1/2}$	Case	Refs
		y_s	b_0	L				
PVP·H$^+$, IrCl$_6^{2-}$ (5% polylysine copolymer); low[IrCl$_6^{2-}$])	Fe^{2+}	1	—	—	2·5	1·7	Lk	31
Nafion, MV·$^+$	O$_2$	1	—	1	—	—	Lk	32
Nafion, PXV·$^+$	O$_2$	1	—	1	—	—	Lk	32
PVP·H$^+$, IrCl$_6^{2-}$ (5% polylysine copolymer); high [IrCl$_6^{2-}$]	Fe^{2+}	1	—	—	2·6	0·4	LSk	31
Poly(lysine·H$^+$),M								
M = Mo(CN)$_8^{3-}$	Co(tpy)$_2^{2+}$	1	—	0	>X_L	0·1	LSk	29
M = W(CN)$_8^{3-}$	Co(tpy)$_2^{2+}$	1	$-\frac{1}{2}$	0	>X_L	0·1	LSk	29
PVP—H$^+$, IrCl$_6^{2-}$	Fe^{2+}	1	>0, but <1	—	—	—	LSk	44
PVP—Ru^{II}(EDTA)	Fe(CN)$_6^{3-}$	1	$\to \frac{1}{2}$ (?)	—	—	—		
Polyhydroxyphenazine	Eu^{2+}	1	—	≠1	—	—	$LSk?$	45

[a] Where appropriate, values are calculated by the reviewer from authors' data.

simply because b_0 is usually ~ 1 mol dm^{-3} and y_s (in these mechanistic studies at least) is $\sim 10^{-3}$ mol dm^{-3} (κ is unlikely to be much above unity, especially if Y carries the same charge as the polymer, and $D_Y \ll D_s$ in most cases.) The result is that the trickle of Y is more likely to be pinned back near the outer interface by the flood of B, than vice versa. In the specific case here,[29] Anson did find that $i_E > i_s$ and $(i_s/i_k)^{1/2} \sim 0.1$ so that the outer 10% of the film was effectively used. Generalising, LE cases are less likely, as evidenced by Table 5, although examples will be given in the applications section of quinone systems where D_E is very low.

The absence of the transport cases LSt_E ('E') and LEt_Y ('S') in Table 5 is due, at least in part, to their rather restrictive requirements, as reflected in the rather small fractions of the zone diagram (Fig. 1) they occupy. Precise conditions are given in the original papers,[10,14,18] but essentially the reaction layers are required to be fairly thin (although greater than molecular dimensions, otherwise 'surface' or 'electrode' cases result) and reasonably comparable (these zones are found near the $X_L = X_0$ line in Fig. 1); given that there are wide variations in the parameters in relation (12) these are comparatively unlikely circumstances. One of these cases, $LRZt_Et_Y$ ('S + E'), is particularly interesting, albeit unusual. It corresponds to a very thin reaction layer located away from the interfaces, its exact position determined by the (necessarily rather similar) fluxes of B and Y. This delicate balance can be tipped towards the outer (inner) interface by increasing b_0 (y_s) very slightly.

Experimental verification of the theory[11] for self-exchange (A/B ≡ Y/Z) within the layer has been sought using the Fe(EDTA)$^{-/2-}$ couple in protonated polylysine films.[47] When the film is thin, the substrate crosses it too rapidly for exchange to occur and the unmediated ('electrode') process dominates under transport (t_Y or 'S') control. As film thickness is increased, so is the opportunity for (charge- or place-) exchange and a second pathway is opened. The thickness at which the transition from i_s to $(i_s + i_E)$ occurs gives access to the exchange rate. The authors found good agreement with their theory, and domination of electron- rather than place-exchange. However, the reviewer is concerned that at the upper end of film thicknesses used (nearly 1 cm in Fig. 6 of ref. 47), since the Levich diffusion layer is $\sim 10^{-3}$ cm, roughness and 'edge' effects in the rotating disc experiment must require attention.

Whilst a number of other papers[48-51] have demonstrated mediated charge transfer processes, notably via rectification effects resulting from the relative values of $E'_{A/B}$ and $E'_{Y/Z}$, they are not sufficiently detailed to allow unambiguous diagnosis.

2.4. Other Mechanistic Studies

Whilst the theoretical treatment of RDE voltammetry is the most developed, other techniques have been used to study faradaic reactions at polymer modified electrodes. Principal among these is cyclic voltammetry which, in its simplest form, has been used to demonstrate mediation to a solution redox couple.[52,53] The 'switching on' of the mediation reaction on conversion of polypyrrole to its oxidised (conducting) form and on illumination has been studied.[54] Use of the standard Nicholson analysis has been made to derive operational rate constants for redox reactions at polypyrrole[55] and some of its derivatives.[56] Use of the mediation reaction to deposit metal particulates/films has been studied by cyclic voltammetry.[26] In the event of slow charge transport, the experiment can be turned on its head and the Y/Z couple used as an 'electron shuttle'.[57,58] The idea here is that Z, produced by reduction of Y at the inner interface, diffuses through the film and reduces otherwise electroinactive A to B. As will be discussed in the applications section, this can be used to enhance the reduction rate of an additional solution redox couple by the A/B mediator system. Effectively, the charge transport (t_E) and kinetic (k) functions are separated here.

Cyclic voltammetry can also be used to study direct reaction of species diffusing through polymer films (the 'electrode' cases): ferrocene permeation through a $Ru(bipy)_3^{2+}$ silane polymer[59] and size discrimination in permeation of species through a polyphenylene film[60] provide examples here.

Microelectrodes have been employed to study the mediated oxidation of $Ru(bipy)_2(CH_3CN)_2^{2+}$ at poly[$Ru(vibipy)_3^{3+}$][61] and to confirm the earlier[34] RDE-based Sk'' diagnosis of Table 4. A thin layer cell configuration led to the same mechanistic conclusion for $Fe(\phi_2phen)_3^{3+}$ reduction by poly[$Os(bipy)_2(vipy)_2^{2+}$].[61] $FeCp_2^{0/+}$ permeation rates (κD_Y) (see Fig. 3) through poly[$Ru(vibipy)_3^{2+}$] under 'electrode' case conditions were determined using the two techniques[61] and pleasing agreement found. In both cases, one is using the Koutecky–Levich relationship

$$1/i_{Lim} = 1/i_F + 1/i_{mt} \tag{13}$$

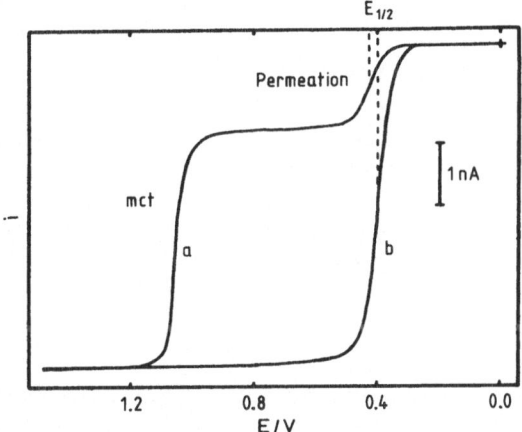

FIG. 3. Steady-state voltammetry of ferrocene at (a) a poly [Ru(vibipy)$_3^{2+}$] coated and (b) a bare (10 μm diameter) Pt microelectrode. The two waves in curve (a) in the vicinity of $E'_{Y/Z}$ and $E'_{A/B}$ represent permeation and mediation processes, respectively, as labelled. (Reproduced by courtesy of Elsevier Science Publishers.)

where the mass transport rate i_{mt} is given by a different expression from that for the RDE in eqn (9); this still retains the essential steady-state behaviour. The relative merits of the three techniques in this context are discussed in ref. 61: microelectrodes give access to rapid kinetics, via high mass transport rates, and the thin layer geometry is useful in the low-current region.

Chronoamperometric determination of charge transport rates has been combined with RDE experiments to study the Mo(CN)$_8^{3-/4-}$ self-exchange process in viologen-functionalised polystyrene films.[62] Analysis of RDE current/voltage waveshape was used to study the Fe(CN)$_6^{3-/4-}$ and Fe$^{3+/2+}$ redox couples at electrodes coated with monolayer films of a viologen polymer.[63] At potentials where only direct reaction was possible, electron transfer was found to be more facile in the former case but slowed in the latter case, a result attributed to electrostatic interactions with the positively charged viologen layer. Mediated reduction of Fe^{3+} (at more negative potentials) proceeded readily.

Optical/spectroscopic measurements have also been used to study mediated charge transfer. Oxalate oxidation by Ru(bipy)$_3^{3+}$ centres in

Nafion[64,65] and polybipyrazine[66] films results in light emission, which has been studied during cyclic voltammetry[64–66] and potential step experiments[65] and for which kinetic parameters have been deduced by digital simulation.[65] Laser interferometry has been used to observe diffusion layer formation at polypyrrole[67] and polycarbazole[68] modified electrodes. Whilst the authors admit that there are still unexplained features, there is some evidence that, in the latter case, the diffusion layers for hydroquinone and for both Fe^{2+} and $Fe(CN)_6^{4-}$ are, respectively, outside and partially inside the film; this would imply 'surface'- and 'layer'-type reactions, respectively.

2.5. Summary

By way of conclusion to this section and introduction to the next, a statement of the requirements for successful mediated charge-transfer-based applications is warranted. Whilst the requirements/performance are dependent on the electrode/reaction properties of each specific system, some general observations are possible. The slower the reaction, the thicker the film required to consume B. However, in order to derive benefit from the additional mediator sites, relatively free diffusion of Y and rapid charge transport are vital. Values of D_Y and D_E in excess of $\sim 10^{-9}\,\text{cm}^2\,\text{s}^{-1}$ are necessary, but the thicker the film the higher this limit becomes.

3. APPLICATIONS

3.1. Introduction

The majority of applications described below are based on mediated charge transfer to solution species. The classification scheme used is based on the end rather than the means, so that where alternative strategies to mediated charge transfer exist they are grouped together to allow their relative merits to be assessed. There are also areas which overlap (electrocatalysis and sensors in particular): to avoid duplication, cross-reference will be made.

3.2. Electrochemical Synthesis

In most cases described here, the general approach has been to take a proven homogeneous redox mediator and immobilise it on the electrode surface. The hope is to combine its virtues with the advantages of electrochemical potential control of mediator concentration, continual rapid regeneration and facile separation of the freqently expensive/toxic mediator. The requirements for success have been discussed in Section 2. Features to look out for are changes in mediator behaviour as a result of polymer immobilisation and the very high local concentration—both being film environmental effects. Environmental effects are the dominant, rather than subsidiary, features in the chiral syntheses described, where one relies not on mediated charge transfer but the provision of specific molecular geometries.

3.2.1. Oxygen reduction

The electrochemical reduction of O_2 is complex and kinetically difficult, except at certain (usually expensive) electrodes, notably Pt. The objective here is therefore to use suitably modified cheaper electrodes, notably carbon. The complexity lies in the possibility of $2e^-$ and/or $4e^-$ reduction pathways to H_2O_2 and H_2O, respectively, the latter either being direct or via the $2e^-$ intermediate. For energy conversion applications $4e^-$ reduction is the goal, preferably not via the high-energy H_2O_2 intermediate. For analytical purposes (see 'Sensors', Section 3.3) either $2e^-$ or $4e^-$ reduction is acceptable provided it is rapid and clean.

The predominance of porphyrin and phthalocyanine systems in Table 6 signals attempts to mimic nature's approach to O_2 reduction, and the Fe species seems rather better than the Co species at $4e^-$ reduction. Perhaps the most elegant of the many monolayer/monomeric multilayer competitors with which these systems should be compared is the di-cobalt face-to-face porphyrin dimer of Collman and Anson[84] ('Co$_2$(FTF4)'): this, together with some close relatives, has a second-order rate constant for $4e^-$ reduction of $\sim 10^5$ mol^{-1} dm^3 s^{-1}. With such a rate constant, comparatively few layers would be very efficient. The work of Wan[70] is of interest, firstly because of the use of copolymer composition to control polymer swelling, and secondly with respect to the variation of i_F (or k'_{ME}) with film thickness. At low coverages, the authors' values of i_s, i_E and i_k place the system just in the Lk ('R') zone. Increasing L should carry one into the LSk zone (see Fig. 1): at the highest coverage quoted (6 nmol cm^{-2}) they find $(i_E/i_k)^{1/2} = X_0/L = 1·6$ and $(i_s/i_k)^{1/2} = (X_L/L) = 0·33$, i.e. optimum thickness. Using their values of κk and Γ, a very respectable modified electrode rate constant (k'_{ME}) of $\sim 6 \times 10^{-2}$ cm s^{-1} is calculated.

TABLE 6
Oxygen Reduction by Polymer Modified Electrodes

Polymer/mediator	Technique(s) used	Medium	Product(s)	Comments	Refs
Poly(methacryl-P)					
P = —NH—ϕ—(Fe^{III}porph)	CV, chronocoulometry	0·05M-H$_2$SO$_4$	H$_2$O	$E_p \sim 0.28$ V (H$^+$/H$_2$, pH 1·3).	69
P = 1-vinyl-2-pyrrolidone copolymer of species above	RDE, RRDE	0·1M-HClO$_4$	H$_2$O (some H$_2$O$_2$ at less negative potentials)	Pyrrolidone component controls swelling/solubility.	70
Nafion, CoTPP	RDE	0·5M-H$_2$SO$_4$	H$_2$O$_2$	Ru(NH$_3$)$_6^{2+}$ 'electron shuttle' to increase effective charging rate. 'Lk' case at high shuttle concentrations.	71
Polypyrrole, CoTPP(SO$_3^-$)$_4$	CV	0·05M-H$_2$SO$_4$	?	$E_p \sim 0.48$ V (H$^+$/H$_2$, pH 1·3).	72
Poly(Fe^{III}TPP·Cl$^-$)	RRDE	0·5M-H$_2$SO$_4$	H$_2$O	('Polymerisation' by heat treatment.) $E_{1/2} \sim 0.22$ V (H$^+$/H$_2$, pH 0·3).	73
Polystyrene, CoPc	CV	6M-KOH	?	$E_p \simeq 0.55$ V (H$^+$/H$_2$, pH 1·7)	74
Polypyrrole, Fe[Pc(SO$_3^-$)$_4$]	RRDE, CV	0·1M-CF$_3$CO$_2$H pH 1·7 pH 6·5 0·1M-NaOH	H$_2$O$_2$ H$_2$O/H$_2$O$_2$ H$_2$O	$E_{1/2} \simeq 0.3$ V (H$^+$/H$_2$, pH 6·5) $E_{1/2} \simeq 0.6$ V (H$^+$/H$_2$, pH 13)	75

Polymer	Method	Electrolyte	Product	Comments	Ref
'Poly(PcM)', M = Fe, Co, Ni (~5% loading)	RRDE	1M-KOH	Mainly H_2O H_2O/H_2O_2	$E_{1/2} \sim 0.7$ V (H^+/H_2, 1M-KOH) (some dependence on loading/molecular mass).	76
Polyxylylviologen	RDE	1M-NaClO$_4$, pH ~ 5	H_2O_2	$E_{1/2} \simeq -0.14$ V (H^+/H_2, pH 5) $\kappa k \sim 10^6 M^{-1} s^{-1}$. (Probable LSk case, with $X_L \simeq 2$ layers.)	77
Poly(styrene sulphonate), polyxylylviologen	RDE	0·2M-$CF_3CO_2^-$/ClO_4^-, pH 0–12	H_2O_2	$E_{1/2} \sim -0.4$ V (SSCE), independent of pH. $k \sim 5 \times 10^5 M^{-1} s^{-1}$ (assuming $\kappa = 1$).	32
Polycyanurviologen	RDE	0·5M-phosphate, pH 7	H_2O_2	$E_{1/2} \sim 0.34$ V (H^+/H_2, pH 7). $k \sim 2 \times 10^6 M^{-1} s^{-1}$.	63
Poly(silylnaphthoquinone)	RDE	0·1M-KCl, pH 7	H_2O_2	$E_{1/2} \sim 0.30$ V (H^+/H_2, pH 7). ($k > 10^5 M^{-1} s^{-1}$)	78, 79
Poly(ethyleneimineanthraquinone)	RDE	0·1M-NaCl, pH 2·3–6·0	H_2O_2	$E_{1/2} \sim 0.15$ V (H^+/H_2, pH 6). $\log[k(M^{-1} s^{-1})] = 0.62$ pH + 0.7 (assuming $\kappa = 1$?) [k(pH 6) $\sim 2.6 \times 10^4 M^{-1} s^{-1}$] ($Lk$ case).	80
Polypyrrole, Co(OAc)$_2$	RRDE	0·05M-H_2SO_4 1M-H_2SO_4	H_2O_2	$E_{1/2} \sim 0.55$ V (H^+/H_2, pH 1·3)	81
Poly(vinylacetic acid), Pt^0	CV		—	$E_p \sim 0.55$ V (H^+/H_2, pH 0)	82
Poly(tetracyanoethylene), M M = Fe, Ni, Cu (oxidation state unspecified).	CV	pH 7	—	$E_p \sim 0.05$ V (H^+/H_2, pH 7)	83

The viologen and quinone systems give H_2O_2 as the primary product. Values of κk (or k, assuming $\kappa = 1$) for viologens immobilised by three different schemes are all $\sim 10^6\,\text{mol}^{-1}\,\text{dm}^3\,\text{s}^{-1}$. The fact that one of these involves only monolayer coverage implies little effect on the reaction through the polymer environment. Variation in $E_{1/2}$ values for the quinones is probably a reflection of E'_{surf} values. (Comparison of $E_{1/2}$ values is not necessarily helpful here, despite conversion to a common reference scale, H^+/H_2 at the appropriate pH, since cases of pH variation[80] and independence[32] have been noted and many studies were restricted to a single pH.)

3.2.2. Other inorganic redox systems

Oxidation of Cl^- to Cl_2 has been effected by monomeric and dimeric bipyridyl ruthenium–oxo complexes ion-exchanged into p-chlorosulphonated polystyrene,[85] via high metal oxidation states, e.g. a Ru^{IV}–Ru^{V} dimeric species is proposed in one case. The problem of Cl^- substitution was overcome by potential cycling and initial currents of $8\,\text{mA}\,\text{cm}^{-2}$ in $5\,\text{mol}\,\text{dm}^{-3}$ NaCl (equivalent to a rate constant of $\sim 2 \times 10^{-5}\,\text{cm}\,\text{s}^{-1}$) were reported.

Two attempts at reduction are noteworthy. In the first (successful) example, polysilylviologen-immobilised Pd^0 particles were used to reduce bicarbonate to formate.[86] Although the rates were slow ($\sim 0.1\,\text{mA}\,\text{cm}^{-2}$ in $\sim 0.05\,\text{mol}\,\text{dm}^{-3}$ HCO_3^-, corresponding to k'_{ME} $\sim 10^{-5}\,\text{cm}\,\text{s}^{-1}$), fair stability ($>10^3$ turnovers per Pd atom) and high current efficiency ($\sim 80\%$) were obtained within $\sim 0.1\,\text{V}$ of the thermodynamic potential for the HCO_3^-/HCO_2^- couple. In the second (unsuccessful) example, N_2 reduction to NH_3 was attempted via N_2 coordination to a series of surface-immobilised Mo/phosphine polymer complexes.[87]

3.2.3. Biochemical Redox Couples

Charge transfer mediation to cytochrome c has been achieved using viologen-[88,89] and ferrocene-[90] based polymer films. In the first case, reduction occurred at $E \simeq E'_{\text{surf}}$ for the $V^{2+}/V^{\cdot +}$ couple, although the silylviologen polymer[88] had much faster kinetics than the vinylviologen polymer.[89] (It should be noted that the mode of operation here, mediation, is distinctly different from that of the unquaternised monomeric bipyridyl system used for ferri/ferro-cytochrome c redox electrochemistry on Au.[91]) In an extension of this work, Pt^0 particles were incorporated in the silylviologen polymer on a glass support to equilibrate the ferri-/ferro-cytochrome c and H^+/H_2 couples.[92] The ferrocene-based mediator has a value of E'_{surf} close to E' for the cytochrome system, so both oxidation and reduction could be carried out.[90] In two of these studies[89,90] 'surface' mechanisms were proposed and the same is probable for the third case:[88] this is a reflection of rapid kinetics[88,90] and unfavourable permeation for the large substrate molecule.

A dopamine-functionalised methacrylate polymer has been used for NADH oxidation.[93] Cyclic voltammetry showed mediation at E'_{surf} (still well positive of $E'_{NAD^+/NADH}$), but at rates that only increased with coverage up to monolayer levels, decreasing thereafter, consistent with a shift to one of the transport-limited cases. Polymer films derived from phenazine methosulphate (PMS) and phenazine ethosulphate[94] also mediated NADH oxidation at E'_{surf}, but this time rather closer to $E'_{NAD^+/NADH}$. RDE data for the PMS mediator indicated first-order behaviour with respect to NADH and coverage, implying an Lk ('R') mechanism. Evaluation of κk (by the reviewer) gave ~300 $mol^{-1} dm^3 s^{-1}$; this moderate rate constant probably explains how the rather large substrate has the opportunity for film permeation. (The same workers also used Meldola Blue for NADH oxidation,[95] but here the films were multilayers of monomer, rather than electrochemically produced polymer.)

In contrast to the above species, a range of inorganic mediators, such as PVF,[96] have been used for ascorbate oxidation. More quantitative studies using coordinated $\sim\sim IrCl_5^{2-}$,[97] and ion-exchanged $Fe(CN)_6^{3-}$ [98] and $Os(bipy)_3^{3+}$ [99] revealed $E_{1/2}$ values at or close to E'_{surf} and values of k'_{ME} greater than 10^{-2} cm s^{-1}. For the iridium system,[97] reaction only occurred over a small part of the layer, as a S or LS case. In the $Fe(CN)_6^{3-}$ system[98] at low loadings, reaction appeared to occur throughout the layer, but on increasing the loading, it occurs only near the outer interface: this is a shift from the Lk to LSk zone resulting from an increase in b_0 (see Fig. 1). As the authors note, it is interesting that the rate constant is increased by a factor of ~500 over the homogeneous solution value. The $Os(bipy)_3^{3+}$ system[99] reacts according to an Sk'' mechanism (X_L ~0·2 nm), and the authors' assumption of a partition coefficient of unity is unnecessary. Increasing ascorbate concentrations (see Fig. 1) would be expected to move the system into the St_E zone, as the authors found, although there were additional complications associated with the Nafion film structure.

Polymeric films derived from methylviologen have been used to effect the reductions of ferredoxin,[100,101] peroxidase[102] and myoglobin.[103] Unfortunately the polymers are not well characterised and the mechanism is not understood, but a conventional Tafel analysis for the ferredoxin case[100] gave an effective heterogeneous rate constant of $\sim 10^{-4}$ cm s^{-1}.

3.2.4. Organic redox reactions

Polyvinyltriphenylamine has been used for the oxidation of carboxylates,[104] such as acetate, where oxidation occurred at E'_{surf}, approximately 0·5 V less positive than at a bare electrode as illustrated in Fig. 4. Increasing catalytic efficiency with Γ up to $\sim 10^{-8}$ mol cm^{-2}, decreasing thereafter, probably implies a shift from the Lk ('R') case ultimately to one of the transport cases. Isopropanol, p-toluic acid and p-xylene have been oxidised by Ru^{IV} centres in PVP-bound bipyridyl complexes,[105] although both rates and stabilities were poor.

The photoassisted reductions of CCl_4 and $CHCl_3$ by PVF have been achieved,[106] although turnover numbers were poor (20 and 40, respectively). Reduction of chloroform and various phenyl derivatives has also been effected by immobilised Ni^0/triphenylphosphine[107] and, to a limited extent, by an immobilised iron porphyrin.[108] A series of 1,2-dibromo species have been reduced to the corresponding alkenes by polynitrostyrene,[109,110] polyvinyl- and polysilyl-viologens,[111] and iron[108] and cobalt[112] porphyrins. In the viologen[111] and nitrostyrene[110] cases, turnover numbers in excess of 10^4 were possible, although only when O_2 and H_2O were rigorously excluded—an important constraint for practical application.

3.2.5. Chiral synthesis

Poly(L-valine) coated electrodes have been used for the reductions of citraconic acid[113,114] and 4-methylcoumarin,[113] and the oxidations of sulphides to sulphoxides[115,116] and a secondary alcohol to a ketone.[117] Variations used by these workers include prior polypyrrole or silane-bonded polypyrrole surface coating[115,116] (to give a bilayer structure) and the use of γ-benzyl-L-glutamate and L-leucine polymers.[116] Whilst optical yields were very variable (1–93%) and fell (often markedly) with use,[116] the approach clearly has some merit. What is not clear, however, is why polymeric films should be any better than monolayer derivatisation, since the chiral part of the film is electroinactive and reaction is thus presumably direct rather than mediated. (Multilayer mediated asymmetric synthesis has been demonstrated, albeit with low

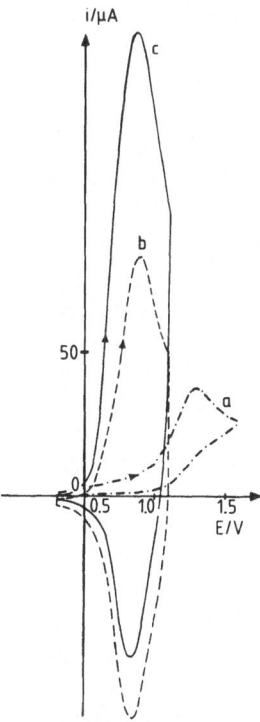

FIG. 4. Use of a polyvinyltriphenylamine modified electrode for the oxidation of acetate in acetonitrile/0·1 mol dm^{-3} n-Bu$_4$NClO$_4$. Cyclic voltammograms (scan rate 50 mV s^{-1}) are for (a) 10^{-3} mol dm^{-3} acetate at bare Pt, (b) the modified electrode ($\Gamma = 2\cdot 8 \times 10^{-7}$ mol cm^{-2}) in background electrolyte and (c) the modified electrode in the solution used for (a). Mediation is indicated by, firstly, diminution of the cathodic peak in (c) as compared with (b) (mediator has been consumed by substrate) and, secondly, by the enhanced anodic current in the vicinity of E'_{surf}. (Reproduced by courtesy of Elsevier Science Publishers.)

(7%) optical purity, in the production of Λ-Co(phen)$_3^{3+}$ from racemic Co(phen)$_3^{2+}$ by Δ-Ru(phen)$_3^{2+}$ immobilised in a clay film on an electrode.[118])

3.3. Sensors

The operational requirement here is a reproducible and stable observable/concentration relationship, with readily discernible changes in the observable over the appropriate concentration range and minimal interference from other species. The three principal ways in which the polymer film can contribute to such a device are to facilitate amperometric or potentiometric detection, or pre-concentration prior to either of these.

Amperometric detection usually involves mediated charge transfer to the target species and is thus intimately related to electrocatalysis—both preparative and analytical applications are readily envisaged for species such as oxygen and biological molecules (see Sections 3.2.2 and 3.2.3). Nitrite has been determined by oxidation at a $Pt/I^-/PVP \cdot R^+$, $IrCl_6^{3-}$ electrode,[119] probably by an LSk ('SR') mechanism according to RDE data. The same system, but without the Ir^{IV} mediator, was used as part of a membrane-covered sensor for nitrite determination in spiked meat extract and environmental water,[120] with linearity of response over the range 4–2000 μmol dm^{-3} and a detection limit of 2 μmol dm^{-3}.

The obvious potentiometric sensor is that for H^+, where a linear pH/voltage correlation is expected on the basis of a Nernstian response. This has been found for poly(1,2-diaminobenzene) films on Pt[121,122] and GC,[122] polyphenol films on Pt and GC[122] and poly('TCNQ') films on Pt[123] over the range $2 \le pH \le 12$. Whilst Heineman[121] attributed the response to amine group protonation, Cheek[122] found the same pH/potential shape at bare and coated electrodes and attributed the response to the underlying surface, with some help from the polymer by way of 'blocking' interfering species. Some support for this is derived from the study on the TCNQ polymer,[123] (see Fig. 5) for which the underlying electrode pH response was initially lost on coating but restored after potential cycling (for discussion of the 'break-in' phenomenon, facilitating electrolyte ingress, see Section 3.4.1 of Chapter 5). Rubinstein has used E_p values for surface-bound couples to determine pH:[124] this was done by recording the E_p for the $Ru(bipy)_3^{2+/3+}$ couple (in Nafion) with respect to a poly(1,2-diaminobenzene) pH-dependent couple on a separate electrode, or on a different part of the same electrode (chloranil or polyaniline). Accuracy was estimated at ±0·1 pH unit. Graphite coated with poly(acrylic acid) has been found to respond to K^+ over the range $1 \le p(K^+) \le 4$,[125] and coated with poly[trialkyl(vinylbenzyl)ammonium] to SCN^- over the range $1 \le p(SCN^-) \le 5$.[125] In the latter case, stability and response time were reasonable although some interference, notably from salicylate, was found.

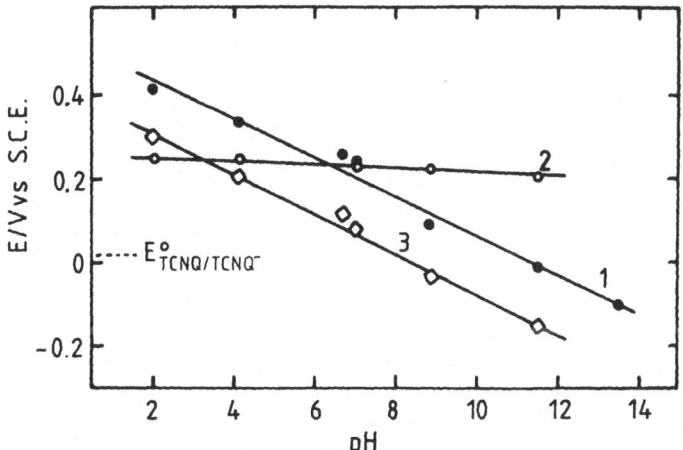

FIG. 5. pH responses of (1) bare Pt, (2) Pt/TCNQ polymer 'as prepared' and (3) Pt/TCNQ polymer following 'break-in' by potential cycling. (Reproduced by courtesy of the American Chemical Society.)

Pre-concentration, based on high partition coefficients for selected ions, prior to voltammetric analysis (cf. anodic stripping voltammetry) has been demonstrated for ion-exchanging films. Accumulation of $Cr_2O_7^{2-}$ in protonated PVP has been shown to occur to the exclusion of excesses of positively charged interferants (Fe^{3+}, Cu^{2+}, Cr^{3+});[126,127] a linear voltammetric current response in the range 10^{-8}–10^{-6} mol dm^{-3} was found. Differential pulse analysis for Cu^{2+} pre-concentrated in sulphonated polystyrene gave a linear current response for bulk concentrations in the 60–300 μmol dm^{-3} range.[128] It has been pointed out that large partition coefficients must be accompanied by rapid equilibration for successful application in certain areas where response time is crucial,[129] i.e. both thermodynamics and kinetics must be favourable. (Chromatographic detection systems, such as the cobalt phthalocyanine carbon paste electrode (CoPc/CPE) electrode for hydrazine,[130] provide an example of such an application.) An alternative to target species pre-concentration is accumulation of the charge resulting from its oxidation/reduction until conveniently measurable. This approach has been demonstrated by bilayer film outer layer 'untrapping' of residual viologen radical sites following exposure to

O_2[131,132] and of photogenerated viologen radical sites following exposure to ethanol under UV irradiation.[132] These are unquantified experiments as yet.

In principle, one could use any physical property suitably correlated with target species activity as an analytical probe. An example is the proposed use of polyfuran film conductivity variation in the presence of moisture as a humidity sensor,[133] although the 'modified electrode' is not being used in an electrochemical cell in this instance.

3.4. Corrosion Protection of Metals

The usual objective here is the suppression of all electrochemical reactions, which is accomplished by the formation of a coherent insulating polymeric film. A range of aniline- and phenol-based polymers have been used to this end on copper[134,135] and iron[136]/steel[137] surfaces. Whilst throwing power and ease of pigment incorporation were not as good as in the competing electrophoretic approach, electropolymerisation of o-alkylphenol[136] was found readily to give stable coatings to which organic overlayers could be bonded.

A more difficult objective is the suppression of corrosion without impeding solution species redox reactions (see also Section 3.5). Silane chemistry has been used to attach multilayers of ferrocene and $\sim\sim Fe(CN)_5^{3-}$ centres to Ni, so that mediated charge transfer to solution $FeCp_2^{+/0}$ was possible without Ni oxidation.[138]

3.5. Semiconductor Electrochemistry

In semiconductor photoelectrochemistry, the principal problem is frequently the useful consumption of photogenerated minority carriers before they cause surface corrosion or undergo recombination. The general approach adopted in many cases is to remove these entities from the surface by electronic or redox conduction for remote mediation to a suitable solution species. (Systems where the film itself is the site of the photochemical process are considered in Section 3.6.)

Photoanodic corrosion of n-type semiconductors has been prevented by mediation via ferrocene polymers (on Si,[139–143] GaAs[144] and Ge[142]), polyxylylviologen/$Fe(CN)_6^{4-}$ (on Si[145]) (see Fig. 6), Ru/bipy complex polymers (on TiO_2,[146] $MoSe_2$[147] and WSe_2[147]), a Ru/phen polymer (on TiO_2[59]) and Nafion/TTF^+ (on Si[43]). Stability increases can be dramatic: for example, under conditions where n-Si passivates rapidly, stable ferrocene mediation to solution reductants such as $Ru(NH_3)_6^{2+}$, $Fe(CN)_6^{4-}$ and I^- ($>10^5$ turnovers) has been reported.[139] Despite this,

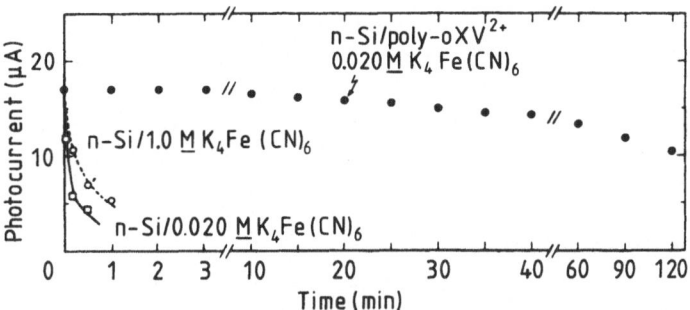

FIG. 6. Improvement in photocurrent stability at n-Si (under 1 mW cm^{-2} 632·8 nm illumination) in aqueous Fe(CN)$_6^{4-}$ solution upon coating with $1·7 \times 10^{-8}$ mol cm^{-2} of poly(o-xylylviologen). Electrode area: 0·053 cm^2; potential: 0·1 V (SCE); solution stirred. In the absence of the polymer, film passivation (via SiO$_x$ formation) occurs in seconds. (Reproduced by courtesy of The American Chemical Society.)

even monochromatic photoelectrode efficiencies are low, e.g. ~5% (at 633 nm) for the n-Si/polyxylylviologen,Fe(CN)$_6^{4-}$ system in aqueous Fe(CN)$_6^{4-}$.[145] Note that in the absence of rapid mediation kinetics (for which the two-dimensional reaction zone concept was recognised at an early stage[141]), the film permeation necessary for reaction may lead to failure through film peeling[43] or photochemical substitution reactions in the polymer,[47] as well as the obvious semiconductor/electrolyte corrosion process. Removal of the photogenerated holes by polypyrrole has been used for stabilisation of n-Si,[143,148-150] n-CdS,[148] n-CdSe[148] and n-GaAs.[148,151] Improvements in stability are again dramatic: currents of ~10 mA cm^{-2} may be passed for many hours at the coated semiconductors in Fe^{2+} or Fe(CN)$_6^{4-}$ solutions, as opposed to only minutes or seconds at the naked surfaces.[148,149] Other polymers which have been used are polyaniline[152] (for n-Si, n-CdS, n-CdSe and n-GaAs) and polyindole[153] (for n-MoSe$_2$), the latter functioning via the blocking of defect sites.[153]

Elaborations on straightforward polymer coatings include polypyrrole overlayers on ferrocene[143] and Au[154] films on n-Si, use of silane chemistry to attach polypyrrole covalently to n-Si,[155] and polypyrrole as a matrix for particulate RuO$_2$ catalyst immobilisation[156,157] (see Section 3.7).

Silyl-[158] and xylyl-[159]viologen polymers have been used in schemes for the improvement of H_2 evolution at p-Si, in conjunction with particulate Pd^0 and Pt^0, respectively. Monochromatic efficiencies of 2–4%[158] and moderate stabilities (limited by heavy metal contamination[158] and viologen photodegradation[159]) were reported. Polyaniline films were found to enhance H_2 evolution at p-Si and p-GaP.[152]

3.6. Photogalvanic Effects

Systems considered here are those where the film, rather than the underlying electrode, is the site of the photo-driven reaction, usually with the chromophore as the polymer-bound reactant. PVP has been used in neutral form to coordinate ZnTPP[160] and $\geq Ru(bipy)_2^{2+}$ [161] and in protonated form to bind Croconate Violet[162] and Rose Bengal.[163] Photocurrents were generally small (less than 5 $\mu m \, cm^{-2}$ (mmol dm^{-3} of solution species)$^{-1}$ even in the most favourable case[161]). Alternative Ru/bipy complex immobilisation procedures[164–167] also produced only small ($\mu A \, cm^{-2}$) currents. Thionine immobilisation in acrylamide[168,169] and merocyanine and rhodamine B immobilisation in ethyleneimine and ethylene oxide[170] polymer films on electrodes has been described. Even 'remarkable'[170] photocurrents were still only at the 10 $\mu A \, cm^{-2}$ level, and power levels were only $\sim 10^{-8}$ W for 10^{-3}–10^{-2} mol dm^{-3} solution species.[169] (Use of a clay matrix for thionine immobilisation[171] was no better, the overall efficiency being $\sim 10^{-4}$%.) These disappointing results are probably due to a combination of low film absorbance (particularly for polychromatic illumination), only partial film participation in the photoelectrochemistry,[165] self-quenching at the high chromophore concentrations found in the film, and enhanced thermal back-reaction.

Bilayers containing $Ru(bipy)_3^{2+/3+}$ and $MV^{\cdot +/2+}$ (in either configuration) were also unsuccessful,[167] but this is probably because of the small fraction of sites sufficiently close to the polymer/polymer interface to react. Photocurrents generated in polypyrrole[172] and a copper phenylacetylide polymer[173] have also been reported.

Quite separate from these effects is the use of a polymer modified electrode to provide the necessary selectivity[174] to discriminate between solution photogenerated products in a photogalvanic cell. For the iron/thionine system, selectivity for the thionine/leucothionine couple on a polymeric thionine coated electrode[175,176] and for the $Fe^{2+/3+}$ couple on a polypyrrole coated electrode[177] have been described.

3.7. Immobilisation of Particulates

Use of polymer films as matrices for the immobilisation of catalyst particles on semiconductor surfaces has already been alluded to in Section 3.5. In the systems previously mentioned, the objective was to use photogenerated minority carriers in a p-type semiconductor (p-Si) to reduce surface attached viologen redox centres, whose function was to act as mediators for H_2 evolution. The problem is that, though thermodynamically favourable for pH < 5, the '$V^{\cdot +}/H^+$' reaction is slow: Pt^{0} [159,178,179] and Pd^0 [158] metal particles were used to speed equilibration. Interestingly, efficiencies for structures with Pd^0 dispersed throughout the film and as the outer layer of a 'sandwich' (see Section 3.9) were similar (~1–4% under 633 nm illumination).[158] In both the Pt^0 [179] and Pd^0 [158] systems, a maximum in H_2 evolution efficiency was observed at pH ~ 4.

Photoelectrochemical O_2 evolution at n-CdS[156] and n-GaP[157] has been catalysed by RuO_2 particles immobilised by polypyrrole. The suspension/evaporation[159] and RuO_4^{2-} ion exchange/reduction[157] procedures used almost certainly give different particle distributions. For n-CdS, the proportion of reaction leading to O_2 production was raised from 1% on the bare electrode to 68% at the polymer/RuO_2 coated electrode.[156] The pressure of trapped O_2 was found to limit stability in the other study.[157]

Both H_2 and O_2 evolution were catalysed by Pt^0 particles trapped in poly(vinylacetic acid) on glassy carbon.[180] Here presumably only those particles in contact with the carbon are active, since the polymer contains no mediator sites.

Carbon particles have been included in very thick polymeric viologen films to increase charging rates;[181] here the particles increase the effective electrode area, i.e. decrease the effective film thickness over which charge must diffuse.

3.8. Display Devices

Whilst many workers have recognised the possibility of using the redox-state-dependent optical properties of polymer films on electrodes as the basis of a display device, comparatively few studies have centred on this application. (It should be noted that these systems are rather different from those based on reduction, precipitation/oxidation, dissolution of monomeric solution phase viologens,[182] where material has to diffuse to/from the surface.) Systems based on surface-bound viologen[183] (0·1, 1), iron phenanthroline[184] (0·16, 0·2),

TTF[185] (—, 0·5) and carbazole[186] (0·018, —) chromophores have been described, where the figures in parentheses represent the important parameters of electrochromic efficiency (mC cm^{-2}) and response time(s), respectively. In one case, a seven-segment display was illustrated.[184] Comparisons of charging rates should be made with caution, for there may be different amounts of polymer present— indeed, there is clearly a trade-off between increasing contrast and decreasing response rate as film thickness is increased. The idea of using conjugated polymers to increase response rate has been explored for polyaniline,[187] polythiophene[188,189] and polypyrrole.[189] For polyaniline, a response time of ~20 ms (illustrated in Fig. 7), excellent stability (>10^6 cycles) and reasonable open circuit memory were achieved.[187] Polythiophene gave a potential-dependent response time in the 30–1000 ms range,[188,189] good stability (>10^3 cycles) and good open circuit memory.[188] Note that, in the absence of mobile 'untrapping' solution species, bilayer assemblies[131] (see Section 3.9) provide a 'latching' facility which could in principle be used to improve open circuit stability. Finally, all these systems are 'passive' (based on light absorption), rather than 'active' (based on emission, e.g. the chemiluminescent systems studied by Bard et al.[64,65]).

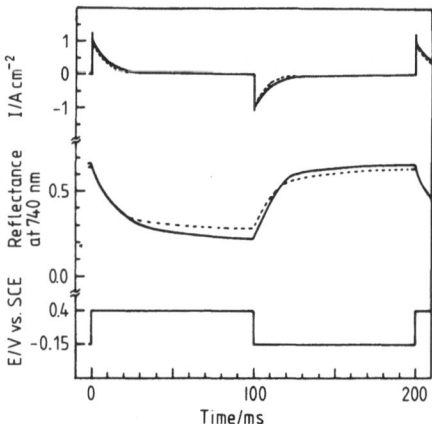

FIG. 7. Electrochemical and optical responses of a polyaniline film to potential steps between −0·15 and 0·4 V (SCE) in 0·1 mol dm^{-3} HCl. The amount of polymer deposited corresponds to a charge of 45 mC cm^{-2}. Full and broken lines show responses initially and after 1·36 × 10^4 cycles. (Reproduced by courtesy of Elsevier Science Publishers.)

3.9. Electronic and Electrochemical Devices

The ability of a surface-immobilised redox couple to act as a poised potential reference electrode has been explored in the case of a PVF film[190] and has already been discussed for some other systems as part of pH sensors in Section 3.3. In this section rather more complex geometrical and chemical combinations which have interesting properties are described. Of these 'structured interfaces', dispersed particulates have already been covered in Section 3.7.

In the simplest arrangement, a thin metal overlayer (sufficiently porous to permit electrolyte passage) is deposited on top of the polymer to produce a metal/polymer/metal' 'sandwich'. Using a series of Fe, Ru, Os/polypyridine complexes, the current/voltage characteristic of such an arrangement has been studied.[191–193] The (steady-state) current is controlled by the concentration gradient and is thus dependent on the potentials of the two metal components both with respect to an external reference (effectively E'_{surf}) and each other. Its maximum value occurs when the metal potentials span E'_{surf} and is given by the charge transport rate discussed in Section 3.4 of Chapter 5. Permeation[26] and H_2 evolution catalysis[194] properties of sandwiches have also been studied.

Alternatively, the outer layer may be another polymer. In this 'bilayer' structure, mediated charge transfer to the outer layer takes place only at the polymer/polymer interface and only over rather narrow potential intervals (effectively, 'conduction bands') in the vicinity of $(E'_{surf})_{inner}$ (for an appropriate couple). This reaction is governed by the thermodynamics and kinetics of the $(O/R)_{inner}/(R'/O')_{outer}$ cross-reaction and the inner film charge transport rate. Initial studies[131,195] demonstrated preparation, 'bilayer' structure and outer film charge transfer mediation (charge 'trapping' and 'untrapping') for a series of Fe/Ru complex and viologen polymer combinations. More recently, Murray's group have carried out[196] and tested[197] an analysis of the voltammetric waveshape: they concluded that, except for cases where the mediated reaction is thermodynamically unfavourable, 'trapping'/'untrapping' is likely to be controlled by inner film charge transport, rather than polymer/polymer interfacial reaction kinetics.

Exploitation of the rectifying nature of the charge transfer processes in 'sandwich' and 'bilayer' assemblies has been used to construct devices possessing diode-like[131,192,195] (Fig. 8) and triode-like[192] current/voltage characteristics. Its use as a 'latching' facility, subject to leakage reactions, for a display device[131] has been mentioned in

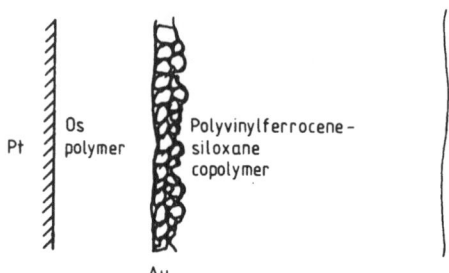

FIG. 8. Schematic representation of a diode-like device using a poly[Os(vipy)$_2$(bipy)$_2^{2+}$] film sandwiched between Pt and Au electrodes. The outer polymer layer is initially placed in a 1:1 FeCp$_2^+$/FeCp$_2^0$ state to act as a reference (as demonstrated by Peerce[190]) by poising the potental of the thin (~1 μm) Au outer film of the sandwich. (Reproduced from reference 192 by courtesy of The Electrochemical Society, Inc.)

Section 3.8. Deposition of a polypyrrole film over an array of three electrodes under separate potential control has been shown to produce a device behaving like a transistor.[198] Whilst switching times do not match those of their solid-state analogues, these types of device are as yet at a very early stage of development.

Complementary to the above electronic properties, the potential dependence of ionic permeability in polypyrrole films has been studied.[199–201] Films sufficiently thick to fill the gaps were deposited on grid electrodes and a.c. impedance used to demonstrate a difference by a factor of 10^3 in ionic permeability between the oxidised (charged) and reduced (neutral) states.[201] Permselectivity for Cl$^-$ over K$^+$ was shown.[201] Response times of this 'ion gate' were improved by using finer grids, allowing use of thinner polymer films.[200]

Whilst stable modification of the surface is usually desired, Miller has shown how instability can be turned to advantage by using it as a means to achieve timed chemical release. Initial work used a polystyrene spine functionalised with isonicotinate units to which dopamine was amide-linked;[202,203] reduction led to amide bond cleavage and dopamine release, analogous to neurotransmitter release in the presynaptic terminal. Enhancement of the same process for glutamate and γ-aminobutyric acid using the MV$^{2+/\cdot+}$ couple as an 'electron shuttle' was subsequently demonstrated,[204] although the fraction of immobilised neurotransmitter released was still small. Alternatively, the species of interest can be electrostatically retained in a polymer film whose charge is variable: ejection of glutamate[and $Fe(CN)_6^{4-}$ as a model] from oxidised (charged) polypyrrole upon reduction to the neutral form demonstrates the point.[205] Whilst glutamate was not released unless a cathodic pulse was applied to the isonicotinate system,[204] there was some loss from the polypyrrole film;[205] however, far greater triggered release was achievable in the latter case. The large numerical value of the Faraday and the ease of measurement/delivery of small charges indicate that this should be a good technique for precise delivery of small amounts of substance.

4. CONCLUSIONS

The theory of mediated charge transfer at modified electrodes would seem to be in fairly good order. The treatments of Albery[18] and Saveant,[12] which include simpler models of others,[3,5,6] (where they overlap) and simplified expressions for several limiting cases have been verified experimentally. The basic model has been elaborated to include more complex reaction schemes by Saveant et al.[11,15-17] Of the experimental techniques, RDE voltammetry is the best established and its steady-state characteristics favour it over cyclic voltammetry. The same steady-state characteristics associated with microelectrodes and thin-layer cells, whilst less widely used in this context, would appear to have much to offer.[61] The unfortunate feature is that comparatively few workers have used the available theories quantitatively—assumptions and convoluted/circular arguments abound. Simple expressions and diagnostic criteria for mechanistic deduction are available and the reviewer encourages their use.

An economic evaluation of polymer modified electrodes for electrocatalytic purposes,[9] including likely capital expenditure, estimated costs of $100–1000 per mole of product. Whilst this indicates viability only for fine, rather than tonnage, chemicals, it is in precisely this area where the fine control/selectivity features of surface-immobilised mediators under electrochemical control are most likely to realise their full potential. Although some workers have tried to catalyse reactions of 'difficult' small molecules, this point is well illustrated by mediated charge transfer to biochemical systems, by chiral synthesis and by the expanding area of electrochemical sensors. Rather few systems have been optimised, but the procedures for this are now available. Selectivity through mediated charge transfer has also been exploited in semiconductor photoelectrochemistry, via physical separation of the photochemical and faradaic reactions.

Perhaps the most exciting area is the recent construction of a range of novel metal/polymer composites, whose geometrical configuration leads to electronic/optical/chemical properties which may be usefully exploited. In the first class, devices mimicking a diode, a triode and a transistor have been produced. Secondly, display devices have been fabricated. In the third class, the principle of an electrode which permits electrochemically controlled release of small doses of chemicals has been demonstrated; this holds great promise for use in biomedical studies.

The global conclusion for Chapters 5 and 6 comes in three parts reflecting the current stage in the evolution of the subject. Firstly, diverse techniques for polymer modified electrode preparation and characterisation are well established. Secondly, theoretical treatments of reactions taking place at this new type of interface have been developed for a range of reaction schemes and the simpler ones have been verified experimentally with model systems. Thirdly, the principles involved in a wide range of applications utilising 'real' systems, based on their chemical, optical and electronic properties, have been demonstrated. The major thrusts of future work are thus expected to be the testing of theoretical treatments for the more complex reaction schemes likely to be encountered in practice, and attempts to realise the considerable promise held by polymer modified electrodes, through optimisation of the principles demonstrated to date.

REFERENCES

1. J. Leddy, A. J. Bard, J. T. Maloy and J. M. Saveant, *J. Electroanal. Chem.*, 1985, **187,** 205.
2. A. J. Bard and L. R. Faulkner, *Electrochemical Methods: Fundamentals and Applications*, Wiley, New York, 1980, p. 92.
3. F. C. Anson, *J. Phys. Chem.*, 1980, **84,** 3336.
4. C. P. Andrieux and J. M. Saveant, *J. Electroanal. Chem.*, 1978, **93,** 163.
5. E. Laviron, *J. Electroanal. Chem.*, 1982, **131,** 61.
6. M. Delamar, M. C. Pham, P.-C. Lacaze and J.-E. Dubois, *J. Electronal. Chem.*, 1980, **108,** 1.
7. C. P. Andrieux, J. M. Dumas-Bouchiat and J. M. Saveant, *J. Electroanal. Chem.*, 1980, **114,** 159.
8. R. W. Murray, *Phil. Trans. R. Soc. Lond. A*, 1981, **302,** 253.
9. D. A. Cox and R. E. W. Jansson, *J. Appl. Electrochem.*, 1982, **12,** 205.

10. W. J. Albery and A. R. Hillman, *Roy. Soc. Chem. Ann. Rep. C.*, 1981, **78,** 377.
11. F. C. Anson, J. M. Saveant and K. Shigehara, *J. Phys. Chem.*, 1983, **87,** 214.
12. C. P. Andrieux, J. M. Dumas-Bouchiat and J. M. Saveant, *J. Electroanal. Chem.*, 1982, **131,** 1.
13. C. P. Andrieux and J. M. Saveant, *J. Electroanal. Chem.*, 1982, **134,** 163.
14. C. P. Andrieux, J. M. Dumas-Bouchiat and J. M. Saveant, *J. Electroanal. Chem.*, 1984, **169,** 9.
15. C. P. Andrieux and J. M. Saveant, *J. Electroanal. Chem.*, 1984, **171,** 65.
16. C. P. Andrieux and J. M. Saveant, *J. Electroanal. Chem.*, 1983, **142,** 1.
17. J. Leddy, A. J. Bard, J. T. Maloy and J. M. Saveant, *J. Electroanal. Chem.*, 1985, **187,** 205.
18. W. J. Albery and A. R. Hillman, *J. Electroanal. Chem.*, 1985, **170,** 27.
19. A. J. Bard and L. R. Faulkner, *Electrochemical Methods: Fundamentals and Applications*, Wiley, New York, 1980, p. 283.
20. V. G. Levich, *Physicochemical Hydrodynamics*, Prentice Hall, Englewood Cliffs, N.J., 1962, p. 68.
21. J. Koutecky and V. G. Levich, *Zh. Fiz. Khim.*, 1956, **32,** 1565.
22. P. C. Lacaze, M. C. Pham, M. Delamar and J.-E. Dubois, *J. Electroanal. Chem.*, 1980, **108,** 9.
23. Y. Ohnuki, H. Matsuda, T. Ohsaka and N. Oyama, *J. Electroanal. Chem.*, 1983, **158,** 55.
24. J. Leddy and A. J. Bard, *J. Electroanal. Chem.*, 1983, **153,** 223.
25. P. V. Kamat and M. A. Fox, *J. Electroanal. Chem.*, 1983, **159,** 49.
26. P. G. Pickup, K. N. Kuo and R. W. Murray, *J. Electrochem. Soc.*, 1983, **130,** 2205.
27. T. Ikeda, R. Schmehl, P. Denisevich, K. Willman and R. W. Murray, *J. Amer. Chem. Soc.*, 1982, **104,** 2683.
28. T. Ikeda, C. R. Leidner and R. W. Murray, *J. Electroanal. Chem.*, 1982, **138,** 343.
29. F. C. Anson, T. Ohsaka and J. M. Saveant, *J. Amer. Chem. Soc.*, 1983, **105,** 4883.
30. F. C. Anson, T. Ohsaka and J. M. Saveant, *J. Phys. Chem.*, 1983, **87,** 640.
31. F. C. Anson, J. M. Saveant and K. Shigehara, *J. Electroanal. Chem.*, 1983, **145,** 423.
32. N. Oyama, N. Oki, H. Ohno, Y. Ohnuki, H. Matsuda and F. Tsuchida, *J. Phys. Chem.*, 1983, **87,** 3642.
33. P. J. Peerce and A. J. Bard, *J. Electroanal. Chem.*, 1980, **112,** 97.
34. T. Ikeda, C. R. Leidner and R. W. Murray, *J. Amer. Chem. Soc.*, 1981, **103,** 7422.
35. C. R. Leidner and R. W. Murray, *J. Amer. Chem. Soc.*, 1984, **106,** 1606.
36. R. H. Schmehl and R. W. Murray, *J. Electroanal. Chem.*, 1983, **152,** 97.
37. N. S. Lewis and M. S. Wrighton, *J. Phys. Chem.*, 1984, **88,** 2009.
38. R. N. Dominey, T. J. Lewis and M. S. Wrighton, *J. Phys. Chem.*, 1983, **87,** 5345.

39. W. J. Albery, M. G. Boutelle and A. R. Hillman, *J. Electroanal. Chem.*, 1985, **182,** 99.
40. K. Shigehara, N. Oyama and F. C. Anson, *Inorg. Chem.*, 1981, **20,** 518.
41. D. J. Harrison, K. A. Daube and M. S. Wrighton, *J. Electroanal Chem.*, 1984, **163,** 93.
42. H. Braun, F. Decker, K. Doblhofer and H. Sotobayashi, *Ber. Bunsenges.*, 1984, **88,** 345.
43. T. P. Henning, H. S. White and A. J. Bard, *J. Amer. Chem. Soc.*, 1982, **104,** 5862.
44. N. Oyama and F. C. Anson, *Anal. Chem.*, 1980, **52,** 1192.
45. O. Haas and H.-R. Zumbrunnen, *Helv. Chim. Acta*, 1981, **64,** 854.
46. C. Degrand and L. L. Miller, *J. Amer. Chem. Soc.*, 1980, **102,** 5728.
47. F. C. Anson, J. M. Saveant and K. Shigehara, *J. Amer. Chem. Soc.*, 1983, **105,** 1096.
48. O. Haas and J. G. Vos, *J. Electroanal. Chem.*, 1980, **113,** 139.
49. N. Oyama, K. Sato and H. Matsuda, *J. Electroanal. Chem.*, 1980, **115,** 149.
50. N. Oyama, Y. Ohnuki, K. Chiba and T. Ohsaka, *Chem. Lett.*, 1983, 1759.
51. S. Geraty and J. G. Vos, *J. Electroanal. Chem.*, 1984, **176,** 389.
52. C. D. Ellis, W. R. Murphy and T. J. Meyer, *J. Amer. Chem. Soc.*, 1981, **103,** 7480.
53. A. B. Bocarsly, S. A. Galvin and S. Sinha, *J. Electrochem. Soc.*, 1983, **130,** 1319.
54. T. Inoue and T. Yamase, *Bull. Chem. Soc. Jap.*, 1983, **56,** 985.
55. A. Diaz, J. M. V. Vallejo and A. M. Duran, *I.B.M. J. Res. Dev.*, 1981, **25,** 42.
56. M. Salmon, M. Aguilar and M. Saloma, *J.C.S. Chem. Comm.*, 1983, 570.
57. M. Fukui, A. Kitani, C. Degrand and L. L. Miller, *J. Amer. Chem. Soc.*, 1982, **104,** 28.
58. C. Degrand, L. Roullier, L. L. Miller and B. Zinger, *J. Electroanal. Chem.*, 1984, **178,** 101.
59. P. K. Ghosh and T. G. Spiro, *J. Electrochem. Soc.*, 1981, **128,** 1281.
60. G. Schiavon, G. Zotti and G. Bontempelli, *J. Electroanal. Chem.*, 1984, **161,** 323.
61. A. G. Ewing, B. J. Feldman and R. W. Murray, *J. Electroanal. Chem.*, 1984, **172,** 145.
62. R. J. Mortimer and F. C. Anson, *J. Electroanal. Chem.*, 1982, **138,** 325.
63. P. Janda, J. Weber and L. Kavan, *J. Electroanal. Chem.*, 1984, **180,** 109.
64. I. Rubinstein and A. J. Bard, *J. Amer. Chem. Soc.*, 1980, **102,** 6641.
65. I. Rubinstein and A. J. Bard, *J. Amer. Chem. Soc.*, 1981, **103,** 5007.
66. P. K. Ghosh and A. J. Bard, *J. Electroanal. Chem.*, 1984, **169,** 113.
67. K. S. V. Santhanam and R. N. O'Brien, *J. Electroanal. Chem.*, 1984, **160,** 377.
68. R. N. O'Brien, N. S. Sundaresan and K. S. V. Santhanam, *J. Electrochem. Soc.*, 1984, **131,** 2028.

69. A. Bettelheim, R. J. H. Chan and T. Kuwana, *J. Electroanal. Chem.*, 1980, **110**, 93.
70. G.-X. Wan, K. Shigehara, E. Tsuchida and F. C. Anson, *J. Electroanal. Chem.*, 1984, **179**, 239.
71. D. A. Buttry and F. C. Anson, *J. Amer. Chem. Soc.*, 1984, **106**, 59.
72. K. Okabayashi, O. Ikeda and H. Tamura, *J. Chem. Soc. Chem. Comm.*, 1983, 684.
73. O. Ikeda, H. Fukuda and H. Tamura, *J. Chem. Soc. Chem. Comm.*, 1984, 567.
74. T. Kikuchi, H. Sasaki and S. Toshima, *Chem. Lett.*, 1980, 5.
75. R. A. Bull, F.-R. Fan and A. J. Bard, *J. Electrochem. Soc.*, 1984, **131**, 687.
76. H. Behret, H. Binder, G. Sandstede and G. G. Scherer, *J. Electroanal. Chem.*, 1981, **117**, 29.
77. P. Martigny and F. C. Anson, *J. Electroanal. Chem.*, 1982, **139**, 383.
78. G. S. Calabrase, R. M. Buchanan and M. S. Wrighton, *J. Amer. Chem. Soc.*, 1982, **104**, 5786.
79. G. S. Calabrase, R. M. Buchanan and M. S. Wrighton, *J. Amer. Chem. Soc.*, 1983, **105**, 5594.
80. C. Degrand, *J. Electroanal. Chem.*, 1984, **169**, 259.
81. O. Ikeda, K. Okabayashi and H. Tamura, *Chem. Lett.*, 1983, 1821.
82. W.-H. Kao and T. Kuwana, *J. Amer. Chem. Soc.*, 1984, **106**, 473.
83. K. Hiratsuka, H. Sasaki and S. Toshima, *Chem. Lett.*, 1979, 751.
84. R. R. Durand, C. S. Bencosme, J. P. Collman and F. C. Anson, *J. Amer. Chem. Soc.*, 1983, **105**, 2710.
85. C. D. Ellis, J. A. Gilbert, W. R. Murphy and T. J. Meyer, *J. Amer. Chem. Soc.*, 1983, **105**, 4842.
86. C. J. Stalder, S. Chao and M. S. Wrighton, *J. Amer. Chem. Soc.*, 1984, **106**, 3673.
87. D. L. DuBois, *Inorg. Chem.*, 1984, **23**, 2047.
88. N. S. Lewis and M. S. Wrighton, *Science*, 1981, **211**, 944.
89. C. M. Elliot and W. S. Martin, *J. Electroanal. Chem.*, 1982, **137**, 377.
90. S. Chao, J. L. Robbins and M. S. Wrighton, *J. Amer. Chem. Soc.*, 1983, **105**, 181.
91. W. J. Albery, M. J. Eddowes, H. A. O. Hill and A. R. Hillman, *J. Amer. Chem. Soc.*, 1981, **103**, 3904.
92. D. C. Bookbinder, N. S. Lewis and M. S. Wrighton, *J. Amer. Chem. Soc.*, 1981, **103**, 7656.
93. C. Degrand and L. L. Miller, *J. Amer. Chem. Soc.*, 1980, **102**, 5728.
94. A. Torstensson and L. Gorton, *J. Electroanal. Chem.*, 1981, **130**, 199.
95. L. Gorton, A. Torstensson, H. Jaegfeldt and G. Johansson, *J. Electroanal. Chem.*, 1984, **161**, 103.
96. M. F. Dautartas and J. F. Evans, *J. Electroanal. Chem.*, 1980, **109**, 301.
97. J. Facci and R. W. Murray, *Anal. Chem.*, 1982, **54**, 772.
98. K.-N. Kuo and R. W. Murray, *J. Electroanal. Chem.*, 1982, **131**, 37.
99. F. C. Anson, Y.-M. Tsou and J. M. Saveant, *J. Electroanal. Chem.*, 1984, **178**, 113.

100. C. D. Crawley and F. M. Hawkridge, *J. Electroanal. Chem.*, 1983, **159**, 313.
101. H. L. Landrum, R. T. Salmon and F. M. Hawkridge, *J. Amer. Chem. Soc.*, 1977, **99**, 3154.
102. V. J. Razumas, A. V. Gudavicius and J. J. Kulys, *J. Electroanal. Chem.*, 1983, **151**, 311.
103. J. F. Stargardt, F. M. Hawkridge and H. L. Landrum, *J. Anal. Chem.*, 1978, **50**, 930.
104. J. C. Moutet, *J. Electroanal. Chem.*, 1984, **161**, 181.
105. G. J. Samuels and T. J. Meyer, *J. Amer. Chem. Soc.*, 1981, **103**, 307.
106. M. F. Dautartas, K. R. Mann and J. F. Evans, *J. Electroanal. Chem.*, 1980, **110**, 379.
107. R. Jasinski, *J. Electrochem. Soc.*, 1983, **130**, 834.
108. C. M. Elliot and C. A. Marrese, *J. Electroanal. Chem.*, 1981, **119**, 395.
109. J. B. Kerr and L. L. Miller, *J. Electroanal. Chem.*, 1979, **101**, 263.
110. J. B. Kerr, L. L. Miller and M. R. Van de Mark, *J. Amer. Chem. Soc.*, 1980, **102**, 3383.
111. K. W. Willman and R. W. Murray, *J. Electroanal. Chem.*, 1982, **133**, 211.
112. R. D. Rocklin and R. W. Murray, *J. Phys. Chem.*, 1981, **85**, 2104.
113. S. Abe, T. Nonaka and T. Fuchigami, *J. Amer. Chem. Soc.*, 1983, **105**, 3630.
114. S. Abe and T. Nonaka, *Chem. Lett.*, 1983, 1541.
115. T. Komori and T. Nonaka, *J. Amer. Chem. Soc.*, 1983, **105**, 5690.
116. T. Komori and T. Nonaka, *J. Amer. Chem. Soc.*, 1984, **106**, 2656.
117. T. Komori and T. Nonaka, *Chem. Lett.*, 1984, 509.
118. A. Yamagishi and A. Aramata, *J. Chem. Soc. Chem. Comm.*, 1984, 452.
119. J. A. Cox and P. J. Kulesza, *J. Electroanal. Chem.*, 1984, **175**, 105.
120. J. A. Cox and P. J. Kulesza, *Anal. Chim. Acta*, 1984, **158**, 335.
121. W. R. Heineman, H. J. Wieck and A. M. Yacynych, *Anal. Chem.*, 1980, **52**, 345.
122. G. C. Cheek, C. P. Wales and R. J. Nowak, *Anal. Chem.*, 1983, **55**, 380.
123. G. Inzelt, J. Q. Chambers, J. F. Kinstle, R. W. Day and M. A. Lange, *Anal. Chem.*, 1984, **56**, 301.
124. I. Rubinstein, *Anal. Chem.*, 1984, **56**, 1135.
125. R. S. Lawton and A. M. Yacynych, *Anal. Chim. Acta*, 1984, **160**, 149.
126. J. A. Cox and P. J. Kulesza, *Anal. Chim. Acta*, 1983, **154**, 71.
127. J. A. Cox and P. J. Kulesza, *J. Electroanal. Chem.*, 1983, **159**, 337.
128. J. Wang, B. Greene and C. Morgan, *Anal. Chim. Acta*, 1984, **158**, 15.
129. M. N. Szentirmay and C. R. Martin, *Anal. Chem.*, 1984, **56**, 1898.
130. K. M. Korfhage, K. Ravichandran and R. P. Baldwin, *Anal. Chem.*, 1984, **56**, 1514.
131. P. Denisevich, K. W. Willman and R. W. Murray, *J. Amer. Chem. Soc.*, 1981, **103**, 4727.
132. K. W. Willman and R. W. Murray, *J. Electroanal. Chem.*, 1982, **133**, 211.
133. T. Ohsawa, K. Kaneto and K. Yoshino, *Jap. J. Appl. Phys.*, 1984, **23**, L663.

134. G. Mengoli, M. M. Musiami, S. Daolio and C. Folonari, *Makromol. Chem.*, 1980, **181**, 1909.
135. G. Mengoli, M.-T. Munari and C. Folonari, *J. Electroanal. Chem.*, 1981, **124**, 237.
136. G. Mengoli, R. Bianco, S. Daolio and M. T. Munari, *J. Electrochem. Soc.*, 1981, **128**, 2276.
137. P. Mourcel, M.-C. Pham, P.-C. Lacaze and J. E. Dubois, *J. Electroanal. Chem.*, 1983, **145**, 467.
138. A. B. Bocarsly, S. A. Galvin and S. Sinha, *J. Electrochem. Soc.*, 1983, **130**, 1319.
139. A. B. Bocarsly, E. G. Walton and M. S. Wrighton, *J. Amer. Chem. Soc.*, 1980, **102**, 3390.
140. J. M. Bolts, A. B. Bocarsly, M. C. Palazzotto, E. G. Walton, N. S. Lewis and M. S. Wrighton, *J. Amer. Chem. Soc.*, 1979, **101**, 1378.
141. N. S. Lewis, A. B. Bocarsly and M. S. Wrighton, *J. Phys Chem.*, 1980, **84**, 2033.
142. M. S. Wrighton, A. B. Bocarsly, J. M. Bolts, M. G. Bradley, A. B. Fischer, N. S. Lewis, M. C. Palazzotto and E. G. Walton, *Adv. Chem. Ser. I*, 1980, **184**, 269.
143. R. E. Malpas and B. Rushby, *J. Electroanal. Chem.*, 1983, **157**, 387.
144. J. M. Bolts and M. S. Wrighton, *J. Amer. Chem. Soc.*, 1979, **101**, 6179.
145. M. D. Rosenblum and N. S. Lewis, *J. Phys. Chem.*, 1984, **88**, 3103.
146. J. M. Calvert, J. V. Casper, R. A. Binstead, T. D. Westmoreland and T. J. Meyer, *J. Amer. Chem. Soc.*, 1982, **104**, 6620.
147. O. Haas, N. Muller and H. Gerischer, *Electrochim. Acta*, 1982, **27**, 991.
148. R. Noufi, D. Tench and L. F. Warren, *J. Electrochem. Soc.*, 1981, **128**, 2596.
149. R. Noufi, A. J. Frank and A. J. Nozik, *J. Amer. Chem. Soc.*, 1981, **103**, 1849.
150. T. Skotheim, I. Lundstrom and J. Prejza, *J. Electrochem. Soc.*, 1981, **128**, 1625.
151. R. Noufi, D. Tench and L. F. Warren, *J. Electrochem. Soc.*, 1980, **127**, 2310.
152. R. Noufi, A. J. Nozik, J. White and L. F. Warren, *J. Electrochem. Soc.*, 1982, **129**, 2261.
153. L. Fornarini, F. Stirpe and B. Scrosati, *J. Electrochem. Soc.*, 1983, **130**, 2184.
154. F.-R. F. Fan, R. L. Wheeler, A. J. Bard and R. N. Noufi, *J. Electrochem. Soc.*, 1981, **128**, 2042.
155. R. A. Simon, A. J. Ricco and M. S. Wrighton, *J. Amer. Chem. Soc.*, 1982, **104**, 2031.
156. A. J. Frank and K. Honda, *J. Phys. Chem.*, 1982, **86**, 1933.
157. R. Noufi, *J. Electrochem. Soc.*, 1983, **130**, 2126.
158. J. A. Bruce, T. Murahashi and M. S. Wrighton, *J. Phys. Chem.*, 1982, **86**, 1552.
159. H. D. Abruna and A. J. Bard, *J. Amer. Chem. Soc.*, 1981, **103**, 6898.
160. H. Kido and C. H. Langford, *J. Chem. Soc. Chem. Comm.*, 1983, 350.

161. T. D. Westmoreland, J. M. Calvert, R. W. Murray and T. J. Meyer, *J. Chem. Soc. Chem. Comm.*, 1983, 65.
162. P. V. Kamat, M. A. Fox and A. J. Fatiadi, *J. Amer. Chem. Soc.*, 1984, **106**, 1191.
163. P. V. Kamat and M. A. Fox, *J. Electrochem. Soc.*, 1984, **131**, 1032.
164. M. Kaneko, A. Yamada, N. Oyama and S. Yamaguchi, *Makromol. Chem. Rapid Commun.*, 1982, **3**, 769.
165. P. K. Ghosh and T. G. Spiro, *J. Amer. Chem. Soc.*, 1980, **102**, 5543.
166. L. D. Margerum, T. J. Meyer and R. W. Murray, *J. Electroanal. Chem.*, 1983, **149**, 279.
167. N. Oyama, S. Yamaguchi, M. Kaneko and A. Yamada, *J. Electroanal. Chem.*, 1982, **139**, 215.
168. R. Tamilarasan and P. Natarajan, *Nature (London)*, 1981, **292**, 224.
169. R. Tamilarasan, R. Ramaraj, R. Subramanian and P. Natarajan, *J. Chem. Soc. Faraday. I*, 1984, **80**, 2405.
170. Y. Morishima, M. Isono, Y. Itoh and S. Nozakura, *Chem. Lett.*, 1981, 1149.
171. P. V. Kamat, *J. Electroanal. Chem.*, 1984, **163**, 389.
172. I. Inoue and T. Yamase, *Bull. Chem. Soc. Jap.*, 1983, **56**, 985.
173. T. M. Abrantes, L. M. Castillo, M. Fleischmann, I. R. Hill, L. M. Peter, G. Mengoli and G. Zotti, *J. Electroanal. Chem.*, 1984, **177**, 129.
174. W. J. Albery and M. D. Archer, *J. Electroanal. Chem.*, 1978, **86**, 19.
175. W. J. Albery, W. R. Bowen, F. S. Fisher, A. W. Foulds, K. J. Hall, A. R. Hillman, R. G. Egdell and A. F. Orchard, *J. Electroanal. Chem.*, 1980, **107**, 37.
176. W. J. Albery, A. W. Foulds, K. J. Hall, and A. R. Hillman, *J. Electrochem. Soc.*, 1980, **127**, 654.
177. A. S. N. Murthy and K. S. Reddy, *Electrochim. Acta*, 1983, **28**, 473.
178. D. C. Bookbinder, J. A. Bruce, R. N. Dominey, N. S. Lewis and M. S. Wrighton, *Proc. Natl. Acad. Sci. U.S.A.*, 1980, **77**, 6280.
179. R. N. Dominey, N. S. Lewis, J. A. Bruce, D. C. Bookbinder and M. S. Wrighton, *J. Amer. Chem. Soc.*, 1982, **104**, 467.
180. W.-H. Kao and T. Kuwana, *J. Amer. Chem. Soc.*, 1984, **106**, 473.
181. P. Burgmayer and R. W. Murray, *J. Electroanal. Chem.*, 1982, **135**, 335.
182. R. C. Cielinski and N. R. Armstrong, *J. Electrochem. Soc.*, 1980, **127**, 2605.
183. H. Akahoshi, S. Toshima and K. Itaya, *J. Phys. Chem.*, 1981, **85**, 818.
184. K. Itaya, H. Akahoshi and S. Toshima, *J. Electrochem. Soc.*, 1982, **129**, 762.
185. F. B. Kaufman, A. H. Schroeder, E. M. Engler and V. V. Patel, *Appl. Phys. Lett.*, 1980, **36**, 422.
186. P. C. Lacaze, J. E. Dubois, A. Desbene-Monvernay, P. L. Desbene, J. J. Basselier and D. Richard, *J. Electroanal. Chem.*, 1983, **147**, 107.
187. T. Kobayashi, H. Yoneyama and H. Tamura, *J. Electroanal. Chem.*, 1984, **161**, 419.
188. Y. Yoshino, K. Kaneto and Y. Inuishi, *Jap. J. Appl. Phys.*, 1983, **22**, L157.

189. K. Kaneto, K. Yoshino and Y. Inuishi, *Jap. J. Appl. Phys.*, 1983, **22**, L412.
190. P. J. Peerce and A. J. Bard, *J. Electroanal. Chem.*, 1980, **180**, 121.
191. P. G. Pickup and R. W. Murray, *J. Amer. Chem. Soc.*, 1983, **105**, 4510.
192. P. G. Pickup and R. W. Murray, *J. Electrochem. Soc.*, 1984, **131**, 833.
193. P. G. Pickup, W. Kutner, C. R. Leidner and R. W. Murray, *J. Amer. Chem. Soc.*, 1984, **106**, 1991.
194. D. J. Harrison and M. S. Wrighton, *J. Phys. Chem.*, 1984, **88**, 3932.
195. H. D. Abruna, P. Denisevich, M. Umana, T. J. Meyer and R. W. Murray, *J. Amer. Chem. Soc.*, 1981, **103**, 1.
196. P. G. Pickup, C. R. Leidner, P. Denisevich and R. W. Murray, *J. Electroanal. Chem.*, 1984, **164**, 39.
197. C. R. Leidner, P. Denisevich, K. W. Willman and R. W. Murray, *J. Electroanal. Chem.*, 1984, **164**, 63.
198. H. S. White, G. P. Kittlesen and M. S. Wrighton, *J. Amer. Chem. Soc.*, 1984, **106**, 5375.
199. P. Burgmayer and R. W. Murray, *J. Amer. Chem. Soc.*, 1982, **104**, 6139.
200. P. Burgmayer and R. W. Murray, *J. Electroanal. Chem.*, 1983, **147**, 339.
201. P. Burgmayer and R. W. Murray, *J. Amer. Chem. Soc.*, 1984, **88**, 2515.
202. L. L. Miller, A. N. K. Lau and E. K. Miller, *J. Amer. Chem. Soc.*, 1982, **104**, 5242.
203. A. N. K. Lau and L. L. Miller, *J. Amer. Chem. Soc.*, 1983, **105**, 5271.
204. A. N. K. Lau, L. L. Miller and B. Zinger, *J. Amer. Chem. Soc.*, 1983, **105**, 5278.
205. B. Zinger and L. L. Miller, *J. Amer. Chem. Soc.*, 1984, **106**, 6861.

CHAPTER 7

Perfluorinated Ionomer Membranes for Use in the Production of Chlorine and Caustic Soda

PETER J. SMITH

General Chemicals Business Group, Imperial Chemical Industries PLC, Runcorn, Cheshire, UK

1. INTRODUCTION

1.1. Historical Perspective to Membrane Cell Technology

The advancement of membrane cell technology has been rapid during the last decade. It has changed in ten years from an interesting laboratory curiosity to the preferred technology of many present-day chlorine/caustic soda producers. Such a scenario is not typical of other industrial processes and because of this it is important to consider the reasons why the dramatic advance in membrane technology has occurred so quickly. This will help to rationalise past advances and it may also provide some insight into future developments.

Caustic soda and chlorine are normally produced by the electrolysis of brine, which is the largest industrial electrolytic process in the world. Vast quantities of chlorine and caustic soda are used as primary feedstocks by many major sections of the world's chemical industry. Chlorine is used to produce a myriad of chlorinated polymers, an extensive series of solvents and many specialised industrial chemicals. Caustic alkali is used in the production of synthetic fibres, in the extraction of aluminium from bauxite and in the fabrication of paper from wood pulp.

As chlorine and alkali are such fundamental building blocks their price has a direct effect on the cost of many bulk chemicals. This means that there is a very great commerical incentive to produce both chlorine and alkali at the lowest possible cost. Any reduction in the price of either is directly transferable to the profitability of commodities such as poly(vinyl chloride). When producing chlorine by traditional technology and to a lesser extent with membrane technology the greatest cost incurred is that associated with the cost of the electricity required to carry out the electrolysis process. This means that the greatest potential opportunity for cost reduction is associated with the electrolysis procedure.

The onset of the first oil crisis in the 1970s made many people acutely aware of the dependence of chlorine production profitability on energy costs. Traditional technologies, which had already been refined for nearly a century, did not appear to be capable of further refinement to improve significantly energy utilisation. However, emerging membrane technology was considered by many chlorine producers as a possible route to lower energy utilisation and the concern about energy costs provided a favourable and sympathetic environment for it to develop.

It was also at this time that a second incident occurred which was to provide a further impetus towards membrane technology. In the early 1970s, quantities of organo-mercury compounds, which had been used as seed preservatives, were discharged into the sea around Minamata, Japan. The organo-mercury compounds were ingested by fish and so introduced into the human food chain and after a period of time mercury started to accumulate in Japanese fishermen. The results were catastrophic and birth abnormalities were soon correlated with mercury ingestion. The occurrence of human birth abnormalities caused a popular upsurge concerning the environmental effects of mercury.

Several years previously the book *Silent Spring* by Rachel Carson[1] had indicated dangers associated with environmental mercury and this coupled with the Minamata disaster precipitated the Japanese Government into action. It was decreed that all Japanese mercury cell chlorine production plants should be converted to non-mercury based technology or cease production. The original date for complete conversion was set for 1979 but this was subsequently extended to 1986. The Japanese government ultimatum provided another very favourable ingredient for creating a sympathetic environment for membrane cell technology.

The only alternative commercial cell process to mercury cells in the early 1970s was diaphragm cell technology. 'Diaphragm' in this review means a non-charged permeable barrier, whereas 'membrane' means a charged semi-permeable barrier. Although diaphragm cell technology is still widely used today, its advancement has been slow compared with that of membrane cell technology. Most commercial diaphragms are made from asbestos, which, like mercury, is considered to have toxicity problems. Several companies have tried to develop a non-asbestos diaphragm but to date none of these has found widespread commercial acceptance.

In addition to the toxicity problems associated with asbestos several other disadvantages are apparent with diaphragm cell technology. The most important limitation is the quality and strength of the alkali produced. In general the product caustic strength is low (~10% w/w) and it is highly contaminated (~10% w/w) with salt. This means that the alkali is suitable only for applications such as paper manufacture. This is to be compared with mercury cell technology which produces strong caustic (~50% w/w) with very low salt content (less than 5 ppm in 50% w/w). If stronger caustic (>30% w/w) is produced in a conventional diaphragm cell equipped with mild steel cathodes, the cell efficiency drops to very low values and problems are encountered with cathode corrosion. All of these factors combined to favour membrane cell technology as the preferred future option.

A final point that is relevant is that it was in Japan that mercury poisoning was encountered and Japan is perhaps unique among other countries in that it has a long history of utilising other commercial membrane processes. This background in traditional membrane technology and the natural acceptance by the Japanese people of new technology put them in a good position to exploit membrane cell technology.

It seems that several important points—concern about rising energy costs, concern about mercury pollution, the inability of diaphragm cells to make strong pure alkali, and the Japanese experience of membranes—gelled together in the 1970s to create a favourable environment for a newly based chlorine technology. When the first membrane materials were developed the industry was ready for them and vast sums of industrial money were made available for their development. Had the first membrane material been discovered at a different time, with a different environment, perhaps the rate of commercial exploitation would have been much slower. It is the interdependence of an industrial need and scientific inventiveness that is responsible for the rapid development of membrane cell technology that has occurred in the last decade.

The close association between strong commercial need and scientific advance has not been without its problems. The development of chlor-alkali membranes has been confined to several large companies, mainly Japanese and American, and the past fierce rivalry, associated with substantial commercial opportunity, between these companies has limited both the supply of membranes to individuals and the publication of information. Policies such as these and the practical difficulties involved in studying chlor-alkali membranes have limited the number of serious academic investigations carried out in the past. However, much high-quality work has been undertaken by the producers and some of this is now gradually starting to be published. Areas concerning membrane composition and membrane structure are jealously guarded by all manufacturers and this creates difficulties in reviewing the current art. Nevertheless, armed with an awareness of the commercial/scientific background and a knowledge of the patent literature it is possible to build up an overview of the state of chlor-alkali membrane development.

1.2. Current Commercial Chlorine Technology

The current technologies employed by chlorine producers are associated with the end use of the electrolysis products. In Europe most chlorine is produced by mercury cells, whereas in North America diaphragm cell technology is in widespread use. This division reflects the use of the product caustic soda—for paper manufacture in the United States and for chemical manufacture in Europe. As different industries require varying qualities of product from their chlorine/caustic soda production facilities it is worth comparing the main features of alternative technologies with those offered by membrane technology.

Basic process outlines for mercury cell technology and diaphragm cell technology are illustrated in Fig. 1. The essential feature of the mercury cell process is that alkali metal ions present in purified brine react with a mercury cathode to form an amalgam. This amalgam is subsequently broken down in a mercury denuder to form products and mercury is recycled to the cell.

There are several important aspects of mercury cell technology.

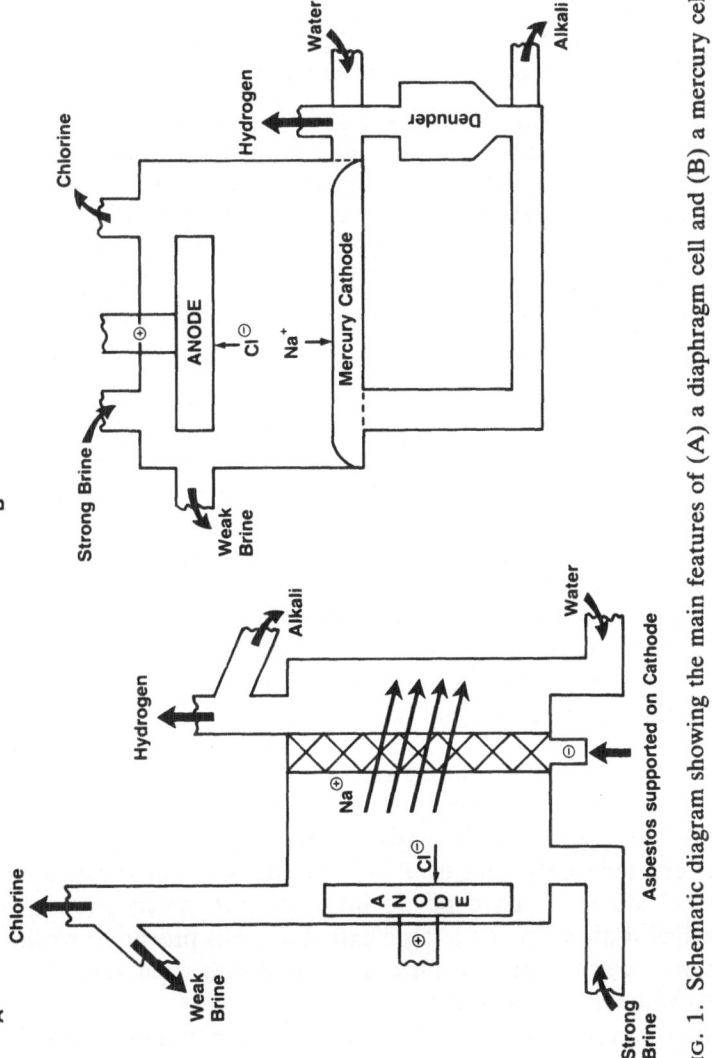

FIG. 1. Schematic diagram showing the main features of (A) a diaphragm cell and (B) a mercury cell.

(i) The thermodynamic reversible voltage for the process is high (*ca* 3·2 V) and this results in a high power utilisation.
(ii) The caustic current efficiency (molar product yield per Faraday of charge passed) is high.
(iii) The product alkali has very little salt contamination.
(iv) Mercury is a dangerous environmental pollutant.

The essential difference between a diaphragm cell and a membrane cell is that in a diaphragm cell there is a hydraulic flow of fluid from anolyte to catholyte. This flow of fluid minimises the transference of alkali from the catholyte and so facilitates the partial separation of electrolysis products. Asbestos is currently the most commonly used diaphragm material.

Some important aspects of diaphragm cell technology are:

(i) The thermodynamic reversible voltage is low (*ca* 2·2 V) and normally results in low electric power utilisation.
(ii) Product caustic soda is normally weak (*ca* 10% w/w) and it is always contaminated with salt ($\sim$14% w/w). Production of concentrated alkali results in very low caustic current efficiency.
(iii) Salt contamination of product caustic can readily be reduced to 1% w/w but alkali containing this level of salt is not suitable for synthetic fibre manufacture.
(iv) Asbestos diaphragms are weak structures and they have poor durability to electric load fluctuations. They are also unsuitable for operation at high current density.
(v) Asbestos is a potential health hazard.

From the foregoing it can be seen that both mercury and diaphragm processes have some very attractive features despite some rather obvious drawbacks. In the early 1970s it was recognised that membrane cell technology might be able to take the best features of both these technologies and combine them without the incurrence of the individual drawbacks. To a large extent this has proved to be the case but some rather specific features associated with membranes have had to be overcome in doing this.

The main principles of a membrane cell are presented in Fig. 2. The anolyte and catholyte chambers are separated by a semi-permeable ion-exchange membrane and there is no bulk fluid flow from anolyte to catholyte. The ion-exchange membrane allows the selective passage of cations and inhibits the movement of negatively charged anions.

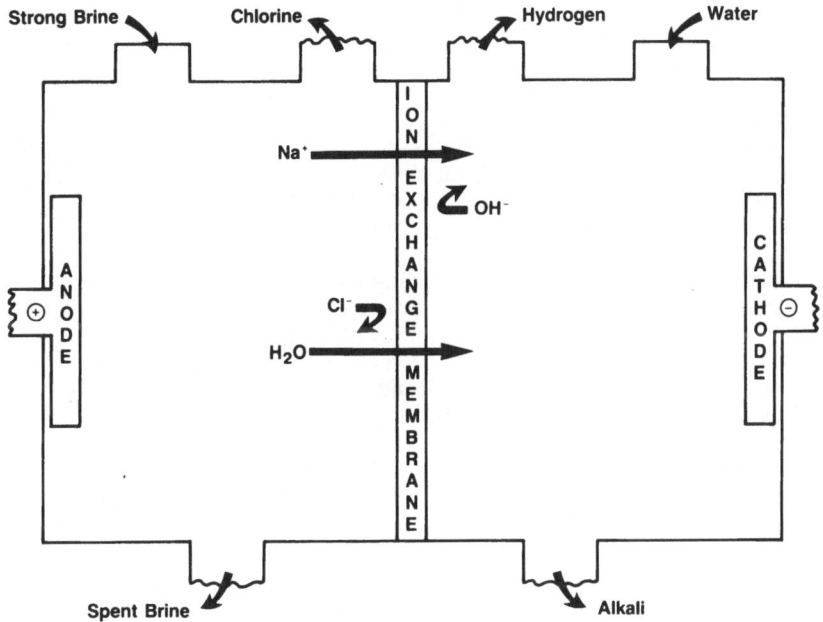

FIG. 2. Schematic diagram showing the essential features of a membrane cell.

Important features of the process are:

(i) The equilibrium thermodynamic voltage is similar (*ca* 2·2 V) to that found in a diaphragm cell.
(ii) Product alkali can be formed at high concentrations (>32% w/w) and alkali of this concentration requires minimal evaporation to give a saleable product.
(iii) The contamination of product alkali by brine is very small (<10 ppm in 50% w/w alkali).
(iv) The ion-exchange membrane is a piece of plastic and it presents no environmental or health problems.
(v) The plastic ion-exchange membrane is mechanically stable and is much better suited to withstanding changes in operating conditions than an asbestos diaphragm.

The essential features of mercury, diaphragm and membrane cell technologies are summarised in Table 1.

Thus it appears membrane cell technology does combine the best

TABLE 1
Comparative Features of Current Commercial Chlor-Alkali Technologies

Features	Mercury process	Diaphragm process	Membrane process
Separator material	Mercury amalgam acts as separator	Asbestos and occasionally starch/PTFE	Cation exchange membrane (plastic sheet)
Anode material	Graphite/metal	Graphite/metal with low-overvoltage coating	Metal with low-overvoltage coating
Cathode material	Mercury	Mild steel	Metal (usually nickel) with coating
Cathode product	Sodium amalgam	Alkali/brine mixture and hydrogen	High-purity alkali and hydrogen
Need for secondary evaporation of alkali	No	Much evaporation needed	Small amount of evaporation needed
Acceptance of change of process conditions	Very good	Poor	Good
Power consumption	High	Low electricity but high steam costs	Medium
Environmental effect	Mercury is a pollutant	Asbestos is a pollutant	None identified

features of the other two alternative technologies whilst alleviating any possible environmental/health related problems. In addition to this it has opened doors to two new and very exciting areas of chlorine production science. The first of these is novel chlor-alkali cell design and the other is the advancement of low-overvoltage cathode systems. The ability of membrane technology to accommodate these areas is a major driving force for the continued advancement of the technology.

The advent of a durable plastic sheet as a cell separator enabled quantum changes to be made in cell design. Cells no longer needed to be large traditional structures and many novel cell designs have been forthcoming to exploit the unique features of ion-exchange membranes. Activated cathodes are not possible with mercury cells as many heavy metal ions generate 'thick butter' on mercury cathodes and so increase the power utilisation. Low-overvoltage cathodes in diaphragm cells were never very successful due to a combination of problems associated with the susceptibility of cathode coatings to poisoning from impure feedstocks. The use of highly purified process fluids in membrane cell technology overcame many of the poisoning problems and present-day activated cathodes have quoted lifetimes in excess of two years.

2. PERFLUORINATED SULPHONATE IONOMERS

Ion-exchange membranes were first used as separators in chlor-alkali electrolysis cells in the late 1950s.[2] However, this did not progress into a commercial electrolysis process as the only membranes that were available were hydrocarbon based materials and these were destroyed very rapidly by chlorinated brine. It was only with the development of a fully fluorinated ion-exchange membrane that the possibility of a commercial process became a reality, and all commercial membranes in current use are fully fluorinated materials.

2.1. Synthesis of Perfluorosulphonate Polymers

During the middle 1960s E. I. Du Pont de Nemours developed a novel series of perfluorosulphonate polymers.[3] These polymers, which were designated XR resins, were originally conceived as elastomers but it was quickly recognised that they represented a significant opportunity for chlor-alkali membranes.[4-6] The synthesis of XR resin is described in detail elsewhere but in essence the reaction can be summarised as follows.[7]

XR resin is formed by the copolymerisation of tetrafluoroethylene with a perfluorovinyl ether compound. The perfluorovinyl ether can be prepared in several ways but the most simple involves the reaction of sulphur trioxide with tetrafluoroethylene to form a sulphone which after rearrangement can react with hexafluoropropylene oxide to generate sulphonyl fluoride adducts. When such adducts are heated with sodium carbonate a sulphonyl vinyl ether is formed. The perfluorovinyl ether

$$CF_2{=}CF_2 + SO_3 \longrightarrow \begin{array}{c} O{-\!-\!-}SO_2 \\ | \quad\quad | \\ CF_2{-\!-\!}CF_2 \end{array} \longrightarrow O{=}CF{-}CF_2{-}SO_2F$$

$$FSO_2{-}CF_2{-}FC{=}O + (n+1)CF_2\overset{O}{\diagdown}CF \\ \qquad\qquad\qquad\qquad\qquad\qquad |\\ \qquad\qquad\qquad\qquad\qquad\qquad CF_3$$

$$\xrightarrow[\text{Pressure}]{\text{Heat}} FSO_2CF_2CF_2(OCF{-}CF_2)_nOCF{-}FC{=}O \\ \qquad\qquad\qquad\qquad\qquad | \qquad\qquad\quad | \\ \qquad\qquad\qquad\qquad\qquad CF_3 \qquad\qquad CF_3$$

$$\text{(A)} \qquad n>1$$

(product A) reacts with tetrafluoroethylene under modest pressure and in the presence of suitable radical initiators to give a high-molecular mass, melt processable, XR resin[8]

$$FSO_2CF_2CF_2(OCFCF_2)_nOCF{-}FC{=}O + CF_2{=}CF_2 \\ \qquad\qquad\quad | \qquad\qquad | \\ \qquad\qquad\quad CF_3 \qquad\quad CF_3$$

$$\qquad\qquad\qquad\qquad\qquad\qquad\qquad\qquad CF_3 \\ \qquad\qquad\qquad\qquad\qquad\qquad\qquad\qquad\; | \\ \longrightarrow {\sim}(CF_2{-}CF_2){-}CFO(CF_2{-}CFO)_n{-}CF_2 \\ \qquad\qquad\qquad\qquad\qquad\;\; | \qquad\qquad\qquad\quad\; | \\ \qquad\qquad\qquad\qquad\qquad\;\; CF_2 \qquad\qquad\;\; F_2OS{-}CF_2 \\ \qquad\qquad\qquad\qquad\qquad\qquad\text{XR resin}$$

As XR resin is melt processable it is easily pressed or extruded into sheet form and the first fully fluorinated chlor-alkali membrane was formed by alkaline hydrolysis of XR resin sheet. This product was subsequently called Nafion* by E. I. Du Pont de Nemours; the discovery of this product was seminal to the development of all future membranes.

2.2. Commercial Exploitation of Perfluorosulphonate Ionomers

The physical properties of XR resin are those primarily associated with the hydrophilic nature of the polymer. Hydrophilicity varies inversely as polymer equivalent weight. This means that a high-equivalent-weight polymer has few ionic side chains and so absorbs very little

---

* Registered Trade Mark of E. I. Du Pont de Nemours.

water. Polymers of low equivalent weight have many pendant side chains and so are more hydrophilic and absorb increased quantities of water. The requirement for a film-forming polymer tends to place a lower limit of about 700 on the equivalent weight and for a simple polymeric structure like XR resin many physical properties are linear functions of the polymer equivalent weight.

Although XR resins are inherently random copolymers, some long-range ordering has been observed with polymers of low fixed charge density. This crystallinity is associated with the different monomer reactivity ratios and crystallinity, as measured by X-ray diffractometry, is a linear function of tetrafluoroethylene content.[9] Crystallinity has also been observed with carboxylate ionomers when the vinyl ether content of the polymer is less than 20 mol %.[10] Polymer crystallinity has been associated with increased membrane selectivity as it is known that crystallinity promotes dehydration of the membrane.[11]

The dependence of catholyte current efficiency (the number of moles of alkali produced per Faraday of charge passed) on the equivalent weight of the XR resin is a linear function whose absolute magnitude depends on the caustic concentration being produced, the current density employed and the previous operational lifetime of the membrane. In general caustic current efficiency increases as polymer equivalent weight increases and decreases as caustic strength increases.[12,13] The maximum caustic current efficiency possible, whilst producing alkali of about 35% w/w, is only 70%. In addition, operation of the membranes at high current density ($>5 \text{ kA/m}^2$) has been found to generate time-dependent transport phenomena.[14]

Although XR resins provided a major advance in membrane lifetime, it soon became apparent that such a simple polymer structure would not provide a successful commercial product. Initially development work centred on trying to combine the good membrane selectivity found with high-equivalent-weight polymers with the good mechanical and reduced voltage characteristics of low-equivalent-weight polymers. This was achieved by laminating films of polymer of different equivalent weight prior to final polymer hydrolysis to the ionic form. Among the first membranes of this type produced by Du Pont was the Nafion 300 series. These membranes were formed by laminating 50 μm of 1500 equivalent weight resin on 100 μm of 1100 equivalent weight polymer. The composite membranes gave improved current efficiencies and were suitable for producing caustic soda of up to about 20% w/w. However, after extensive study in the mid-1970s it was concluded that they were unsuitable for development to a commercial stage.[15]

The next phase of development extended the idea of a dual functional membrane but this time it was by modification of the water absorption characteristics of one surface of a single perfluorosulphonate film. In addition to higher cation specificity this eliminated the need for lamination and also increased the mechanical integrity of the product. In order to produce a thin barrier layer capable of producing high selectivity it was proposed to convert surface sulphonic acid ion exchange sites into less acidic sulphonamide cation exchange sites.[16,17]

Prehydrolysed XR resin was single-surface-reacted with ammonia to form a surface layer of sulphonamide groups. These surface groups were then further reacted with a variety of amines including methylamine and ethylenediamine. The ethylenediamine modified membranes exhibit some crosslinking whereas this is not possible with the ammonia or monoamine derivatives which form pendant groups at the sulphonyl site.

The measured caustic current efficiency of sulphonamide modified Nafion membranes was much in excess of that found with previous laminates of varying equivalent weight. It was now possible to produce 30% w/w caustic soda at a current efficiency in excess of 90% compared with the 70% value normally obtained with XR resins. Nevertheless sulphonamide modified membranes were not without their problems as transport properties were still not invariant with time and the polymers were prone to exhibiting odd memory effects (that is, performance was dependent on past membrane operating history). In addition, some membranes had to be extensively wetted-out prior to cell installation (by placing in water at 80°C for several hours) and some membranes had an amine smell.

In spite of the drawbacks of sulphonamide membranes it was still possible to achieve caustic current efficiency in excess of 88% for more than one year while producing 28% w/w NaOH at $3 \cdot 1 \text{ kA/m}^2$. The average power consumption was 2860 kW h/tonne NaOH and this appeared a very reasonable value considering the age of the technology in the late 1970s. Further development of sulphonamide membranes was curtailed soon after these data were obtained due to the increasing promise shown by perfluoro membranes which used carboxylate ion exchange groupings.

2.3. Physical Properties of Perfluorosulphonate Ionomers

The period of intense commercial interest in the development of pure and amine surface modified sulphonate ionomers lasted for about eight years. During this time much fundamental work was carried out by membrane producers but little of this was published in the open literature. It was only several years later, about 1980, when most industrialists had lost primary interest in sulphonic ionomers, that academic studies started to appear in the open literature.

The sorption characteristics of sulphonate ionomers were among the first properties to be reported.[18] The sorption temperature dependences of the membranes in the acid and salt forms were studied and details of the kinetics of sorption obtained. Equilibrium properties of membranes immersed in aprotic solvent have also been measured,[19] as have selectivity coefficients for hydrogen ion exchange with alkali metal atoms.[20,21] Interest in Nafion was not limited to America and in the late 1970s the first papers concerning Nafion started to appear in the Russian literature.[22,23] One paper has attempted to relate equilibrium sorption data with factors that must be considered when designing novel membranes.[24]

The diffusion characteristics of sulphonate ionomers have also been reported. Early work by Du Pont provided information regarding the dependence of sodium hydroxide transfer rate on equivalent weight.[25] The rate of permeation of sulphuric acid through Nafion has been investigated.[26] Prompted by the difficulty in interpreting the meaning of bulk electrolyte transfer diffusion coefficient, several workers have studied self-diffusion of alkali cations in Nafion.[27] Self-diffusion coefficients for sodium and caesium ions in Nafion membrane immersed in organic solvents are reported.[28] A recent review by Yeager provides more detailed information about diffusional properties of Nafion.[29]

Several studies concerning selectivity and resistivity of perfluorosulphonate ionomers have been reported. The electrical conductivity of simple perfluoro ionomers immersed in acid and various alkali metal hydroxide solutions was measured.[30,31] The dependence of d.c. ohmic drop on current density and alkali concentration at operating cell temperatures has been reported.[32] As would be expected, membrane resistance increases with both caustic concentration and applied current density. Model investigations using a well mixed flow cell have shown that membrane ohmic drop is a linear function of current density until about $12 \, kA/m^2$ and there is no indication of concentration polarisation effects.[33] This would mean that there is no mass transfer limitation to operating well mixed cells at a current density below the limiting current density value of about $12 \, kA/m^2$. Other authors have found different results.[34]

The variation of sodium ion transport number with process parameters has been determined by several different techniques. The true sodium ion transport number found in a conventional chlor-alkali cell can be measured using a steady-state electrolysis cell:

Titanium coated anode/$NaCl(m_1)$/membrane/$NaOH(m_2)$/nickel
$$\text{cathode}$$

In this type of cell, electrolyte and water are generally fed continuously to the cell and the sodium ion transport number is obtained by determining the number of moles of alkali produced per Faraday of charge passed.

Calculation of the number of moles of sodium hydroxide produced, in a given time interval, requires a knowledge of the sodium hydroxide concentration and an accurate measure of the exit catholyte flow rate. If the exit catholyte flow rate is not known accurately catholyte liquor must be collected over a long period of time (at least 16 h for flow rates $<25 \, cm^3/h$) to minimise errors in flow measurement. This means that a rapid measurement of transport number is not possible with the steady-state electrolysis procedure. However, average transport numbers of high precision are obtainable by the method.

In order to determine transport numbers more rapidly, methods involving radioactive traces were initiated.[35–37] Several automated techniques are now in existence and using these methods sodium ion and water transfer numbers can be measured in less than one hour. There is only one drawback associated with such a rapid measurement of transport number and this is inherent in any rapid measurement. The average value of the measured transfer number is an average over a very short time period. If this short average is time-invariant the measurement can be related to steady-state values, but if the membrane has not attained steady state before the rapid measurement is completed the experimental values will not be characteristic of final operational values obtained during long-term electrolysis.

It is the author's experience that steady-state values of transport properties are very rapidly attained with sulphonate ionomers and because of this, rapid radioactive tracer techniques are very useful for measuring cation and water transfer number for perfluorosulphonate polymers. However, many carboxylate ionomers take a considerable period of time, in some situations several days, before reaching steady-state values; for this reason caution is needed when interpreting transport numbers obtained from radioactive tracer studies.

The variation of sodium ion transport number with process conditions for several sulphonate ionomers has been reported by different workers.[17,35-40] In broad terms the following observations can be made. Temperature has only a small effect on cation transport numbers,[37] current density generally results in an increase in transport number,[14] transport number decreases with decreasing brine strength, catholyte chloride levels decrease with increasing current density,[41] and transport number exhibits a non-linear dependence on caustic concentration. It is perhaps the non-linear dependence of cation transport number on external alkali concentration which is the most interesting.

The variation of sodium transport number with external sodium hydroxide concentration for a typical 1200 equivalent weight sulphonate ionomer operating at $3\,kA/m^2$ and 90°C is shown in Fig. 3. Transport number does not decline monotonically with increasing alkali concentration. Between 0 and 20% w/w NaOH, transport number falls but after this there is an increase, followed by a further decline. This type of behaviour is quite typical of both sulphonate and carboxylate ionomers and is what distinguishes them from conventional crosslinked hydrocarbon ion exchangers.

There is not a universally accepted explanation of this non-classical variation of transport number with increasing electrolyte concentration. Explanations involving mechanisms for hydroxyl ion tunnelling,[42] increases in local fixed ion concentrations,[43] and frictional interactions between migrating water and hydroxyl ions[44] have all been suggested to play some part in determining the complicated behaviour pattern. Whatever the detail is, the phenomenon is certainly intimately related to polymer structure.

Many of the modern analytical techniques have been employed to investigate the microstructure of sulphonate ionomers. The nature and environment of sorbed sodium ions and water in sulphonate ionomers have been studied by multi-nuclear magnetic resonance.[45-49] Variable-temperature Raman spectroscopy has also been used to probe the supermolecular structure[50] and the infrared spectra of several sulphonate ionomers have been measured.[51-55] Electron microscopy studies

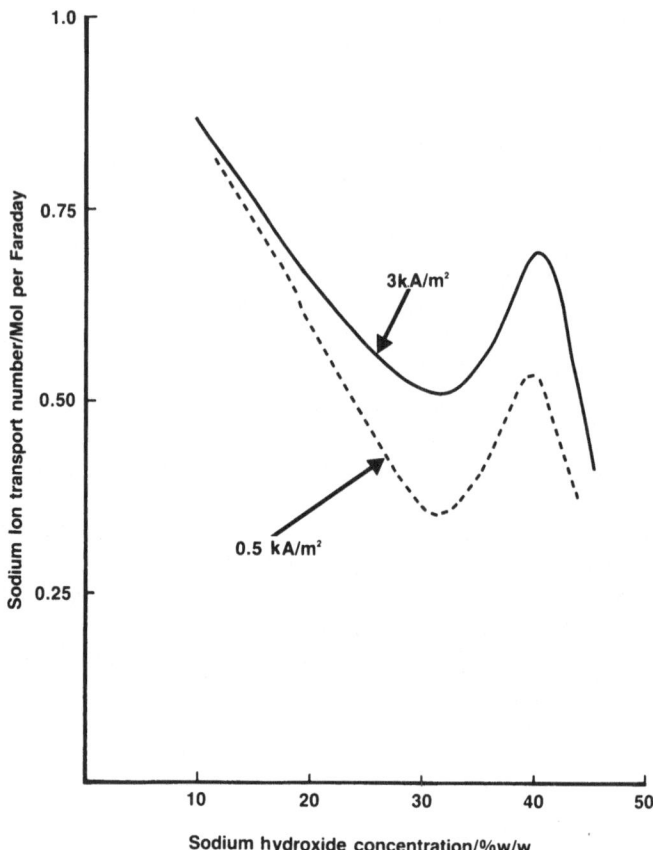

FIG. 3. The dependence of sodium ion transport number on sodium hydroxide concentration for a typical sulphonate ionomer, equivalent weight 1200. NaCl 230 g/litre, 90°C.

have been undertaken,[56,57] as have measurements of mechanical and dielectric relaxation.[58] Extensive small-angle X-ray scattering (SAXS) and wide-angle X-ray scattering (WAXS) studies have been reported.[57–60] Neutron scattering experiments have also been carried out.[61]

The molecular morphology of sulphonate ionomers is substantially different from that encountered with conventional styrene/divinylbenzene ion exchangers as the hydrophilic ion exchange sites and the hydrophobic polymer backbone phase separate to generate a clustered morphology.[57] These clusters, which are of average dimension less than 50 Å diameter, play a significant role in controlling both the mechanical properties and electrochemical specificity and conductivity. As clusters are also found with carboxylate ionomers, a description of the relationship between cluster morphology and membrane performance is left to Section 5.

3. PERFLUORINATED CARBOXYLATE IONOMERS

The history of the discovery and initial development of perfluorocarboxylate ionomers for use as chlorine cell separators is fraught with uncertainty: it has been, and still is, the subject of much patent litigation and the author therefore offers no opinion as to patentability of these developments.

As was the case for perfluorosulphonate ionomers, simple perfluorocarboxylate ionomers are also formed by homopolymerisation, but in this case a perfluorovinyl ether monomer containing a group convertible to a carboxylic moiety is polymerised with tetrafluoroethylene. In general terms the sulphonate and carboxylate monomers are very similar:

$$CF_2{=}CFO{-}(CF_2{-}\underset{\underset{CF_3}{|}}{C}FO)_n{-}(CF_2)_m{-}SO_2F$$

Typical sulphonate monomer

$$CF_2{=}CFO{-}(CF_2{-}\underset{\underset{CF_3}{|}}{C}FO)_n{-}(CF_2)_m{-}C\overset{O}{\underset{OCH_3}{\diagdown}}$$

Typical carboxylate monomer

Many of the more recent carboxylate membranes are formed with the inclusion of a third non-ion exchange monomer.

Subsequently to polymerisation, membranes of different equivalent weight are often laminated, in some cases with additional reinforcement in the form of a net placed between them.

The basic steps which are needed in the fabrication of any industrial membrane are shown in Table 2; each of these areas possesses its own challenges and difficulties.

TABLE 2
Stages Involved in Developing Perfluorocarboxylate Ionomer Membranes

Monomer Synthesis	Copolymerisation	Extrusion	Lamination	Surface modification
Tetrafluoroethylene ⟶	Polymers of different equivalent weight	Film manufacture	Inclusion of reinforcement	Surface treatment to aid performance
Carboxylate perfluorovinyl ethers ⟶				
Additional monomers ⟶				

3.1. Synthesis and Polymerisation of Perfluorinated Carboxylate Monomers

Organic fluorine chemistry is an interesting but difficult area of science, and because of this there is only a limited number of industrial organisations involved in the production of fluorochemicals. Therefore most work concerning the development of perfluorocarboxylate monomers has been carried out by a very few companies.

Most perfluorovinyl ether monomers are formed by the reaction of hexafluoropropylene oxide with a bifunctional compound, followed by subsequent vinylisation by heating with a suitable catalyst.[62-66] Although the patent literature exemplifies and claims many different possible monomers it is likely that only certain specific monomers are used by the membrane producers. Synthetic routes to two monomers which have found commercial application are presented to illustrate the stages of production.

Stages involved in preparing typical perfluorovinyl ether monomers are shown in Scheme 1 when $n = 0$ the product is methyl perfluoro-5-oxa-6-heptenoate and when $n = 1$ the product is methyl perfluoro-5,8-dioxa-6-methyl-9-decenoate.

Although there are many other possible reaction schemes to produce other monomers, the scheme above is sufficient to illustrate the complexity of the chemistry and the number of stages involved to produce a single monomer. More exhaustive information is available in the patent literature.[67,68]

The polymerisation reaction between the vinyl ether monomer and tetrafluoroethylene can be carried out by either solution polymerisation or emulsion polymerisation.[8,10,69,70] During both polymerisation processes the reaction temperature is usually modest (60°C) and the total pressure is regulated to about 30 kg/cm^2 (3 MPa). Non-aqueous polymerisation is normally carried out in an inert halogenic solvent, e.g. Arcton 113, and the initial radical source is often provided by the decomposition of azobisisobutyl nitrile (AIBN). The molecular mass of a typical copolymer structure is between 10 000 and 100 000 molecular weight units.[10,70,71]

The ratio of active side chain to polymer backbone is determined by the reactivity ratios of the perfluorovinyl ether and tetrafluoroethylene and the monomer feed ratio. Examples of monomer activity ratios for polymerisation of both sulphonate and carboxylate perfluorovinyl ethers with tetrafluoroethylene are given in Table 3. Suitable control

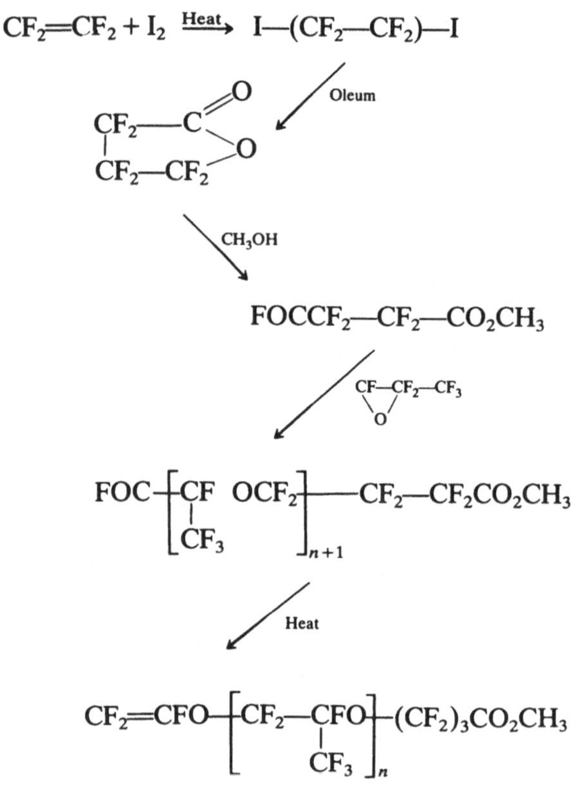

$$CF_2{=}CF_2 + I_2 \xrightarrow{\text{Heat}} I{-}(CF_2{-}CF_2){-}I$$

$$\begin{array}{c} CF_2{-}C{\diagup}^{O}_{\diagdown O} \\ | \hspace{1.2cm} | \\ CF_2{-}CF_2 \end{array} \xleftarrow{\text{Oleum}}$$

$$\downarrow CH_3OH$$

$$FOCCF_2{-}CF_2{-}CO_2CH_3$$

$$\swarrow \begin{array}{c} CF{-}CF_2{-}CF_3 \\ \diagdown O \diagup \end{array}$$

$$FOC{-}\!\!\left[\begin{array}{cc} CF & OCF_2 \\ | & \\ CF_3 & \end{array}\right]_{n+1}\!\!{-}CF_2{-}CF_2CO_2CH_3$$

$$\downarrow \text{Heat}$$

$$CF_2{=}CFO{-}\!\!\left[\begin{array}{c} CF_2{-}CFO \\ | \\ CF_3 \end{array}\right]_n\!\!{-}(CF_2)_3CO_2CH_3$$

SCHEME 1

TABLE 3
Monomer Activity Ratios, r

Sulphonate		Carboxylate	
Monomer	r	Monomer	r
$CF_2{=}CF_2$	8	$CF_2{=}CF_2$	7
$CF_2{=}CFOCF_2{-}\overset{\overset{CF_3}{\|}}{C}FO{-}(CF_2)_3SO_2F$	0·08	$CF_2{=}CFOCF_2{-}\overset{\overset{CF_3}{\|}}{C}FO{-}(CF_2)_3CO_2CH_3$	0·09

of monomer feed ratios allows high-molecular-mass polymers of varying ion-exchange capacity to be formed.[72-74]

As was noted earlier with perfluorosulphonate ionomers, crystallinity increases with decreasing perfluorovinyl ether component. However, as polymer crystallinity increases, the membrane tensile elongation reduces and this makes the membrane mechanically weak. In order to increase tensile elongation while maintaining low perfluorovinyl ether content, polymerisation is carried out in the presence of a third non-active monomer. The function of this monomer is to attach to the main polymer chain and inhibit crystal packing.[10,75] Diagrammatically this can be represented as:

$$CF_2{=}CF_2 + CF_2{=}CFOCF_2{-}\overset{\overset{\displaystyle CF_3}{|}}{CF}O{-}(CF_2)_3{-}CO_2CH_3$$

$$+ CF_2{=}CF{-}O{-}(CF_2)_3{-}CF_3$$

$$\downarrow \text{AIBN/pressure}$$

$$-CF_2{-}CF_2{-}CF_2{-}\underset{\underset{\displaystyle CF_3}{|}}{\underset{\displaystyle (CF_2)_3}{|}}{\underset{\displaystyle O}{|}}{CF}{-}\underset{\underset{\underset{\underset{\underset{\displaystyle CO_2H}{|}}{\displaystyle CF_2}}{|}}{\underset{\displaystyle O}{|}}}{\underset{\displaystyle CFCF_3}{|}}{\underset{\displaystyle CF_2}{|}}{\underset{\displaystyle O}{|}}{CF}{-}CF{-}$$

The inclusion of this third non-active monomer has a tremendous influence on the physical properties of the resulting ionomer. The flexibility implemented by the third monomer aids polymer phase separation and this has a large effect on both polymer conductivity and cation specificity.

3.2. Fabrication of Perfluorinated Membranes

Membrane manufacturers are always trying to optimise three things—mechanical strength, cation specificity and power dissipation. The thickness of a membrane has little influence on cation specificity but it has a pronounced effect both on mechanical properties and ohmic resistance. Reduction in membrane thickness yields major energy savings but the penalty for this is usually low mechanical strength. This can be alleviated to some extent by a reinforcing material.

An ingenious and unique method of membrane reinforcement has been developed by the Asahi Glass Company Ltd, Yokohama, Japan.[76] It relies on incorporating small polytetrafluoroethylene (PTFE) particles into the polymer mix. The PTFE is not immediately obvious to the eye and at correct loading levels it does not reduce significantly the surface area of the membrane. It confers mechanical rigidity to the whole membrane while the active polymer remains a highly flexible non-crystalline structure. This combination of flexible and rigid polymers allows for low voltage, high cation specificity and good mechanical properties when the membranes are operated in well defined ways. Operation of PTFE fibre reinforced membranes outside these well defined areas can generate problems.

As the carboxylate polymers described in Section 3.1 are all melt processable they can be formed into films by either conventional moulding or extrusion methods. Polymers which contain PTFE fibres do not extrude or mould well: in order to overcome this, membranes made from such polymer mixes must be formed by controlled stretching techniques.[77] There is much art involved in producing a laminated membrane sheet and the problems are even more complex if a reinforcing gauze is to be included between the polymer layers.[78-80]

Many commerical membranes incorporate some form of mechanical reinforcement to enable them to be handled and installed without damage in commercial electrolysers. The nature of the reinforcement can take several forms: inert gauze inclusion,[78] a combination of inert and sacrificial reinforcement, surface gauze adhesion and PTFE fibre inclusion.[76] The positioning of a typical reinforcement material in a membrane is shown in Fig. 4.

All these methods of reinforcement improve the mechanical properties of finished membranes significantly and it is normally found that under a defined set of operating conditions the inclusion of the supporting material has little influence on the electrochemical performance of the membrane. However, as the reinforcement material is of a different composition from the ionomer, perfect wetting/bonding is rarely achieved and the non-perfect wetting/bonding can result in a membrane which is prone to delamination when operated outside defined regions.

The inclusion of inert gauzes or surface-adhered nets can influence

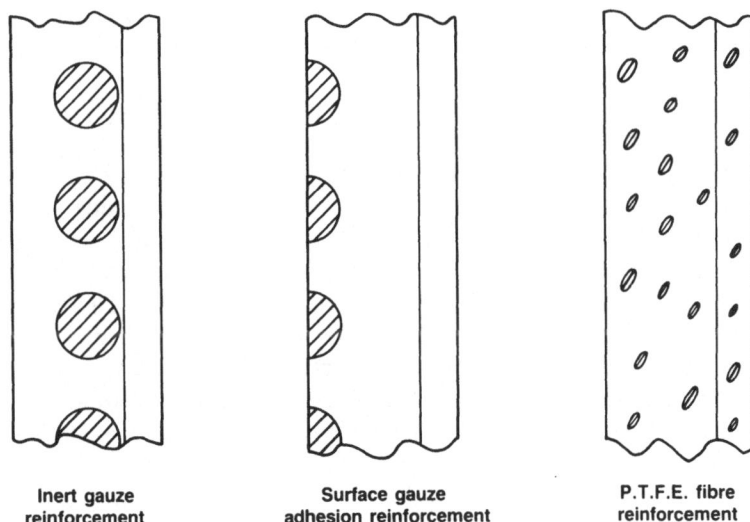

FIG. 4. Schematic diagram showing typical methods used to reinforce commercial membranes.

the topography of the membrane surface and an undulating or corrugated surface may pose certain problems for sealing the membrane in commercial cells. A pocketed surface may also produce high cell voltages by trapping gas when the electrodes are close to the membrane surface. Operation of surface-adhered reinforced membranes at small membrane–cathode gaps can generate problems as product gas absorbs on to the non-wetted reinforcement material, and then spills over on to the active membrane surface and increases the effective operating membrane current density.

Due to the above effects it is the author's firm view that reinforcement material of all kinds should be avoided if at all possible. Careful design of the size and shape of a commercial cell can effectively eliminate the need for membrane reinforcement.[81]

3.3. Surface Topography of Perfluorinated Ionomers

Until about four years ago little attention was paid to the topography of membrane surfaces. However, this has now changed, and the most important advance in recent years has been the recognition of the vital role played by a membrane surface in determining both cation specificity and overall cell power performance. The recognition of the key role of the membrane surface has meant that all present-day commercial membranes have their surface topography modified prior to cell operation.

During operation of a chlor-alkali membrane cell, gas bubbles adhere to the surface of the membrane. The extent of adhesion depends on the nature of the absorbing gas (hydrogen, or chlorine), the membrane–electrode gap, the degree of electrolyte mixing present and the nature of the membrane surface. At small electrode–membrane distances gas adhesion can be pronounced and this can result in inhibited current flow and increased cell voltage. The effect is associated with the hydrophilicity of the membrane surface; it is not simply gas entrapment by a poor membrane reinforcement.

The phenomena occurring when the membrane–electrode gap is reduced were extensively investigated by workers at the Asahi Glass Company Ltd, and they concluded that gas adhesion could be minimised by increasing the hydrophilicity of the membrane surface.[82] The effect of increasing the surface hydrophilicity is impressive and results for a typical fibre reinforced carboxylate ionomer are illustrated in Fig. 5.

The methods by which surface hydrophilicity can be increased are described in an extensive series of patent applications.[83–87] One of the principal ways to increase surface hydrophilicity is to deposit on the surface of the membrane a porous non-electrode inorganic material such as titanium dioxide, silicon dioxide, silicon carbide or zirconia, which provides the membrane with a non-smooth surface. Although the inorganic material used must be chosen with regard to its chemical stability in the process fluids, the main feature of the surface coating is physical topography. The rough surface engendered by the inorganic particles makes gas adhesion difficult and gas is easily removed from the membrane surface.

The generation of a hydrophilic non-gas-adhering surface can be achieved without the use of a discrete porous non-electrode inorganic layer. Simple abrasion of the surface of the ionomer with sandpaper or other abrasive generates a microscopically rough surface and this too aids gas release properties.[88] It is also possible to impregnate a surface with aluminium dust and then remove it by alkali to generate a non-smooth surface.[89] There are other possible methods of producing rough surfaces e.g. ion etching.[90]

Mechanical abrasion techniques and chemical impregnation/removal procedures yield excellent rough surfaces subsequent to treatment but

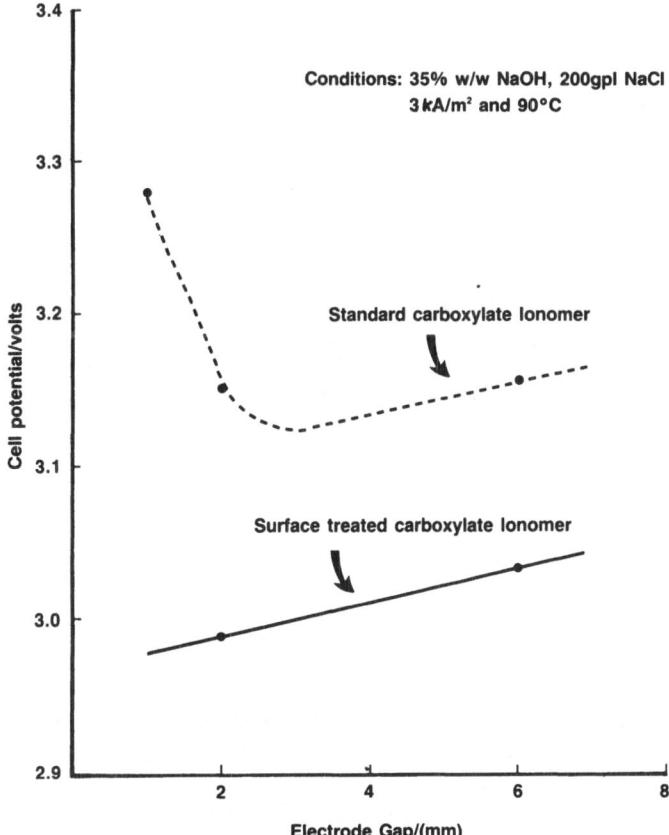

FIG. 5. The effect of surface roughening on cell voltage as a function of electrode gap for a typical carboxylate ionomer.

it is the author's experience that the long-term stability of these treatments is inferior to that of the adhered porous non-electrode layer technique. However, longevity may be counterbalanced by the simplicity of the procedures involved and the low capital cost needed to carry out the treatment.

The generation of carboxylate membranes with excessive surface roughness is not always beneficial and in some instances it can be detrimental to the overall performance of the membrane. It is known that a surface coating of a non-electrode porous medium which covers only approximately 10% of the membrane area is sufficient to bring about significant voltage reductions without decrease in membrane specificity.[91] Excessive roughening of the catholyte face of a carboxylate polymer can result in a large decrease in membrane selectivity and it is claimed that for optimum current efficiency the catholyte surface of a commercial membrane must be smooth.[92]

Hence there is a balance between low voltage caused by modest surface roughening, and poor selectivity caused by excessive roughening. With careful control of the degree of roughening it is possible to obtain significant power savings without loss of selectivity. The most likely reason for the reduction in membrane selectivity is that pockets of concentrated alkali are formed in the interstices between the coating particles.

3.4. Physical Properties of Perfluorocarboxylate Ionomers

The commercial importance of perfluorocarboxylate ionomers as chlor-alkali cell separators has resulted in very few samples of membrane being supplied to academic institutions. This has meant that the study and characterisation of these materials has been generally limited to industrial organisations and in view of the uncertainty associated with patents, very little of this information has been published in the open literature. However, during the last few years some structural and transport properties of representative polymers have started to emerge slowly.

Data regarding the water and electrolyte sorption characteristics of simple perfluorocarboxylate ionomers have been reported.[11,24,82,93,94] The observed electrolyte and water contents of carboxylate polymers are normally much smaller than these found with the corresponding sulphonate polymers and this low sorption power is related to the much reduced acidity of the carboxylate grouping. It means that for a given equivalent weight the conductivity of the sulphonate ionomer is greater than that of the corresponding carboxylate polymer. However, by increasing the ion-exchange capacity of the carboxylate system, increased conductivity can be achieved.

The specific conductivity of carboxylate ionomers depends markedly on temperature and electrolyte concentration.[11] Conductivity declines rapidly with increasing electrolyte concentration, and apparent activation energy for ionic conductance increases quickly with increasing electrolyte concentration. The variation of ionic conductance with external electrolyte concentration has been explained in terms of membrane dehydration.[10] At low temperature and high sodium hydroxide concentration, contact ion-pairs form more rapidly with carboxylate ionomers than with sulphonate polymers.

The diffusional properties of perfluorinated polymers have been reported by Suhara and Oda.[11] The dependence of ionic conductivity on ion exchange capacity and external electrolyte concentration at 25°C and 90°C is reported. Data concerning the diffusion of anions are correlated with measured electrical conductance and a definite correlation between specific conductance and diffusion coefficient is proposed.

Very few structural investigations concerning perfluorocarboxylate ionomers have been published but a good exposition of the subject has been given by Hashimoto et al.[95]

4. MIXED PERFLUORINATED SULPHONATE/CARBOXYLATE IONOMER MEMBRANES

Although membranes formed totally from carboxylate ionomers are capable of providing low voltage and high selectivity they are not problem-free. The weak acidity of the carboxylic functionality (pK_a approx. 2–3) means that they are more susceptible to acid protonation than sulphonic acid membranes (pK_a approx. 1) and, as many chlorine producers use acid addition to regulate anolyte oxygen levels, this means that more operational control is often needed when using carboxylate ionomers. In view of this, certain membrane manufacturers have developed and patented an extensive series of membranes based on a composite structure of sulphonate and carboxylate ionomers.

The main products which are used commercially are either laminates of a sulphonate and carboxylate polymer (Du Pont) or a carboxylate surface modified sulphonate ionomer (Asahi Chemical Industries and Tokuyama Soda Co Ltd). Attempts have also been made to form a membrane by terpolymerisation of tetrafluoroethylene with sulphonate and carboxylate monomers[96,97] but none of these has been very successful. Membranes have also been made from blends of melt processable carboxylate and sulphonate ionomers but the performance of these membranes left much to be desired.[98]

Very successful high-performance membranes have been derived by carboxylate modification of the catholyte surface of conventional sulphonate polymers.[99,100] An illustrative example of this type of procedure is shown schematically below:

$$\sim(CF_2-CF_2)_m-CFO[CF_2-CFO]CF_2-CF_2-SO_2F$$
$$\qquad\qquad\qquad |$$
$$\qquad\qquad\qquad CF_3$$

$$\xrightarrow{\text{Reaction}} \sim(CF_2-CF_2)_m CFO[CF_2-CFO]CF_2-CO_2-CH_3$$
$$\qquad\qquad\qquad\qquad |\qquad\qquad |$$
$$\qquad\qquad\qquad\qquad CF_2\qquad\;\; CF_3$$

The important points are that the basic polymer structure is maintained while the pendant side chain length is reduced by one —CF_2— unit. This requires the chemical transformation of either the sulphonic acid site[99,100] or a modified sulphonic acid site.[101–105]

Membranes formed by surface modification techniques eliminate the need to laminate polymers of different functionality and they are often mechanically very robust. They can be reinforced by conventional techniques and it is possible to treat the membrane surfaces to aid gas release properties. High performance is not restricted to carboxylate modified sulphonate ionomers and very high cation selectivity is also possible from laminated sulphonate/carboxylate polymers.[106]

5. STRUCTURE/TRANSPORT RELATIONSHIPS IN PERFLUORINATED IONOMER MEMBRANES

5.1. Basic Structural Features of Perfluorinated Ionomers

Pre-hydrolysed sulphonate and carboxylate polymers exhibit crystallinity, which is not radically reduced upon hydrolysis. Studies of the effect of tensile draw on the polymer structure indicate a periodicity corresponding to molecular chain alignment but the available evidence is not sufficient to conclude that 'chain folding', as is found with polyethylene, does occur.[107]

Structural studies of hydrolysed polymers by techniques such as X-ray diffractometry, electron microscopy,[56] NMR spectroscopy,[45] Mössbauer spectroscopy[108,109] and dielectric relaxation[110] have shown that the predominant structural feature is the existence of ionic clustering.[111,112] These ionic clusters or domains, which are formed by the self-association of pendant ionic groupings, are also found in hydrocarbon ionomers.[113] The size and morphology of the cluster are controlled by the counterbalancing of two opposing features: the energy released by cluster collapse, and the energy required to deform polymer chains to generate cluster collapse.

Several structural models have been proposed to account for cluster formation in both hydrocarbon ionomers and the perfluoro ionomers. The simplest is a two-phase model where ionic clusters are considered to be formed as discrete entities distributed in a polymer matrix which contains few other ionic groupings.[114] The clusters do not contain polymer chains and they are regarded as being composed entirely of hydrophilic sites. This model has been extended to cover systems where the ionic aggregates are dispersed in a fluorocarbon matrix of intermediate non-clustered ionic activity.[107,115] A second structural model is the core shell model, in which the ionic cluster (the core) is surrounded by a shell of inert fluorocarbon polymer (the shell) and core–shell units are dispersed throughout the matrix. The geometry of the core–shell particles can be spherical[116,117] or lamellar.[118]

At present there is no agreement about which model best represents the structure of ionomers, but in view of the very low dielectric constant found with perfluoro ionomers it is unlikely that the core–shell model has extensive application. Although there is a lack of agreement about which cluster model best describes molecular structure, there is agreement that the morphology of the cluster is critical in determining final membrane transport properties.

The most detailed information about cluster size and geometry has been provided by SAXS and WAXS investigations. Measurements of this type provide a value for an average Bragg spacing which can be related to average cluster size. Measured values of cluster size depend on the polymer ion-exchange capacity, the cation form and the quantity of absorbed water present in the polymer.

At constant humidity, the average cluster size of sulphonate and carboxylate ionomers, of similar structural form, increases with increasing polymer ion-exchange capacity. The variation of cluster size with ion-exchange capacity reflects the corresponding increasing flexibility of the polymer matrix. The more flexible ionomer requires less energy to deform the polymer backbone and hence larger ionic clusters are more easily formed.

The average cluster size is not markedly affected by changing the cation form, although different cation forms cause significant changes in the mechanical properties of the polymer. Increased temperature causes only a small decrease in cluster size and ionic clusters still persist even at temperatures above the melting point of the ionomer. The decreased cluster size at elevated temperature is associated with increased bulk polymer mobility.

Cluster size is particularly sensitive to polymer water content: it reduces quickly as absorbed water content declines. This means that average cluster size depends on the external concentration of electrolyte with which the polymer is in contact; concentrated solutions favour small clusters whilst dilute solutions promote larger clusters.

Several attempts have been made to determine average cluster size from solvent absorption studies. When it is assumed that the clusters are spherical and that they form a cubic array, equations for average cluster size, the number of ion exchange sites per cluster and the average number of water molecules in a cluster can be derived. Calculations of this type suggest that as ion-exchange capacity decreases, the cluster size, the number of ion-exchange sites per cluster and the water uptake by each ion-exchange site also decrease. Similar calculations have been applied to membranes in varying cationic forms; as cation mass increases the cluster diameter and absorbed water per exchange site decrease whilst the number of exchange sites per cluster increases. Explanations for these effects have been given in terms of cation/exchange site bonding properties.[59]

Cluster morphology calculations have also been carried out for ionomers of different internal water content. As the polymer absorbs more water, the average cluster size, the number of exchange sites per cluster and the number of water molecules per exchange site all increase. The increase in the number of exchange sites per cluster has been interpreted as suggesting that the cluster does not simply form by expansion of a dehydrated unit but by a combination of expansion and continuous reorganisation of the exchange sites so that there is a reduced number of clusters in a fully hydrated membrane.

The average size of clusters observed with sulphonate and carboxylate ionomers of similar structure has been determined. It was noted that, in general, carboxylate ionomers generate much smaller ionic clusters. This is most probably associated with the smaller electrostatic energy release upon cluster formation by the low-acidity carboxylate grouping. The small cluster size found with carboxylate ionomers is present in both wet and dry polymers.

The ionic clusters in both sulphonate and carboxylate ionomers must be connected, otherwise the polymers would not facilitate ionic

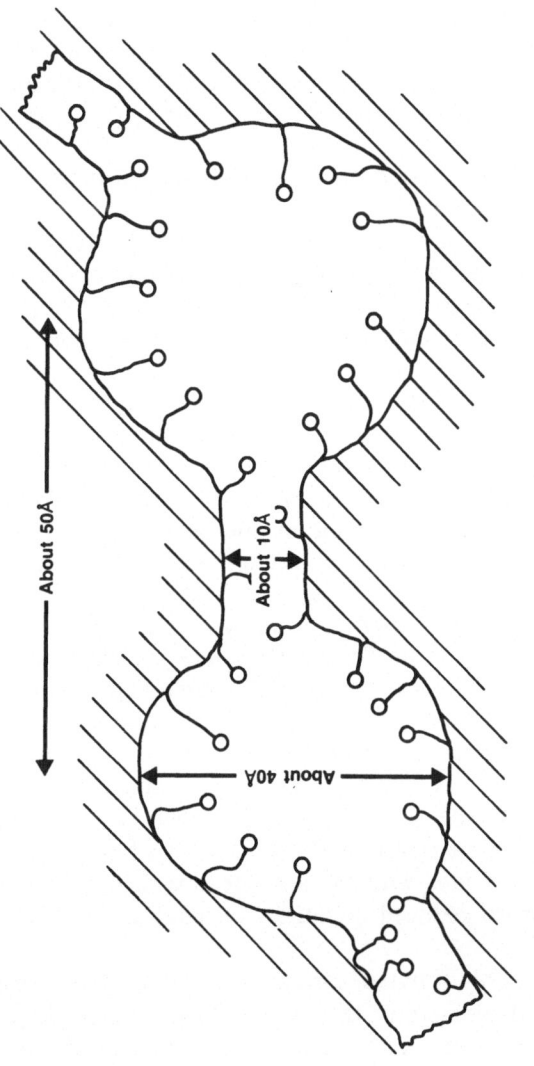

FIG. 6. A schematic representation of possible cluster-channels found in typical perfluorinated ionomers.

conduction and hydraulic flow. Hydraulic permeability studies and diffusion transfer rates through the ionomers have been used to estimate the size of the cluster-connecting, rate-controlling structural feature. This information, coupled with the cluster morphology, provided the basis for the Gierke cluster-network model. This model is shown schematically in Fig. 6. The average cluster diameter is about 40 Å, the cluster separation is about 50 Å and the dimensions of the connecting pathways are estimated at about 10 Å.

5.2. Experimental Correlations Between Polymer Structure and Transport Properties

Data concerning the average cluster dimensions of a series of perfluorosulphonate ionomers immersed in water have been published, as has the efficiency of these membranes as chlor-alkai cell separators. A qualitative indication concerning the influence of cluster size on membrane efficiency can be inferred from a plot of separator efficiency as a function of cluster diameter. This is shown in Fig. 7, but it must be stressed that this graph is only qualitative since the precise size and nature of a cluster in an operating membrane will be different from, but related to, that measured when the membrane is immersed in water. In spite of this limitation the general trend is readily visible—small cluster formation correlates well with high separator cation specificity.

A corresponding investigation concerning estimation of the average cluster size for a series of perfluorocarboxylate ionomers has recently been presented.[119] As was found with sulphonate ionomers, the efficiency of the polymers as chlor-alkali cell separators was highly dependent on the size and distribution of the ionic clusters. The dependence of sorbed alkali content on cluster diameter for this series of polymers was also measured and this is shown in Fig. 8. The data displayed in Fig. 8 show clearly that the quantity of electrolyte sorbed by the polymer correlates with the average size of the ionic clusters. Small clusters absorb reduced quantities of alkali and so favour high cation specificity.

This study of perfluorinated carboxylate ionomers is particularly noteworthy as it demonstrates that initial cluster morphology can be varied by factors other than ion-exchange capacity. Initial cluster size can be modified by the incorporation of a third monomer in the polymer matrix. This additional monomer, although having no ion-exchange functionality, reduces the matrix crystallinity and so aids

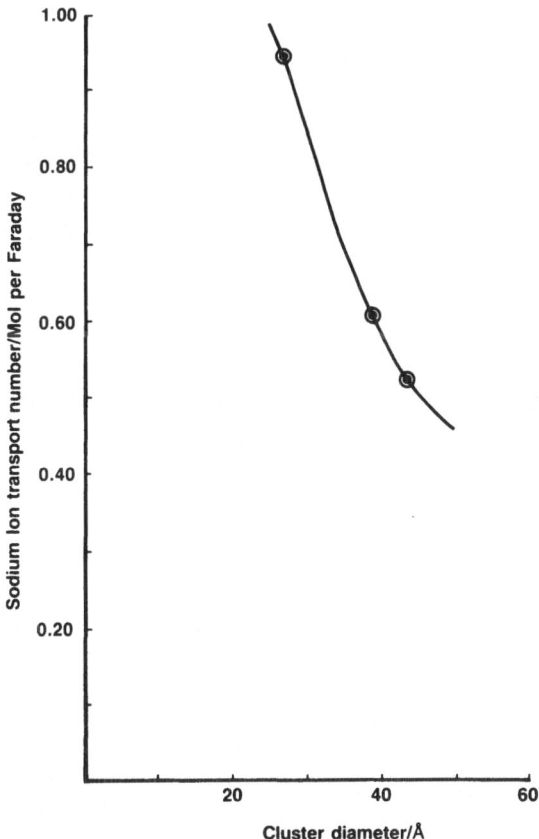

FIG. 7. The variation of sodium ion transport number with cluster size for a typical sulphonate ionomer.

more extensive phase separation. During the polymerisation process increased phase separation favours the generation of a large number of small clusters and if the subsequent polymer hydrolysis is carried out correctly the large number of small clusters persists in the ionomer. If the hydrolysis reaction is carried out in a high-swelling medium the highly flexible polymer chains reorientate to give clusters of increased size.

The structural effect of non-active termonomers means that it is possible, with due care, to synthesise polymers with morphology

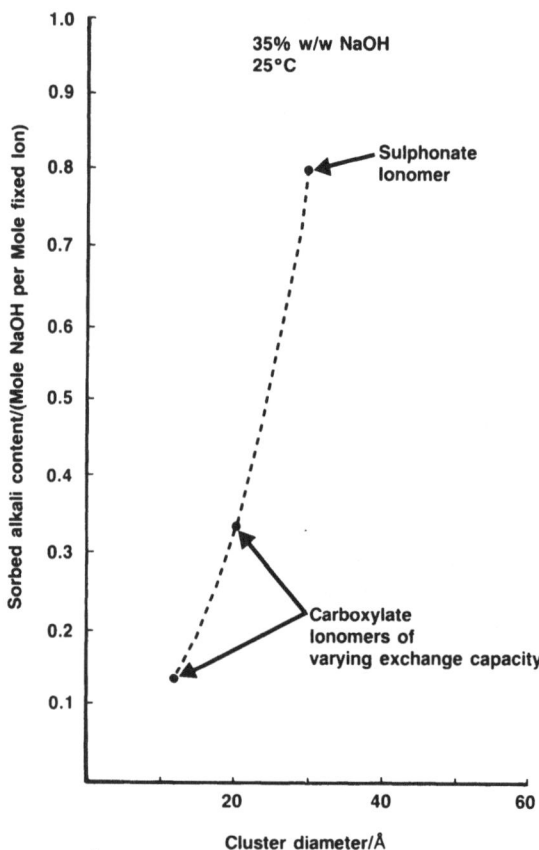

Fig. 8. The dependence of sorbed alkali concentration on cluster size for a typical carboxylate ionomer.

capable of yielding high cation specificity while retaining many ionic groupings to facilitate good conductivity. The relationship between bulk ionomer conductivity and termonomer content for a typical perfluorocarboxylate membrane is shown in Fig. 9. The ionic conductivity of the bulk polymer increases rapidly with increasing termonomer content. This is to be associated with reducing polymer crystallinity and the subsequent increased phase separation between hydrophilic ionic sites and the hydrophobic polymer matrix.

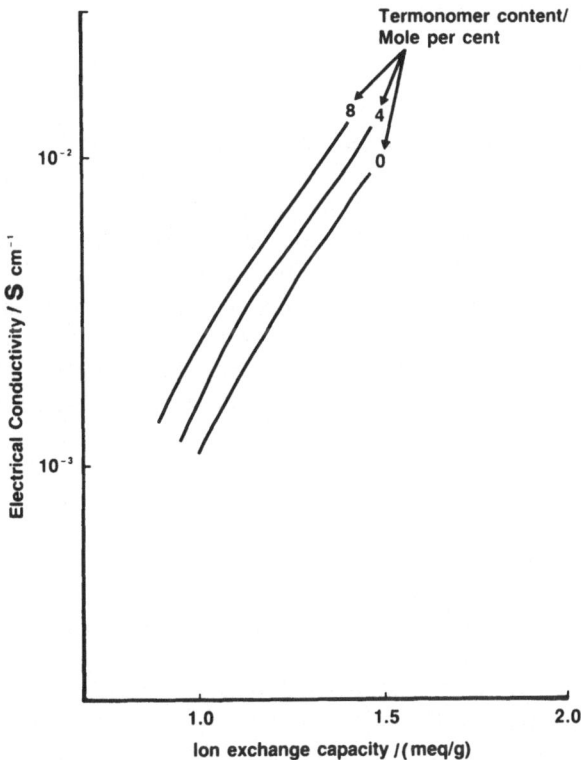

FIG. 9. The effect of non-active termonomer on the electrical conductivity of typical carboxylate ionomers.

The empirical relationships outlined above illustrate the strong dependence which exists between ionic cluster size and the resultant efficiency of the polymer as a low-voltage–high-efficiency chlor-alkali separator. Favourable membranes are formed from relatively flexible polymers which require minimum work to deform polymer chain segments to generate small ionic clusters distributed regularly throughout the matrix. In addition to polymer backbone flexibility, the length of the pendant side chain on to which the ion-exchange group is attached is also important in determining average ionic cluster size. Polymers which include small pendant side chains are capable of forming smaller clusters and it is known that the cation specificity of such polymers is better than those formed with longer side chains.

5.3. Quantitative Relationships Involving Ionic Transport

5.3.1. Gierke cluster-network theory

For any given ionomer, immersed in a particular solution, there is a definite interconnected distribution of ionic clusters such as that illustrated in Fig. 6. Water and electrolyte are not absorbed to any major extent by the hydrophobic matrix and because of this the imbibed liquor is restricted to the ionic clusters and pathways which connect them. Electrostatic repulsion effects exclude anions from the surface of the cluster and the interconnecting chanels, but in the centre of the cluster, which may be effectively screened, large quantities of electrolyte may be accommodated. Thus the ionomers absorb relatively large quantities of liquor but as this electrolyte is located in very special sites, the transfer rate across the membrane is very slow. Transfer of anions from the cluster centre is severely inhibited by the need to overcome significant electrostatic repulsion effects at the cluster surface and in the connecting channels. It is this large electrostatic energy barrier, produced as a direct result of the polymer morphology, which is considered to explain the high cation specificity afforded by perfluorinated ionomers in concentrated solution.

The qualitative picture previously described can be turned into a quantitative predictive model by applying absolute reaction rate theory. A detailed discussion of how this has been done has been provided by Gierke and it is sufficient to say that the theory allows accurate prediction of separator specificity and polymer diffusivities.[59]

5.3.2. Phenomenological theories of membrane transport

It is possible to attempt to calculate sodium ion transport number without referring to molecular structure. In this approach the calculation is often based on concentrated solution theory[120] and the Stefan–Maxwell transport equations.[121] A detailed exposition of this approach has been given by Bennion and coworkers.[122] The method is mathematically inclined and the theory has very little predictive value.

A theory for rationalising the observed caustic current efficiencies found with two-layer laminated membranes has been proposed by Krishtalik.[123,124] An explanation of the high observed current efficiency found with a membrane containing a more selective cathode site is given in terms of ionic diffusion of sorbed alkali. Possible concentration profiles of sorbed alkali in laminate membranes are also discussed.

6. PRESENT AND FUTURE TRENDS IN PERFLUORINATED IONOMER MEMBRANE CELL TECHNOLOGY

Present world installed annual capacity for chlorine production is about 40 million metric tons.[125] Only about 2–3% of this now makes use of membrane technology, but there are signs that the percentage is about to grow rapidly. The world's largest membrane cell plant (*ca* 280 000 metric tons per annum of caustic soda) was recently opened by Akzo Zout Chemie and capacity of about 0·4 million metric tons was announced during 1983. In Japan it is estimated that a further 1·3 million ton capacity will be converted to membrane technology by 1986.

Membrane cell technology has come of age—about one million tons capacity installed, and a further one million planned. This situation heralds the start in a new phase of membrane cell technology development. Full-scale plant installations will allow people to gather experience of commercial cell operation and there may now be a period of consolidation in the industry. The important features needed for safe and steady commercial cell operation will be learned and the wish for continually improving power utilisation will begin to assume less importance.

It is likely that the direction of membrane improvement will also change in the next five years. We now have membranes which provide high cation specificity and low ohmic drop and it is unlikely that these factors will be refined further. What is more likely to happen is that effort will concentrate on developing membranes which are easy to use and are tolerant to upset conditions. Membranes which can operate with brine containing magnesium and calcium may also be possible in the next five years.

ACKNOWLEDGEMENT

I would like to thank ICI (Mond) PLC for permission to publish this review.

REFERENCES

1. R. Carson, *Silent Spring,* Hamish Hamilton, London, 1963.
2. Ionics Inc., US Patent 2 636 851.

3. Anon., *X-R Perfluorosulphonic Acid Membranes*, Du Pont New Product Bulletin, October 1969.
4. W. G. F. Grot, G. E. Munn, and P. N. Walmsley, 'Perfluorinated Ion Exchange Membranes', Paper presented at *141st Meeting of the Electrochemical Soc., Houston, Texas, May 1972*.
5. F. B. Leitz, M. A. Accomazzo and S. A. Michalek, 'Development of Electrochemical Hypochlorite Generators', Paper presented at *141st Meeting of the Electrochemical Soc., Houston, Texas, May 1972*.
6. F. G. Will, *J. Electrochem. Soc.*, 1979, **26**, 36.
7. E. I. Du Pont de Nemours Inc., US Patent 3 909 378.
8. E. I. Du Pont de Nemours Inc., US Patent 3 528 954.
9. M. Seko, In *Perfluorinated Ionomer Membranes, ACS Symposium Series*, **180**, American Chemical Society, Washington DC, 1982.
10. M. Ukihashi, *Chemtech*, February 1980, 118.
11. M. Suhara and Y. Oda, Paper presented at *161st Meeting of the Electrochemical Soc., Montreal, Canada, May 1982*.
12. H. L. Yeager, In *Perfluorinated Ionomer Membranes, ACS Symposium Series*, **180**, American Chemical Society, Washington DC, 1982.
13. P. J. Smith and T. L. Jones, *J. Electrochem. Soc.*, 1983, **130**, 885.
14. S. G. Cutler, *Polymer Physics Series*, Vol. 2, Academic Press, London, 1977.
15. E. H. Price and D. E. Maloney, Paper presented at the *Chlorine Institute's 21st Plant Manager Seminar, Houston, Texas, February 1978*.
16. E. I. Du Pont de Nemours Inc., US Patent 3 969 285.
17. C. J. Hora and D. E. Maloney, 'Nafion membranes structured for high efficiency chlor-alkali cells', Paper presented at *Electrochemical Soc. Meeting, Atlanta, Georgia, October 1977*.
18. T. Takamastsu, M. Hashiyama and A. Eisenberg, *J. Appl. Polym. Sci.*, 1979, **24**, 2199.
19. M. Lopez, B. Kipling and H. L. Yeager, *Anal. Chem.*, 1976, **48**, 1120.
20. H. L. Yeager and A. Steck, *Anal. Chem.*, 1979, **51**, 862.
21. A. Steck and H. L. Yeager, *Anal. Chem.*, 1980, **52**, 1215.
22. N. I. Men'shakova, V. I. Kubasov and L. I. Krishtalik, *Electrokhimiya*, 1981, **17**(2), 275.
23. N. I. Men'shakova, V. I. Kubasov and L. I. Krishtalik, *Electrokhimiya*, 1978, **14**, 356.
24. K. Kimoto, *J. Electrochem. Soc.*, 1983, **130**, 334.
25. T. Berzins, Paper presented at the *152nd Electrochemical Soc. Meeting, Atlanta, Georgia, October 1977*.
26. A. Narebska and R. Wodzki, *Angew. Makromol. Chem.*, 1979, **80**, 105.
27. H. L. Yeager, B. Kipling and R. L. Dotson, *J. Electrochem. Soc.*, 1980, **127**, 303.
28. H. L. Yeager and B. Kipling, *J. Phys. Chem.*, 1979, **83**, 1836.
29. H. L. Yeager, in *Perfluorinated Ionomer Membranes, ACS Symposium Series*, **180**, American Chemical Society, Washington DC, 1982, Chapter 4.
30. R. S. Yeo and J. McBreen, *J. Electrochem. Soc.*, 1979, **126**, 1682.

31. R. S. Yeo, J. McBreen, G. Kissel, F. Kulesa and S. Srinivasen, *J. Appl. Electrochem.*, 1980, **10**, 741.
32. T. Berzins, Paper presented at the *Electrochemical Soc. Meeting, Atlanta, Georgia, October 1977.*
33. P. J. Smith and K. G. Moss, ICI Mond Division Internal Report.
34. Asahi Chemical Co. Ltd, British Patent 1 543 249.
35. S. F. Burkhardt, Paper 440 presented at the *Electrochem. Soc. Meeting Atlanta, Georgia, October 1977.*
36. H. L. Yeager, B. O'Dell and Z. Twardowski, Papers 413, 414 presented at the *Electrochemical Soc. Meeting, Minneapolis, Minnesota, May 1981.*
37. P. J. Smith and T. L. Jones, *J. Electrochem. Soc.*, 1983, **130**, 885.
38. R. L. Dotson, R. W. Lynch and G. E. Hillard, in *Proc. Symp. Ion Exchange*, ed. R. S. Yeo and R. P. Buck, The Electrochemical Society, Pennington, NJ, 1981.
39. C. H. A. Kruissink, *J. Mem. Sci.*, 1983, **14**, 331.
40. O. P. Romashin, M. M. Fioshin, R. G. Erenburg, E. F. Ryabov, V. L. Kubasov and L. I. Krishtalik, *Elektrokhimiya*, 1979, **15**, 653.
41. N. I. Men'shakova, L. I. Krishtalik and V. L. Kubasov, *Elektrokhimiya*, 1977, **13**, 1732.
42. K. A. Mauritz, J. K. Branchick, C. L. Gray and S. R. Lowry, *Polym. Prepr., ACS Divn Polym. Chem.*, 1980, **20**, 122.
43. M. Suhara and Y. Oda, in *Proc. Symp. Ion Exchange*, ed. R. S. Yeo and R. P. Buck, The Electrochemical Society, Pennington, NJ, 1981.
44. T. R. E. Kressman and F. L. Tye, *Trans. Faraday Soc.*, 1959, **55**, 1441.
45. R. A. Komoroski and K. A. Mauritz, *J. Amer. Chem. Soc.*, 1978, **100**, 7487.
46. R. A. Komoroski, *ACS Adv. Chem. Ser.* I, 1980, **187**, 155.
47. R. Duplessix, M. Escoubes, B. Rodmacq, F. Volino, E. Roche, A. Eisenberg and M. Pinieri, *ACS Symposium Series*, **127**, American Chemical Society, Washington DC, 1980, p. 469.
48. M. Wong, J. K. Thomas and T. Nowak, *J. Amer. Chem. Soc.*, 1977, **99**, 4730.
49. K. Erdmann and A. Narebska, *Acad. Pol. Sci. Ser. Sci. Chem.*, 1979, **27**, 589.
50. A. Neppel, I. S. Butler and A. Eisenberg, *Canad. J. Chem.*, 1979, **57**, 2518.
51. C. Heitner-Wirguin, *Polymer*, 1979, **20**, 371.
52. M. Falk, *Canad. J. Chem.*, 1980, **58**, 1495.
53. K. A. Mauritz and S. R. Lowry, *J. Amer. Chem. Soc.*, 1980, **102**, 4665.
54. M. Falk, in *Perfluorinated Ionomer Membranes*, ACS Symposium Series, **180**, American Chemical Society, Washington DC, 1982, Chapter 8.
55. L. Y. Levy, A. Jenard and H. D. Hurwitz, *J. Chem. Soc., Faraday I*, 1980, **76**, 2558.
56. J. Ceynowa, *Polymer*, 1978, **19**, 73.
57. T. D. Gierke, Paper presented at *152nd Meeting of the Electrochemical Society, Atlanta, Georgia, October 1977.*
58. S. C. Yeo and A. Eisenberg, *J. Appl. Polym. Sci.*, 1977, **21**, 875.

59. T. D. Gierke and W. Y. Hsu, *J. Mem. Sci.*, 1983, **13,** 307.
60. A. Narebska, W. F. Bakew and V. I. Geraslmov, *Roczniki Chemii Ann. Soc. Chem. Polonvrum*, 1974, **48,** 1761.
61. M. Pineri, R. Duplessix and F. Volino, in *Perfluorinated Ionomer Membranes, ACS Symposium Series,* **180,** American Chemical Society, Washington DC, 1982, Chapter 12.
62. E. I. Du Pont de Nemours Inc., US Patent 3 282 875.
63. E. I. Du Pont de Nemours Inc., US Patent 3 301 893.
64. E. I. Du Pont de Nemours Inc., Laid-open Japanese Patent Application 53—132519.
65. Asahi Glass Co. Ltd, Laid-open Japanese Patent Application 52—59111.
66. Asahi Glass Co. Ltd, Laid-open Japanese Patent Application 52—83417.
67. Asahi Chemical Industries Co. Ltd, Laid-open Japanese Patent Application 55—160030.
68. M. Seko, Paper presented at the *Electrochemical Society Meeting, Minneapolis, Minnesota, May 1981.*
69. US Patent 4 126 588.
70. Asahi Chemicals Industries Co. Ltd, Laid-open Japanese Patent Application 55—160008.
71. Asahi Glass Co. Ltd, US Patent 4 126 588.
72. Asahi Glass Co. Ltd, US Patent 4 138 373.
73. Asahi Glass Co. Ltd, British Patent 1 522 877.
74. Asahi Glass Co. Ltd, British Patent 1 523 047.
75. Dow Chemicals Inc., European Patent Applications 0 041 732 and 0 041 735.
76. Asahi Glass Co. Ltd, US Patent 4 218 542.
77. Imperial Chemical Industries PLC, European Patent Application 0—086595.
78. E. I. Du Pont de Nemours Inc., US Patent Application 317 280.
79. E. I. Du Pont de Nemours Inc., US Patent Application 138 681.
80. Asahi Glass Co. Ltd, Laid-open Japanese Patent Application 55—139842.
81. Imperial Chemical Industries PLC, European Patent Application 40 920.
82. M. Nagamura, H. Ukihashi and O. Shiragami, Paper presented at *Symposium on Electrochemical Membrane Technology, A. I. Chem. Winter Meeting, Florida, March 1982.*
83. Asahi Glass Co. Ltd, European Patent Application 61 594.
84. Asahi Glass Co. Ltd, European Patent Application 61 080.
85. Asahi Glass Co. Ltd, Laid-open Japanese Patent Application 56—14999.
86. Asahi Glass Co. Ltd, Laid-open Japanese Patent Application 57—126979.
87. Asahi Glass Co. Ltd, Laid-open Japanese Patent Application 57—158231.
88. Asahi Chemical Ind., Laid-open Japanese Patent Application 56—145927.
89. Oronzio de Nora Imp., European Patent Application EP 50188A.
90. Permelec Electrode Company Ltd, UK Patent Application 2119405A.

91. Asahi Chemical Ind. Co. Ltd, Laid-open Japanese Patent Application 56—116891.
92. E. I. Du Pont de Nemours Inc., British Patent 2091297A.
93. P. J. Smith, 'Equilibrium and transport properties of perfluorinated membranes immersed in concentrated electrolyte at elevated temperatures', Paper presented at *SCI Symposium, Runcorn, England, April 1983*.
94. H. L. Yeager, Z. Twardowski, L. M. Clarke and B. O'Dell, *J. Electrochem. Soc.*, 1982, **129**, 324.
95. T. Hashimoto, M. Fujimura and H. Kawai, *Perfluorinated Ionomer Membranes ACS Symposium Series*, **180**, American Chemical Society, Washington DC, 1982, Chapter 11.
96. Asahi Chemical Industries, Laid-open Japanese Patent Application 55—14148.
97. Asahi Glass Co. Ltd, Laid-open Japanese Patent Application 52—36589.
98. Asahi Chemical Industries Co. Ltd, British Patent 1 497 749.
99. Tokuyoma Soda Co. Ltd, US Patent 4 200 711.
100. Asahi Chemical Industries Co. Ltd, US Patent 4 151 053.
101. Tokuyoma Soda Co. Ltd, Laid-open Japanese Patent Application 54—64090.
102. Asahi Chemical Industries, Laid-open Japanese Patent Application 53—125986.
103. E. I. Du Pont de Nemours Inc., Belgian Patent 866 122.
104. Tokuyoma Soda Co. Ltd, Laid-open Japanese Patent 54—21478.
105. Asahi Chemical Industries Co. Ltd, Laid-open Japnese Patent 54—6887.
106. S. M. Ibrahim, E. H. Price and R. A. Smith, Paper presented at the *25th Chlorine Plant Operation Seminar, Atlanta, Georgia, February 1982*.
107. M. Fujimura, T. Hashimoto and H. Kawai, *Macromolecules*, 1981, **14**, 1309.
108. R. Rodmacq, M. Pineri, and J. M. D. Coey, *Rev. Phys. Appl.*, 1980, **15**, 1179.
109. C. Heitner-Wirguin, E. R. Bauminger, A. Levy, F. Labensky de Kanter and S. Ofer, *Polymer*, 1980, **21**, 1327.
110. R. S. Yeo and A. Eisenberg, *J. Appl. Polym. Sci.*, 1977, **21**, 875.
111. A. Eisenberg and M. King, in *Ion Containing Polymers*, Vol. 2, *Physical Properties and Structure*, ed. R. S. Stein, Academic Press, New York, 1977.
112. A. Eisenberg, *Macromolecules*, 1970, **3**, 147.
113. R. Longworth and D. J. Vaughan, *Polym. Prepr., ACS Divn Polym. Chem.*, 1968, **9**, 525.
114. S. L. Cooper, D. F. Caulfield and C. L. Marx *Macromolecules*, 1973, **6**, 344.
115. W. Y. Hsu, T. D. Gierke and C. J. Molnor, *Macromolecules*, 1983, **16**, 1945.
116. W. J. Macknight, W. P. Taggart and R. S. Stein, *J. Polym. Sci., C*, 1974, **45**, 113.
117. J. Kao, R. S. Stein, W. J. Macknight, W. P. Taggart and G. S. Corgill III, *Macromolecules*, 1974, **7**, 95.

118. E. J. Roche, R. S. Stein, T. P. Russel and W. J. Macknight, *J. Polym. Sci., Polym. Phys. Ed.*, 1980, **18,** 1497.
119. O. Shiragami, H. Ukihashi and M. Miyake, Paper presented at *SCI Symposium, Runcorn, England, April 1982.*
120. J. Newman, *Electrochemical Systems*, Prentice-Hall, Englewood Cliffs, NJ, 1973, Chapter 12.
121. J. O. Hirschfelder, C. F. Curtiss and R. B. Bird, *Molecular Theory of Gases and Liquids*, Wiley, New York, 1954, p. 714.
122. Papers 392, 393, 394 and 395 presented at *Electrochemical Soc. Meeting. St Louis. Missouri, May 1980.*
123. L. I. Krishtalik, *Elektrokhimiya*, 1979, **15,** 734.
124. L. I. Krishtalik, *Elektrokhimiya*, 1979, **15,** 438.
125. R. D. Varjian and D. E. Hall, *J. Electrochem. Soc.*, 1984, **131,** 374C.

Index

Activation energies, 206
Activation volume, 56
Alternating copolymer, 28–9
Alternating current techniques, 208
Amorphous polymers, 29–32, 42, 63
 flexibility and flow, 29–30
Amorphous region transport, 16
Amorphous structure, 24, 51
Amperometric detection, 274
Anion, 1, 63, 64
Anode, 2, 5
Anodic polarisation, 12
Arrhenius expression, 53
Atactic form, 32
Auger electron spectroscopy (AES), 210, 217–24
Azobisisobutyl nitrile (AIBN), 311

Batteries, 4, 6–7, 13, 14, 52, 58, 67, 71, 90, 96–7
 see also under specific types of battery
Beer–Lambert law, 211
Bilayer assemblies, 281
Biochemical redox couples, 270–2
Block copolymer, 29
Branched polymer, 26
Butler–Volmer equation, 12

Cathode, 2, 5
Cathodic polarisation, 12
Cation, 1

Cationic transport number, 17
Caustic soda production, 293–334
CE scheme, 249–50
Cell
 constant, 13
 discharge, 74
 potential, 78
Chain
 folding, 38–9
 length, 30–1
 reaction, 27
 structures, 24
Charge
 carriers, 1
 reaction, 95
 transfer, 5, 251
  coefficient, 12
 transport rates, 196
Chiral synthesis, 272–3
Chlor-alkali
 cell separator, 324, 327
 membrane, 296
 cell, 316
 technologies, 300
Chlorine
 production, 293–334
 technology, 296–301
Chromatographic detection, 275
Chronoamperometry, 194–207
Chronocoulometry, 194–207
Chronopotentiometry, 207
cis structure, 34
Cluster morphology calculations, 322

Cluster size effects, 322, 324–7
Complex plane representation, 18
Conductance, 2, 3
Conducting polymers, 1, 90–6
Conductivity, 3, 40, 48, 59, 62
  measurement, 17–21
Constant frequency technique, 17
Constant phase elements, 18
Copolymerisation, 26
Corrosion, 8
  metals, protection of, 276
Cottrell equation, 195
Creep, 30, 43
Crosslinking, 26, 43
Crystallinity, 32–8, 40, 41
Crystallisation, 37–8, 43
Current collectors, 3
Cyclic voltammetry, 61, 190–4, 264, 266

Daniell cell, 8, 10
Defect conduction, 15
Degree of crystallinity, 36
Degree of polymerisation, 24
Diaphragm cell technology, 295–301
Dielectric relaxation, 320
Differential pulse voltammetry, 208
Differential scanning calorimetry, 31, 41
Diffusion
  coefficient, 55
  Fick's first law of, 13
Discharge
  characteristics, 76–7, 79, 82, 83
  curves, 84
  reaction, 87, 93, 94
Disordered sub-lattice motion, 16
Display devices, 279–80
Divalent units, 24
Donnan domains, 187
Dopants for polyacetylene, 69–71
Doping, 2
Double layer capacitance, 17
Dynamic bond percolation (DBP) theory, 57

EDAX, 210, 217–24
Electrical measurement techniques, 58–61
Electroactive materials, 67–101
  present status, 96–7
Electrochemical cells, 4
Electrochemical devices, 2, 23, 281
Electrochemical polymerisation, 152–75
Electrochemical reactors, 4–7
Electrochemical synthesis, 266–73
Electrochemistry, basic concepts of, 71
Electrode, 2
  blocking, 17
  non-blocking, 17, 59
  reactions, 71
  reversible, 7–9
  see also Polymer modified electrodes
Electrode/electrolyte interphase, 5, 7–13
Electrode/electrolyte laminations, 61
Electrolytes, 2, 3, 7, 16–17, 24
  liquid, 51, 58
  solid, 31–2, 42, 48, 63
Electrolytic cells, 4, 5
Electromotive force, 7
  concentration, and, 10–11
  temperature, and, 10–11
Electron carrier, 2
Electron energy loss spectroscopy (EELS), 225
Electron holes, 1
Electron microscopy, 320
Electron spin resonance (ESR), 215, 216
Electronic conductivity, 19–21
Electronic devices, 281
Electronically conducting polymers, 1–2
Electrostatic binding, 125
Ellipsometry, 216
End groups, 24, 28
Energy density, 78–9
Equivalent circuit, 18
Ethylene oxide, 28

INDEX

Exchange current density, 12
Extent of crystallinity, 36, 38

Faradaic reactions, 248, 264
Faraday transition, 15
Faraday (unit), 9
Fast-ion conductors, 3
Fibrils, 39, 85
Fick's first law of diffusion, 13
Film thickness, optimisation, 247–8
Flexibility, 29–30
Flow, 29–30, 43
Free volume theory, 53–5, 57, 58
Friction coefficient, 210
Fringed micelle model, 34, 38

Galvanic cells, 4, 5
Gel polymer electrolyte, 45
Gibbs–Helmholtz equation, 10
Gibbs phase rule, 41
Gierke cluster-network model, 324
Gierke cluster-network theory, 328
Glass transition temperature, 30–3, 61
Glyme series, 51
Graft copolymer, 29

Half-cell
  potential, 8
  reaction, 5
Homopolymer, 29

Impedance, 18
  low-frequency, 59
  measurements, 208
  spectrum, 17, 59
Initiation, 27
Inorganic redox systems, 270
Insulating polymers, 1
Intercalation cathodes, 15
Intercalation electrodes, 17, 24
Internal resistance, 8
Interphases, 5, 7, 14

Ion carriers, 2
Ion conducting polymers, 1, 2, 3, 15–16, 41, 45–66, 326
  theory, application of, 55–7
Ion exchange membranes, 301
Ion exchange polymers, 130, 141
Ion gate, 282
Ion pairs, 1
Ion transport, 52–8, 55, 328
Ionic clusters, 322
Ionic permeability, potential dependence of, 282
Ionomers, 46
Isotactic form, 32

Kinetic zone diagram, 246
Koutecky–Levich equation, 250, 264
Koutecky–Levich plots, 256, 260, 261

Lamellae, 39
Large-scale integrated (VLSI) circuits, 15
Laser interferometry, 266
Lattice defects, 15
Layer cases, 245–7, 251, 261, 262
Lead-acid accumulator, 13
Lead-acid battery, 5, 14
Leclanché cell, 14
Levich diffusion layer, 263
Levich equation, 250
Limiting current density, 13
Linear polymer, 24, 29
Lithium batteries, 8, 61, 62

Macromolecular structure, 24–9
Macromolecules, 23
Maltese Cross, 39
Mechanical strength, 63
Mediated charge transfer
  analysis of, 244–50
  basic model and concepts, 242–3
  experimental test of, 250–64
  theoretical treatments of, 242–66
Melting point, 34, 41

Membrane cell technology, 293–6, 329
Membrane transport theories, 328
Mer, 25
Mercury cell technology, 296–301
Microelectrode, 264
Mixed conductor, 2
Molecular size effects, 23
Molecular weight, 24
Monomer molecules, 26
Monomer/redox centre observation, 211–16
Monomer units, 25
Morphological observations, 210–11
Mössbauer spectroscopy, 320

n-doping, 69
n-type conduction, 1
NADH oxidation, 271
Negative electrode, 5, 6
Nernst equation, 10
Network structures, 24
Neutron scattering experiments, 308
Normal pulse voltammetry, 208
Nuclear magnetic resonance (NMR), 50, 225, 320
Number-average rmm, 24

Ohmic drop, 8, 14
Ohm's law, 8
Oligomers, 25
Open circuit voltage, 7
Optical measurements, 265–6
Optical microscopy, 38–9, 210
Organic redox reactions, 272
Overpotential, 11
  activation, 11, 12
  concentration, 11, 12
Overvoltage, 11
Oxygen reduction by polymer modified electrodes, 267–70

p-doping, 69
p-type conduction, 1
Particulate immobilisation, 279
Partition coefficient, 187

Perfluorinated carboxylate ionomers, 309–19
Perfluorinated carboxylate monomers, synthesis and polymerisation of, 311–13
Perfluorinated ionomer membranes, 293–334
  basic structural features of, 320–4
  electrical conductivity, 327
  fabrication, 313–15
  present and future trends in cell technology, 329
  sorbed alkali concentration, 326
  structure/transport relationships in, 320–8
  surface topography of, 315–18
Perfluorinated sulphonate/carboxylate ionomer membranes, 319–20
Perfluorinated sulphonate ionomers, 301–9
Perfluorocarboxylate ionomer membranes
  physical properties of, 318–19
  stages involved in developing, 310
Perfluorosulphonate ionomers
  commercial exploitation of, 302–4
  physical properties of, 305–9
Perfluorosulphonate polymers, synthesis of, 301–2
Permittivity, 59
Permselectivity, 282
pH effects, 274
Phase transitions, 31
  β–α first-order, 16
Photoanodic corrosion, 276
Photogalvanic effects, 278
Polarisation
  concentration, 17
  development of, 6
  effects, 79
  state of, 11
Polyacetylene, 2, 67–9
  anode, 86–7, 89–90
  as electrode material, 72–4
  cathode, 74–86, 89–90
  n-doped, 69–74, 80–6
  p-doped, 69–80

Polyaniline, 93, 94, 95, 280
Polycarbazole, 266
Polyelectrolytes, 46
Poly(ethylene glycol), 28
Poly(ethylene oxide), 2, 16–17, 32, 40–2, 47, 51, 56–8, 61, 64
Polymer complexes, 49–66
  preparation of, 49
Polymer modified electrodes, 103–239, 241–91
  applications, 266–83
  behaviour of model redox couples, 250–64
  cyclic voltammetric behaviour, 183
  electrochemical characterisation, 180–224
  electrochemical polymerisation, 152–75
  Faradaic reactions, 248, 264
  kinetic parameters, 190–209
  objectives and limitations, 180–1
  polymerisation/coating procedures, 176–80
  pre-formed polymers, 107–45
    coordinatively functionalised after coating, 127
    modified by ion exchange, 130
    modified by non-coordinative/ion-exchange strategies, 141–5
    post-coating functionalisation of, 126–45
    pre-functionalised polymers, 107–26
  preparation of, 106–80
  prepared from completely formed polymers, 108
  silane-based, 145–80
  simple models of, 181–2
  simultaneous coating/polymerisation methods, 145–80
  thermodynamic parameters, 182–90
  waveshape effects, 185
Polymer structure, 23–43
Polymeric insertion electrode, 64
Polymeric salt, 45

Polymerisation, 26, 34
  chain-addition, 27–8
  degree of, 24
  electrochemical, 152–75
  step-growth, 26, 28
Poly(p-phenylene), 90–2
Poly(phenylene oxide), 190
Polypropylene, 32, 33
Poly(propylene oxide), 33, 42, 46, 47, 57
Polypyrrole, 2, 90, 92, 208, 264, 266, 278, 280, 282, 283
Polyquinolines, 91, 92
Polysiloxanes, 61
Polytetrafluoroethylene (PTFE), 314
Polythiophene, 90, 280
Poly(L-valine), 272
Poly(vinylacetic acid), 210
Polyvinylferrocene (PVF), 180, 185, 187, 193, 211, 251, 271, 281
Poly(4-vinylpyridine)(PVP), 126, 141, 184, 185, 193, 208, 275, 278
Polyvinyltriphenylamine, 272
Positive electrode, 6
Positive plate, 5
Post-coating functionalisation of pre-formed polymers, 126–45
Potential dependence of ionic permeability, 282
Potential step methods, 194–207
Potentiometric detection, 274
Pre-formed polymers, 107–45
  coordinatively functionalised after coating, 127
  modified by ion exchange, 130
  modified by non-coordinative/ion-exchange strategies, 141–5
  post-coating functionalisation of, 126–45
Pre-functionalised polymers, 107–26
Primary battery, 4, 67
Propagation, 27
Proportionality constant, 2

Raman spectroscopy, 217
Random copolymer, 28
Recycling characteristics, 84

Reference temperature, 55
Resistance, 2
Rotating disc electrode (RDE), 209
　voltammetry, 250, 264, 265, 271, 283
Rotating ring disc electrode (RRDE), 209
Rubber-like behaviour, 30

Sandwich assemblies, 281
Scanning electron microscopy (SEM), 38–9, 210
Secondary battery, 4, 67
Self-discharge, 7, 21
Self-exchange theory, 263
Semiconductor electrochemistry, 276–8
Semicrystalline polymers, 48
Sensors, 273–6, 281
Shelf life, 7
Siemen (unit), 2
Silanes, 145–80
Small-angle X-ray scattering (SAXS), 308, 321
Sodium ion transport number, 325
Sodium–sulphur battery, 14
Solid state battery, 7, 79
Solid state cells, 13–17
Solvating polymers, 46, 48, 52, 63
Spectroscopic techniques, 209–24, 265–6
Spherulites, 38–9
Standard electrode potential, 8–10
Standard Gibbs free energy, 9
Standard hydrogen electrode (SHE), 8
Steady-state diffusion equations, 245
Stefan–Maxwell transport equations, 328
Stereoregular polymer, 28
Structural repeat unit (sru), 25
Structure–property relationships, 40
Sub-molecular units, 216–17

Superionic conductors, 3
Syndiotactic form, 32

Tacticity, 32–3
Tafel relationship, 12
Temperature coefficient of emf, 11
Temperature effects, 10–11, 15, 37, 38, 54, 55, 62
Termination, 27
Thermal history, 38
Thermomechanical analysis (TMA), 31
Time domain experiments, 61
trans structure, 34
Transport control, 257
Transport number, 13
Transport properties, 54
Triple ions, 1

Vacancies, 15
Variable frequency technique, 18
Viscosity, 30
Voltammetric analysis, 275

Wagner polarisation cell, 19
Walden's rule, 256
Waveshape effects, 185
Weight-average rmm, 25
Wide-angle X-ray scattering (WAXS), 308, 321
Williams–Landel–Ferry (WLF) equation, 55

XR resins, 301–4
X-ray diffraction patterns, 34
X-ray diffractometry, 320
X-ray photoelectron spectroscopy (XPS), 210, 217–24

Zeolites, 24